JOURNEY
OF THE MIND

Praise for *Journey of the Mind*

"*Journey of the Mind* . . . [is] rich in illustrative stories from the origin of life to the evolution of *Homo sapiens*—to the murder of George Floyd. A daring book and an absorbing read."
    —Peter Sterling, University of Pennsylvania Perelman School of Medicine
                    and coauthor of *Principles of Neural Design*

"An engaging and surprisingly accessible account of matter's journey to mind. . . . An encouragingly optimistic reminder of how our shared experience and common understanding of reality far surpass our superficial differences."   —Douglas Rushkoff, professor of media theory and
                    digital economics at CUNY/Queens and best-selling author of
                            Team Human and *Throwing Rocks at the Google Bus*

"No one can accuse Ogi Ogas and Sai Gaddam of lacking literary courage. [T]hey offer an imaginative tale of who we are and how we came to be that will keep you reading into the wee hours until the last page."
—Jeffrey Lieberman, former president of the American Psychiatric Association,
            chairman of the Department of Psychiatry at Columbia University,
                                        and author of *Shrinks*

"Ogi Ogas and Sai Gaddam deftly convert sophisticated ideas and concepts into accessible and compelling language, unraveling the mystery of how thinking went from a single-celled organism to the magnificence of the human mind."
              —Annie Duke, winner of the World Series of Poker Tournament of
                        Champions and best-selling author of *How to Decide*

"This book is stunning in its range and reach. . . . Whether or not you agree with the authors' views, you will be led to think about the material in interesting ways and will be led to appreciate the marvelous power, complexity, and beauty of the mind."
              —Stephen Kosslyn, former John Lindsley Professor of Psychology in
                    Memory of William James and dean for the social sciences at
Harvard University, and founder and chief academic officer of Foundry College

ALSO BY OGI OGAS AND SAI GADDAM

A Billion Wicked Thoughts

JOURNEY OF THE MIND

HOW THINKING EMERGED FROM CHAOS

Ogi Ogas and
Sai Gaddam

W. W. NORTON & COMPANY
Celebrating a Century of Independent Publishing

For information about permission to reproduce selections from this book, write to
Permissions, W. W. Norton & Company, Inc., 500 Fifth Avenue, New York, NY 10110

For information about special discounts for bulk purchases, please contact
W. W. Norton Special Sales at specialsales@wwnorton.com or 800-233-4830

Manufacturing by Lakeside Book Company

Production manager: Anna Oler

Library of Congress Cataloging-in-Publication Data

Names: Ogas, Ogi, author. | Gaddam, Sai, author.
Title: Journey of the mind : how thinking emerged from chaos / Ogi Ogas and Sai Gaddam.
Description: First edition. | New York, NY : W. W. Norton & Company, [2022] |
Includes bibliographical references and index.
Identifiers: LCCN 2021050489 | ISBN 9781324006572 | ISBN 9781324006589 (epub)
Subjects: LCSH: Philosophy of mind.
Classification: LCC BD418.3 .O33 2022 | DDC 128/.2—dc23/eng/20220103
LC record available at https://lccn.loc.gov/2021050489

ISBN 978-1-324-05057-5 pbk.

W. W. Norton & Company, Inc., 500 Fifth Avenue, New York, N.Y. 10110
www.wwnorton.com

W. W. Norton & Company Ltd., 15 Carlisle Street, London W1D 3BS

1 2 3 4 5 6 7 8 9 0

For Agastya, Meera, and Priyanka

And for Tofool

Chaos isn't a pit. Chaos is a ladder. . . . The climb is all there is.

—Littlefinger, *Game of Thrones*

Contents

STAGE III

MODULE MINDS

STAGE IV

SUPERMINDS

JOURNEY
OF THE MIND

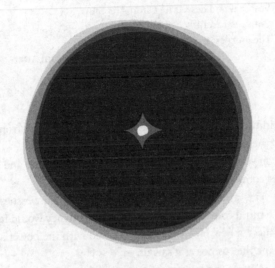

Initial Value Problem

Do not imagine that the Way is short;
Vast seas and deserts lie before His court.
Consider carefully before you start;
The journey asks of you a lion's heart.
    —*The Conference of the Birds*, Farid ud-Din Attar

In the beginning, fourteen billion years ago, existence arose from nonexistence and the universe commenced.

Four billion years ago, give or take, life arose from nonlife and the evolution of species commenced.

A billion years after that, purpose arose from purposelessness and the journey of the mind commenced. Eventually, the journey would forge a god out of godlessness, a new breed of mind endowed with the power and disposition to reshape the cosmos as it saw fit.

This book retraces the journey of the mind from the aimless cycling of mud on a dark and barren Earth until the morning a mind woke up and declared to an indifferent universe, "I am aware of *me*!"

The chapters ahead visit seventeen different living minds, ranging from the simplest to the most sophisticated. First up is the tiniest organism on Earth, the humble archaeon, featuring a mind so minuscule that you would be forgiven for questioning whether it's a mind at all. From there, our itinerary will take us forward through a series of increasingly brawny intellects. We will sojourn with amoeba minds, insect minds, tortoise minds, and monkey minds, until we arrive at the mightiest mind to ever grace our solar system . . . one that may be something of a surprise.

Each chapter highlights a new mental challenge thrown down in a mind's path by chaos, the purposeless churn of physical matter, before revealing the mental innovation that surmounted it. This gauntlet of challenges begins with one of the most perplexing of all: *How did a mind emerge from mindlessness?*

You will learn how each new innovation led to an even more daunting challenge that spurred minds to become smarter still. You will see how bacteria make situation-specific decisions without the benefit of a single neuron, let alone a brain. You will discover why the housefly's surprisingly intelligent mind marks nature's boundary between minds that are unquestionably nonconscious and those indisputably conscious. You will come to appreciate how monkeys rely upon hope, rage, and awe to chart a course through life. Each new form of thinking is explained in plain language without any mathematical equations, though curious readers can find additional details and references in the endnotes.

This book is motivated by three goals. First, to help you appreciate the hidden connectedness of all minds. Amoeba minds and human minds are linked through an unbroken continuum that parallels the one linking the ancient Roman town of Londinium with the twenty-first-century London megalopolis.

The second aim of retracing the mind's three-billion-year pilgrimage is to obtain new answers to very old questions. Why do we exist? Where are we all headed? Is there a hidden relationship between chaos and purpose? Is there a cosmic role for love in the universe? For decency?

The third goal is to explain the physical basis of the "Big Three" forms of thinking: consciousness, language, and the Self. How do minds—unlike pebbles and dust and sunspots—boast the ability to *experience* things? You may have heard that mortal consciousness is the greatest locked-room mystery in science, an unsolved puzzle that may never be unriddled. The journey ahead suggests otherwise. By progressing through the sequence of innovations that led to sentience, step-by-step, this narrative offers an incremental account of *why* and *how* consciousness appeared in the universe. You will learn how consciousness works and come to see new strains of consciousness in places you never expected. You will learn, too, how language was constructed atop the architecture of consciousness and why a hairy, hooting biped with a fondness for rocks became the first beast to compose poetry, rather than a soft-feathered flyer with a melodious song. And, finally, you will acquire a deeper understanding of the greatest innovation in biological minds: the human Self.

Such grandiose finales often have modest origins. The journey of the mind begins eons before the first human mew, in a bleak and unthinking void without a trace of purpose in sight. The old fable of Genesis got one thing right, for in this boundless reign of chaos the first mind strove to cleave the light from the darkness . . .

Stage I
Molecule Minds

CHAPTER ONE / **FIRST MIND**

Purpose

We're building something here, Detective. We're
building it from scratch. All the pieces matter.
                                    —Lester Freamon, *The Wire*

1.

A long, long time ago, somewhere in the dark and the deep, volcanic vents
blasted through the crust beneath the ocean and spewed out sulfur and minerals in a furious maelstrom of pressure and heat. Out of this infernal brew,
the first living organisms began to multiply.

Or maybe they oozed out of asteroid-infused mud. Or seeped from the
cracks of sunbaked clay. Or bubbled out of wet pockets of rock galvanized
by lightning. Or wriggled out of freshwater pools on volcanic islands. Or—
another serious contender—perhaps they hitched a ride across the heavens
on a meteor.

The origins of life on Earth remain one of science's most enduring mysteries. Nobody has a particularly persuasive theory regarding the identity or
circumstances of the primal chemistry that jump-started metabolism and
reproduction, the two fundamental hallmarks of life. Making the task of
recombobulating life's firstborn even more knotty is the fact that conditions
on primeval Earth were radically different than now.

If you were somehow transported back to the day that the first audacious
speck of life xeroxed itself, you would find that the day was a brisk ten hours
long, the Earth completing an entire rotation during the length of a modern
commuter's workday. A dim chilly red sun barely penetrated a hazy sky the
color of brickdust. During the brief night, the moon loomed low and enormous in the sky like some pale fairybook beast. An oxygen-less atmosphere
drenched with poison churned over a cold and sludgy ocean.

In this exotic and speculative realm, some combination of molecules
interacted with some other combination of molecules and sparked the jour-

ney of life—an event whose staggering import scientists dryly mask with the term "abiogenesis." Regarding the identity of this sacred web of chemical permutations, all is contested conjecture. Well, almost all. One hypothesis has risen above the rampant guesswork and attained something close to consensus among scientists. The hypothesis is this:

> *The chemical system of newborn life—whatever it might have been—was swaddled in a tiny ball of fat.*

We should all sing hymns of thanksgiving to this wobbly little globe. It proved to be the cornerstone character in two cosmic narratives: the journey of life and the journey of Mind. (Biologists have a ready term to refer to *all species that ever were*: "life." Neuroscientists lack a corresponding term for *all minds that ever were*, so this book shall employ "Mind" for the task.)

Certain large molecules containing fatty acids—*lipids*, in the language of chemistry—possess a special property. They automatically self-assemble into a membrane. Their physical nature is to link together into an elastic wall that bends back on itself to create a sphere. You've witnessed this process anytime you've noticed a bubble emerge from soapy water. Soap bubbles contain molecules similar to those found in the membranes of living organisms—and similar, perhaps, to those in the primeval membranes that originally cordoned off life from not-life, thereby constructing a private room where the story of biology could unfold in fragile safety.

The establishment of a distinct physical boundary around metabolizing and self-replicating chemical processes inaugurated something marvelous. A *body*. A physical configuration whose constituents were diligently laboring to preserve their configuration's existence.

By separating the stuff performing all the indispensable tasks of self-preservation from all the other stuff in the world—by dividing *me* from *not-me*—an ancient shroud of lipids fulfilled an essential precondition for the emergence of Mind. Every mind needs a border partitioning the physical processes of its *body* from its *environment*, where chaos maintains its indomitable reign.

The inception of a living body introduces the first of four deep principles of Mind that will help us make sense of Mind's ascendance from mindlessness to sentience: the *embodied thinking principle*. According to the embodied

THE BIRTH OF BODY

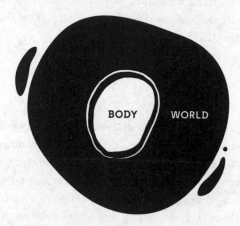

thinking principle, it is not possible to separate a mind from its body, any more than we can separate a city from its buildings. In order to understand any particular form of thinking, we must always take into account the configuration of the physical *stuff* that the thinking is incarnated within, even when much of that stuff may not appear to be directly involved with manufacturing thoughts.

Every mind has a body. Yet, no matter how complex its breathing, feeding, and excreting, a body alone is not a mind. Innumerable generations of membrane-enveloped organisms nursed a vibrant biochemical metabolism and were perfectly alive yet remained totally mindless. That's because these pioneering species lacked something else indispensable for the genesis of thought.

They needed something to think with.

2.

A mind is a physical system that converts sensations into action. A mind takes in a set of inputs from its environment and transforms them into a set of environment-impacting outputs that, crucially, influence the welfare of its

THINKING ELEMENTS

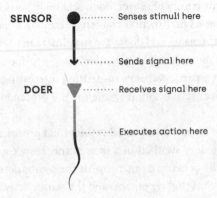

body. This process of changing inputs into outputs—of changing sensation into useful behavior—is *thinking*, the defining activity of a mind.

Accordingly, every mind requires a minimum of two *thinking elements*:

- A *sensor* that responds to its environment
- A *doer* that acts upon its environment

Some familiar examples of *sensors* that are part of your own mind include the photon-sensing rods and cones in your retina, the vibration-sensing hair cells in your ears, and the sourness-sensing taste buds on your tongue. A sensor interacts with a *doer*, which *does something*. A doer performs some action that impinges upon the world and thereby influences the body's health and well-being. Common examples of doers include the twitchy muscle cells in your finger, the sweat-producing apocrine cells in your sweat glands, and the liquid-leaking serous cells in your tear ducts.

A mind, then, is defined by what it *does*, rather than what it *is*. "Mind" is an action noun, like "tango," "communication," or "game." A mind responds. A mind transforms. A mind acts. A mind *adapts* to the ceaseless assault of aimless chaos.

The identities of the sensors and doers of Earth's first mind are lost in deep time and may never be recoverable. These primal thinking elements may have been simple, perhaps single molecules that changed their shape when they contacted a photon of light. However, it's more likely that the original sensors and doers were ungainly contraptions composed of clumsy chains of free-floating molecules that did not reliably respond to stimuli or consistently exhibit a desired behavior.

Together, this trinity of sensor, doer, and body launched the journey of Mind by solving its first mental challenge: *creating a mind out of mindlessness*. Cast apart, sensors and doers are inanimate matter, flecks of vagrant chemical junk. But when you assemble a sensor, doer, and body so that they interact in a particular way, something interesting happens. To illustrate this physical revolution, it's time for the protagonist of our story to step on stage.

The simplest hypothetical mind is the proud proprietor of a single sensor and a single doer swathed in a membrane. How could a single sensor and doer usefully benefit an organism? If the sensor detects some resource in the environment (light, perhaps) and the sensor activates the doer when it detects the resource and if the doer affords some kind of locomotion

---

SIMPLEST POSSIBLE MIND

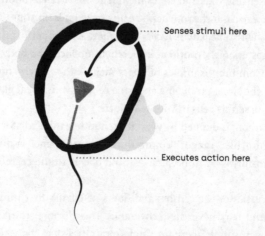

Senses stimuli here

Executes action here

(propulsion, perhaps), then this super-simple mind could advance toward the resource.

Behold *purpose*, an unprecedented development in the universe!

Purpose is a special class of physical activity, different from all other physical activity that existed before the first mind thought its first thought. A moon orbits a planet because the laws of gravity command it. But a moon with purpose might very well throw off the shackles of its planetary master and soar off into the freedom of open space. Whereas physics obeys rigid laws without aim, a mind pursues an aim without following rigid laws. Physics *ensues*. A mind *adapts*.

Purposeful phenomena are best explained using different mathematics than purposeless phenomena. Imagine predicting the path of an ant as it walks through a forest. It dodges rocks, clambers over twigs, crawls along the side of stumps. Its path is complex and nonlinear, and if physicists attempted to derive an equation that accounted for the ant's trajectory using the same mathematical tools they employ to account for the trajectory of an asteroid, tornado, or electron, they would labor in vain. If they instead recognized that the ant's behavior was *purposeful*—the ant was trying to get home to its anthill—then they could predict the ant's trajectory by assuming that at any given moment, the ant will *choose* the path offering the easiest route home. Predicting the ant's motion requires a different kind of mathematical model than any found in the physical sciences: a model describing the thinking processes responsible for the ant's decision-making. One of the aims of this book is to guide your intuitions about the odd principles of purpose so you can better understand consciousness, language, and the Self.

The first thinking on Earth, whatever it may have been, surely served a specific purpose. And whatever that purpose was, it was surely an advantage in the pitiless contest for survival. Even if the first mind's sensor was feeble and its doer clumsy, such a configuration would move itself closer to food (or away from danger) with a probability better than chance and thereby engage the engine of natural selection. A photosynthesizing organism that could move toward light or away from darkness would outcompete an identical organism that floated about aimlessly like a feather on a pond.

Earth's first mind remains entirely speculative, to be sure, and is the only mind on our journey that we cannot examine directly. Sadly, no organism with one sensor and one doer roams the microscopic wilds of twenty-first-century Earth. So let's visit the next best thing. Let's examine the simplest *living* mind in nature and see how it compares to the simplest *hypothetical* mind.

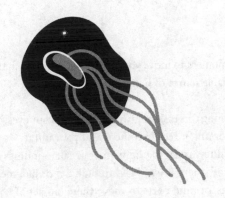

CHAPTER TWO / **ARCHAEA MIND**

Targeting

We sit in the mud, my friend, and reach for the stars.
—*Fathers and Sons*, Ivan Turgenev

1.

Please take a moment to move yourself one centimeter in the direction of the brightest available source of light.

Done?

Accomplishing this task required the use of your eyes. Your muscles, too. But most important, it required the use of your mind.

You needed to survey the light in your surroundings (an act of perception), determine where it was brightest (an act of judgment), and then will your legs, arms, or hind parts to move there (an act of volition). In a word, you needed to *think*.

Accomplishing this task solves the first challenge that every mind must address. Let's call it the targeting problem: How do you locate a desired objective in your environment and move yourself toward it? On the face of it, the thinking you just engaged in would appear to require a fair bit of mental wiring. Some kind of visual processing circuitry to apprehend the light, you might guess, as well as judgment circuitry that evaluates where the light is brightest, and motor-control circuitry that formulates and executes the plan to move your body toward the light in question. This sounds fairly complicated. And yet, this complex feat of targeting is accomplished by the most rudimentary mind in nature.

Archaea (pronounced "ar-KEY-uh") are wee little beasties. Like bacteria, archaea are single-celled microbes with a cell membrane but no nucleus. Not only do they lack neurons, some archaea are so diminutive that their entire body can squeeze into the nucleus of a human neuron. The smallest known cellular organism is an archaeon: *nanoarchaea* are one-fifth the length of an *E. coli* bacterium.

Archaea form one of the three great domains of earthly life. The other two are bacteria and eukaryotes, the latter of which includes all known species of animals, plants, and fungi. But eukaryotes and bacteria are the youngbloods in the family. Biologists believe that archaea's branch of life extends the furthest back in time.

Archaea are found in every habitat on the planet, yet they spent most of the past century hiding in plain sight. Whenever a biologist happened to notice an archaeon on a microscope slide it was misclassified as a bacterium. Science finally recognized that archaea were our long-lost relatives in 1977. That's when an obscure biologist analyzed the RNA of a methane-producing microbe. He discovered that its genetic code diverged from that of every known branch of life. It was a bit like delving into your family tree and discovering that your great-great-great-grandmother didn't emigrate from Germany like everyone thought, but Jupiter.

Though in general archaea look and behave a lot like bacteria, they employ physiological processes unique to their kind that enable them to feed upon sources of nutrition that are indigestible or even lethal to other organisms, including arsenic, ammonia, gypsum, petroleum, and uranium. This allows them to thrive in inhospitable habitats such as volcanic vents, polar ice caps, boiling hot springs, toxic-waste dumps, and subterranean rock a mile beneath the ocean crust.

Haloarchaea, for instance, dwell in water with concentrations of salt so extreme that they are toxic to almost all other life. Haloarchaea derive their energy from the sun. Not through plant-style photosynthesis using green chlorophyll. Instead, haloarchaea conduct their own brand of photosynthesis using a purplish molecule known as bacteriorhodopsin. (The Great Salt Lake in Utah is often tinged with a pinkish-purple hue around its edges from haloarchaea blooms.) Since the sun is its source of "food," the haloarchaeon must ensure it can access a reliable supply of sunlight. That's where the world's simplest living mind comes in.

Introducing Archie, a model of a haloarchaeon:

ARCHIE

A model is a simplified representation of a complex phenomenon. A useful model allows you to focus on the important stuff and ignore everything else. A map of the Boston subway is a useful model. It allows you to swiftly chart your course from Copley station to South Station, because the map omits distracting real-life details like the bustling crowds of commuters, the twisty, screechy track around Boylston, and the train's broken-down air-conditioning.

All the models in this book are designed to help you focus on the important stuff—namely, the ideas most useful for making sense of the Big Three (consciousness, language, and the Self). In Archie's case, the most important idea is that no matter where you put him, he will always move toward the brightest source of light.

It might not seem like a sun-basking microorganism would require the ability to think to survive. But as you demonstrated a moment ago, to accomplish what a haloarchaeon does, a human being would certainly need to think. Sure, the single-celled haloarchaeon doesn't sport a visual cortex or a motor cortex, let alone a brain. But *some* kind of mental apparatus must survey its environs, judge the direction of the brightest source of light, and command its body to move there.

What constitutes thinking in a creature so piffling that it managed to elude microscope-brandishing scientists for more than three hundred years?

ARCHIE'S MOVEMENT

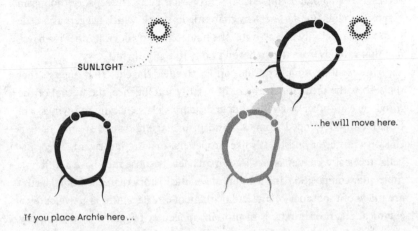

SUNLIGHT

...he will move here.

If you place Archie here ...

2.

Archie's mind consists of just four thinking elements: two sensors and two doers. Each sensor is connected to one doer, in two parallel sensor-doer pairs.

Archie is an example of a *molecule mind*, the first stage of thinking on our journey. All the thinking elements in molecule minds consist of individually

ARCHIE

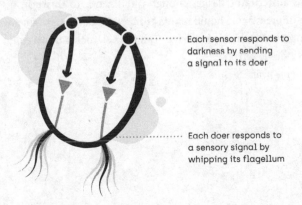

Each sensor responds to darkness by sending a signal to its doer

Each doer responds to a sensory signal by whipping its flagellum

identifiable molecules. The membrane of an actual haloarchaeon is traversed by a shape-shifting molecule known as a sensory rhodopsin. This molecule serves as the haloarchaeon's visual sensor—its "eye." The sensory rhodopsin responds to light (or darkness) by changing its shape, which triggers a cascade of molecular activity that activates the haloarchaeon's doer. Its doer is a *flagellum* that extends from the archaeon's body like a bullwhip.

Almost all the simplest minds on Earth wield flagella. That suggests that the lash is the primal instrument of earthly intelligence, the ur-tool of the mind that all other anatomical doers (such as claws, teeth, and wings) are functionally descended from. A handful of flagella extend from a haloarchaeon's posterior and thrash like propellers, driving it forward. These flagella (tec$ically known as *archaella* in archaea, because they have a different molecular composition and a different genetic history than bacterial flagella) are made of thousands of molecular filaments. A flagellum is activated by a chemical chain of hundreds of protein molecules triggered by the sensory rhodopsin. That means that the entire sensor-doer pathway in a haloarchaeon consists of fewer than ten thousand molecules. By comparison, scientists have estimated that a single human neuron contains somewhere in the ballpark of fifty billion molecules of protein alone.

When one of Archie's "eyes" detects darkness (much like the night sensor on a streetlamp), it sends a signal to its linked "propeller," which begins to whip Archie forward. The dimmer the light, the stronger the signal. The stronger the signal, the stronger the propulsion, which means that Archie's flagella whip their hardest in total darkness.

Archie is barely more sophisticated than the simplest possible mind, yet despite his austerity he flashes enough intelligence to thrive in his chaotic world, an unpredictable liquid realm buffeted by eddies and shadows where he must obtain sunshine to survive. The four elements of his mind all contribute to a single purpose:

Seek out the light.

ARCHIE'S MOVEMENT EXPLAINED

SUNLIGHT

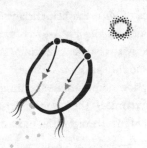

1.
Initially, the **left-side sensor** detects more darkness because it is farther from the light, so it sends a stronger signal than the right-side sensor, which makes the **left-side door** whip faster than the right-side doer, which makes Archie move forward and turn to the right...

2.
The **left-side sensory signal** starts to weaken as it gets closer to the light, which makes the two flagella closer in strength, which makes Archie slow down and turn less sharply...

3.
Eventually, the light source is located directly between the two sensors, which makes them send **signals of equal strength**, which makes **the flagella whip with equal strength**, which makes Archie go straight forward. Since the sensory signals continue to weaken, Archie slows down as he nears the light...

4.
Finally, the light is so close and strong that both sensors shut off, which shuts off the flagella, which makes Archie halt.

3.

Imagine you visit a park with a basketball court. Pick out ten strangers at random, invite them onto the court, and toss them a ball. They stand around blinking at you. Is this a basketball game? No. Not even if all ten people, through an uncanny stroke of serendipity, happened to be professional athletes employed by the Boston Celtics.

A game is not defined by the identity of the players. A game is defined by how the players *interact* with one another. If your ten randomly selected strangers formed into two teams of five and began to dribble and pass the ball around the opposing team as they attempted to hurl the ball through the hoop, well then. Now you've got a basketball game.

It's the same with Mind. A mind is not defined by the identity of the physical stuff inside an organism. A mind is not defined by its neurons or, in Archie's case, by its molecules. A mind is defined by how its thinking elements *interact*. A man who has been dead for an hour has the same physical stuff in his brain he had the hour before, but no thinking is going on because his neurons are no longer interacting. A corpse is like ten people standing around a basketball court: no game, no mind.

Both a game and a mind are examples of what mathematicians call a *dynamic system*. A dynamic system is defined by *change*. A given system's dynamics consist of the way the system's activity changes over time. Whenever you read the term "dynamics," you should think of "change," "activity," or "action." The word "mind" is an action noun *because* the mind is a dynamic system. Understanding a mind's dynamics is the key to understanding how it thinks, which is why dynamic systems theory is the most useful branch of mathematics for investigating Mind.

Though the size of a mind's sensors and doers, their location in the body, and their chemical identity undoubtedly influence a mind's dynamics—this is what the embodied thinking principle advises us—these features by themselves don't constitute a mind any more than the height of the players, their position on the court, and the color of their jerseys constitute a basketball game. A game consists of jump shots and blocks and free throws and fouls—that is to say, the *action*. A mind consists of the activity of its thinking elements as they move around, change their shape, send out signals, and react to other thinking elements. We'll call this activity-dictated conception of the mind the *basketball game principle*, the second key principle in the journey of Mind.

The basketball game principle holds a very important lesson for how we should think about thinking. If you were watching an actual basketball game, you'd probably focus on the ball. After all, the ball drives all the action and accounts for all the scoring. This seems to suggest that if you simply traced out the trajectory of the ball over the course of a game, you'd acquire a pretty good understanding of the game. Not so. A basketball game consists of the activity of *all* ten players, even the ones who aren't touching the ball.

As every coach knows, if you want to understand how a particular player scored, you need to know what all the other players were doing. During a given play, a player might be cutting toward the basket, blocking out an opponent, setting a screen for a teammate, or performing many other actions that influence the flow of the game. All these individual actions are happening simultaneously in real time and each player's activity influences every other player's activity. If one player leaps up to block an anticipated shot, her opponent might react by passing the ball to someone else, or faking a shot before taking the real shot, or leaning into the blocker in an attempt to draw a foul. Meanwhile, one of the blocker's teammates might sprint down the court in the opposite direction in the hope of a fast break. In a basketball game, everything is happening everywhere all at once. According to the basketball game principle, then, if we wish to understand the dynamics of a game—or a mind—we must learn to think *holistically*.

There is no isolated or compartmentalized thinking activity within Archie's mind. Thinking occurs everywhere at the same time. Archie's mind is nothing like a computer program, which is the opposite of a dynamic system. Archie does not follow a functional sequence of: perception *then* judgment *then* action. In fact, Archie has no distinct "perception circuit" that creates a representation of the light. Nor does he possess a "navigation circuit" that judges which direction to go. There's not even a "motor-control circuit" that executes his navigational plan. It's simply not possible to say, "*Here* and *now* is where Archie's mind forms an image of the light" or "*Here* and *now* is when his mind makes the decision to move." Instead, Archie's "eyes" and "propellers" all interact continuously and simultaneously in real time to move him toward his desired target through ceaseless loops of blurred causality.

You might imagine that even though archaea's thinking is simple and holistic, that eventually the journey of Mind must cross some decisive bridge separating primitive forms of thought (like Archie's sun-seeking) from magical forms of thought (like consciousness, language, and the Self). Not so. The basketball game principle proclaims: *the magic was already there from the very start.*

In physical matter's transition from purposelessness to purpose, it wasn't the *elements* of mind that were enchanted. It was their *arrangement*. The holistic mental dynamics solving the archaea's targeting problem are generated by a special configuration of ordinary molecules. A more complex configuration of the same ordinary molecules generates the holistic mental dynamics responsible for the Big Three.

The relevance of a mind's configuration becomes apparent if we make a single tiny change to Archie's wiring and thereby transform him from a creature of light into a creature of darkness.

4.

Here is Cross-Wired Archie:

CROSS-WIRED ARCHIE

Cross-Wired Archie has the exact same thinking elements as his predecessor, in the exact same locations. The *only* thing that has changed is that all of his sensor-doer connections are now wired *across* his body, instead of running parallel to one another. There's been no change at all in the substance of Archie's mind. No new thinking elements, no shuffling around existing elements, no new wires. The only change we've made is swapping the *ends* of the wires.

Let's place Cross-Wired Archie in the same spot as before. How does he behave now?

CROSS-WIRED ARCHIE'S MOVEMENT EXPLAINED

SUNLIGHT

1.
Initially, the **left-side sensor** detects more darkness because it is farther from the light, so it sends a stronger signal than the right-side sensor, which makes the **right-side doer** whip faster than the left-side doer, which makes Archie move forward and turn to the left...

2.
As Archie turns, **both sensors** move away from the light, which makes both send stronger signals, which makes **both flagella** whip harder, which makes Archie swim faster and continue to turn to the left...

3.
As Archie continues to turn away from the light, **both sensors** eventually send signals of equal strength, which makes Archie stop turning as he continues to swim faster and faster...

4.
Finally, Archie reaches his maximum velocity as he swims directly away from the light, deeper and deeper into the darkness.

If you alter a mind's dynamics, you alter its purpose. Cross-Wired Archie's behavior—avoiding light and seeking darkness—is the exact opposite of Parallel-Wired Archie's. Parallel-Wired Archie was a Light-*Seeker*. But Cross-Wired Archie has become a Light-*Shirker*.

The divergent behaviors of Parallel-Wired Archie and Cross-Wired Archie illustrate how the evolution of the earliest minds likely unfolded. Slight changes in the location or activity of a single molecular thinking element can lead to dramatic transformations in a molecule mind's behavior. (Similarly, replacing a short player with a tall player or a great shooter with a great defender can lead to dramatic changes in a basketball game.) During the molecule mind stage of the journey of Mind, minor tweaks to molecular thinking elements produce major innovations that empower molecule minds to better adapt to their world.

Calculus, comedy, and kitesurfing might seem impossibly remote destinations from the unassuming archaeon, but even the lengthiest of journeys begins with a single purposeful step—or twitch of the flagellum.

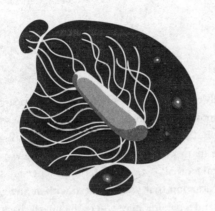

Decision-Making

Exploration is in our nature. We began as wanderers,
and we are wanderers still.

—*Cosmos*, Carl Sagan

1.

Bill Belichick is widely regarded as the most talented American football coach
alive. He's won the most playoff games in the history of the sport. He's been
to twelve Super Bowls, the World Cup of American football, winning eight
of them. One of the main tactics powering his winning ways is what Belichick calls "situational football"—the notion that you need to select the right
play for the specific situation that you are faced with at any given moment.
Some scenarios demand a passing play, where the quarterback hurls the ball
through the air to a receiver. Other scenarios demand a running play, where
the quarterback hands the football to a running back who tries to bulldoze
his way through the defense. The challenge, of course, is correctly deciding
which action the circumstances call for. Coach Belichick has mastered this
challenge so well that he is considered a football genius.

Nobody would consider bestowing the epithet of "genius" upon a lowly
bacterium. Despite this snub, bacteria have mastered what we might call "situational food-ball."

In 1972, biologists noticed that bacteria exhibited two distinct types of
locomotion. The first consists of rapid darts straight ahead. They dubbed this
type of movement *running*. The second consists of haphazard, erratic jostling.
This was dubbed *tumbling*.

After further investigation, it became apparent that bacteria run toward
food and tumble about when they arrive at the food. They were somehow
deciding to switch from running to tumbling or back again according to the
conditions they encountered. But how can a microscopic molecule mind
make context-dependent decisions?

RUNNING VS. TUMBLING

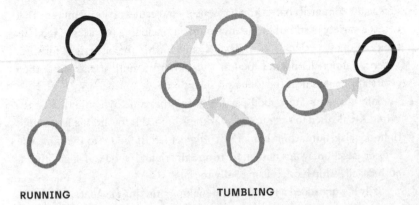

RUNNING   **TUMBLING**

As it turns out, it's possible to transform Cross-Wired Archie into a situational decision-maker without adding any new thinking elements or modifying any of his connections. To see how, please make the acquaintance of Sally the salmonella:

SALLY

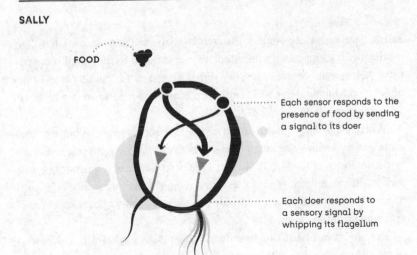

FOOD

Each sensor responds to the presence of food by sending a signal to its doer

Each doer responds to a sensory signal by whipping its flagellum

Salmonella enterica is a bacterium. It is one of the most common causes of food poisoning in humans and is the pathogen responsible for typhoid fever. About 155,000 people die every year from salmonella-related disease. Like archaea, a salmonella bacterium has sensory molecules on its membrane that detect a variety of stimuli in its environment, including particles of food. Its food sensors send signals to its flagella, which whip the bacterium forward. When a salmonella notices food, it *runs* toward it until it gets close, then erratically *tumbles* near the food as it slurps it up.

Sally steers toward a food particle using the same dynamic that Cross-Wired Archie used to steer toward darkness. If food is toward the left, Sally's right-side flagellum whips harder, turning her left. If food is to the right, the left-side flagellum whips harder, turning Sally right. If food is straight ahead, both flagella whip hard, causing Sally to *run*.

Sally has not added a new decision-making thinking element. Her configuration of sensors and doers is identical to that of Cross-Wired Archie, who does not choose whether to run or tumble. So how is Sally able to decide when her circumstances call for running and when they call for tumbling—and how does she switch between the two?

The answer lies within an upgrade to Archie's doers.

2.

Sally's flagella operate slightly differently than Archie's. Archie's flagella, recall, always whipped with an intensity proportional to the signal received from their sensors. Stronger sensory signal = stronger whipping. No sensory signal = no whipping. But Sally's enhanced flagella can perform two distinct actions: *whipping* and *twitching*.

When one of Sally's flagella receives a signal from a sensor, it behaves the same way as Archie's flagella, namely, it *whips* with a strength proportional to the intensity of the signal. But let's see what happens when no signal is arriving at a flagellum. Without any incoming stimulation, Sally's flagellum *twitches* intermittently at an irregular pace. It's as if Sally's doers have a low and random "idling rate."

But how does a twitching flagellum furnish Sally's mind with the power of "situational food-ball"? Whenever Sally detects a distant food particle, she

SALLY'S DOER

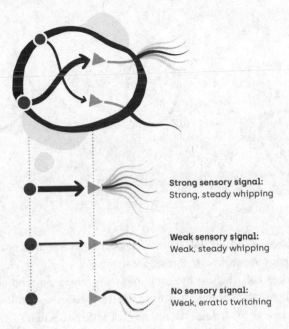

Strong sensory signal:
Strong, steady whipping

Weak sensory signal:
Weak, steady whipping

No sensory signal:
Weak, erratic twitching

reacts in the same manner as Archie, with her "eyes" activating her "propellers," prompting Sally to run toward the food.

But when there is no food nearby, something unusual happens. Her flagella go into "idling mode." They start to twitch randomly. Since each fla-

SALLY RUNS

SALLY TUMBLES

gellum is now quivering at its own independent and irregular interval, Sally jounces about unpredictably. She begins to *tumble*.

How does Sally benefit by tumbling when there's no food close by? It allows Sally to aim her food sensors in different directions until she detects a new particle of food. Since she has no idea which direction might be most fruitful, a random search works just fine. As soon as she notices a distant whit of nutrition, her sensors light up, her flagella whip in harmony, and she scurries for her supper.

By making a small change to Archie's doers, while keeping everything else the same, Sally gained a powerful new skill: situational decision-making, a talent that enables salmonella to track down our intestinal cells and slip inside, triggering a nasty case of gastroenteritis.

3.

One of the chief problems that every organism must address is how to get the resources it needs to stay alive. Every living thing, no matter how enormous or tiny, must find *food*. Before the first minds came on the scene, the earliest microscopic critters probably dwelled in a nutritional Garden of Eden where food was abundant and easy to access. Perhaps they harvested minerals or electrons from the rocks they clung to, perhaps they floated around pools or oceans soaking up compounds dissolved in the water, perhaps they simply luxuriated in the sun's rays and converted solar energy into metabolic energy. For a long time, life would have been passive and uncompetitive.

As the food supply eventually began to shrink from endless generations of feeding or competition from ever-growing numbers of rivals, the first sensors and doers likely appeared, making it possible to *search* for food. The first mental revolution in the journey of Mind was probably a transformation from the sedentary lifestyle of tiny mindless organisms to a hunting-and-gathering lifestyle of tiny thinking organisms.

Once the first mind was able to discriminate between light and darkness, or food and toxins, the lazy days of Eden were over. The hunting-and-gathering lifestyle introduced new challenges for the fragile young minds, none bigger than the problem of choice. The capacity for targeted movement introduced a dilemma that all minds must constantly grapple with. *Should I keep exploring for more promising opportunities—or exploit what I've already found?*

This is the exploration dilemma.

A dilemma is a mental challenge where a perfect solution is not possible because there's an unavoidable trade-off. An organism confronting the exploration dilemma must decide whether to enjoy the resources previously obtained or spend resources on a quest for new resources—and risk finding none.

Our own species has come up with a variety of maxims that capture the uncertainty inherent to the exploration dilemma. "Fish or cut bait" is a common admonition in business, giving voice to the question of whether you should continue to invest money and time in an opportunity that hasn't paid off yet and may never pay off (that is, continue fishing)—or cut your losses and move on (that is, cut the bait off your line so you can pack up your fishing rod and go fish somewhere else). "A bird in the hand is worth two in the bush" provides concrete guidance for managing the exploration dilemma by suggesting one should consume a single serving of food rather than go hunting for two servings of food that you may not be able to acquire. The exploration dilemma is expressed in its most naked form in the ultimatum "Take it or leave it."

Archie's implicit solution to the exploration dilemma is to whip in the direction of the brightest source of light—that is, to *explore*—then stop whipping once he gets close enough to the light to soak up its nourishing rays—switching over to *exploiting*. Sally, in contrast, handles the exploration dilemma by shifting between two modes of behavior: *running* for exploration (moving toward a place where there might be food) and *tumbling* for exploitation (staying put upon reaching food so she can eat it).

The mind's solution to the exploration dilemma embodies a dynamic that distinguishes minds from other physical systems: decision-making. A mind makes choices. The ability to *choose* one course of action over another is the ultimate source of agency in the universe: unlike protons, rivers, or black holes, a mind can *decide* to explore rather than exploit—or to exploit rather than explore.

Sally illustrates the simplest biological dynamic of situational decision-making, and it's somewhat counterintuitive. Sally does not possess a "Coach Belichick molecule" that issues an authoritative command to run or tumble. There is no explicit site in Sally's mind where she "recognizes" there is no food nearby. We cannot identify a specific place in her mind where she "decides" to switch from running to tumbling nor a precise moment when such a decision occurs. This is not a matter of our having incomplete information about Sally's mind. We can identify *absolutely everything* happening inside Sally. There is simply no answer to the question "What part of Sally's mind is responsible for situational decisions?" because her decision-making process is holistic, involving all her thinking elements working together in unison.

As we continue onward, the environments of more advanced organisms will steadily expand in size, scope, and complexity, goading their minds to develop increasingly sophisticated tactics for contending with the exploration dilemma, not just in the pursuit of food, but in the pursuit of sexual partners, nesting spots, and entertainment, and in the avoidance of toxins, predators, and stress. At its heart, the exploration dilemma presents a profound philosophical quandary that speaks not only to the essence of the human condition but to the condition of all thinking beings.

Should we settle for *this*? Or risk it all in the hope for something better?

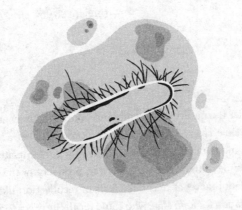

CHAPTER FOUR / **BACTERIA MIND II**

Memory

I've got a more graceful solution to the memory
problem. . . . I use habit and routine to make my life
possible.

—Leonard Shelby, *Memento*

1.

When we ponder the nature of remembrance, our mind may gravitate toward
fond memories, like a romantic trip with an amorous partner, a hard-fought victory, or a sweet moment with our child. Or we may contemplate *Jeopardy!*-style
facts, like the name of the sixteenth American president or the capital of Bahrain. Or we might marvel at celebrated feats of recollection like those of Kim
Peek, the inspiration for *Rain Man*, who could recall the exact contents of twelve
thousand books and the location of every ZIP code in the United States.

These examples can make us instinctively conceive of memory as a kind
of hard drive filled with files that we retrieve and store, somewhat like the
Pixar film *Inside Out*. In it, a girl's memories are visualized as glowing orbs
stacked tidily upon enormously long shelves. The blue orbs are sad memories.
Yellow orbs are happy memories. Green orbs are disgusting memories. The
widespread impression of human memory as a kind of cloud-based file storage from which we upload and download facts and experiences is undoubtedly influenced by our immersion in the Internet Age. But memory does not
work like this in any living creature.

Like all the mental innovations in the journey of Mind, memory first arose
to solve a specific challenge. We might call this challenge the gradient problem. On the face of it, the gradient problem doesn't seem to have anything to
do with memory. In fact, it naturally presents itself as a navigation problem.

When we examined Sally and her reaction to food, we presumed that the
stimuli her sensors detected were singular and discrete—that is to say, we presumed Sally's mind treated her environment as if it contained *one* particle of

VARIABLE CONCENTRATION OF FOOD

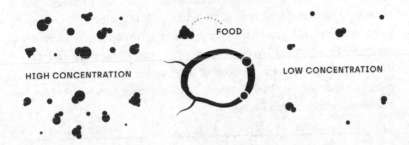

HIGH CONCENTRATION   FOOD   LOW CONCENTRATION

food at a time. In reality, bacteria are usually immersed within hazy clouds of food particles of ever-shifting densities.

Piloting through such a nebulous environment presents a novel stumbling block. If Sally attempted to navigate through a fog of food particles, she would end up twisting and careening about in an inefficient manner as each new movement of her body brought a different morsel near her sensors and triggered another reorientation. She would spend more time twirling and jiggling than eating.

One strategy for solving this navigational challenge would be to ignore the individual crumbs and instead attempt to determine where the greatest *concentration* of crumbs lies. This requires a major mental shift from thinking in terms of discrete points to thinking in terms of continuous gradations. Instead of searching for the nearest *particle*, in other words, search for the nearest *region* with a high density of particles. This is exactly what *E. coli* does.

Like salmonella, *Escherichia coli* is a rod-shaped bacterium important to both scientists and our intestines. Many strains of *E. coli* live symbiotically inside our guts, aiding with digestion and protecting our digestive tract from malicious bacteria. Regrettably, several strains of *E. coli* are pathogenic, which is why it occasionally moonlights as the villain in a national food recall. *E. coli* has long served as a kind of guinea pig for microbiologists. The bacterium can be cultured easily and inexpensively in a laboratory setting and is consequently one of the most studied organisms in science. Running and tumbling were first discovered and fully described in *Escherichia coli*.

Like salmonella, *E. coli* has sensor molecules that detect a variety of stimuli, including sugar, salinity, acidity, temperature, amino acids, and toxins. *E. coli* runs toward high densities of food, then tumbles when it arrives. To illustrate this, here is an actual path taken by an *E. coli* bacterium in a petri dish where the concentration of food steadily increases to the right. Though the bacterium's herky-jerky path is far from optimal—it is so minuscule that its body is constantly jostled by molecules, like bicycling in a blizzard—it reliably moves itself into the region with the highest available concentration of food.

E. COLI LOCOMOTION IN FOOD GRADIENT

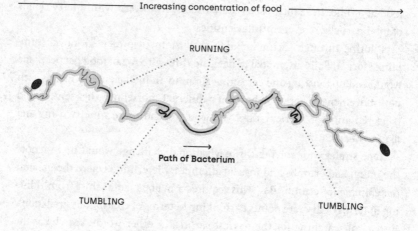

Adapted from Waite et al.

So how does a bacterium manage to steer itself through an ever-morphing cloud of nutrition? One method might be to compare the density of edibles at two different locations on *E. coli*'s body. That's what larger single-celled organisms such as amoebas do. They compare what their sensors detect at one end of their body to what their sensors detect at the other end and calculate which side senses the higher concentration.

But *E. coli* are too small to perform this trick. Amoebas are one thousand times larger than *E. coli*. There's simply not enough distance between the front and rear tips of *E. coli* to effectively assess the relative density of food. Fortunately, another approach is available to the bacterium, one that makes use of an indirect source of information.

E. coli can harness time.

Instead of comparing food concentrations at two different *locations* on its body at the *same* moment in time, a bacterium compares concentrations at the *same* location on its body at two different *moments* in time. Thus, a spatial problem becomes a temporal problem. If a bacterium detects more food nearby than it did previously, that means it's moving into a higher concentration. Great! Keep going! If it detects less food than previously, that means it's moving into a lower concentration. Oops! Better stop!

To evaluate food gradients using this time-based approach, an *E. coli* mind must somehow store information about the amount of food it detected in the recent past.

In other words, it must form a memory.

2.

Please exchange greetings with Eska the *E. coli*:

ESKA

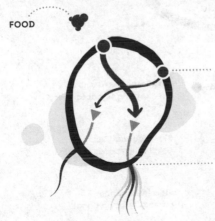

FOOD

Each sensor responds to the presence of food by sending a signal to its doer

Each doer responds to a sensory signal by whipping its flagellum

Eska has the same four-element configuration as Sally. Her two sensors respond to food. Her two doers *whip* when activated by a sensor signal and *twitch* with no signal. Thus, Eska *runs* toward food and *tumbles* when no food's nearby. So far, so familiar.

But Eska trots out her own upgrade, another molecular tweak. Until now, all the sensors we've encountered exhibit a response proportional to the strength of the stimulus. Strong light evokes a strong signal, weak light evokes a weak signal. But this is not how Eska's sensors behave. After they encounter a stimulus—like food—they steadily lose their responsiveness. They send a weaker and weaker signal to their doer like the fading light from a dying ember. This is known as *habituation*.

When Eska's sensors detect food, they react the same way that Archie's and Sally's sensors did. They fire off a signal to their flagellum proportional to the strength of the stimulus. Lots of food = strong signal. But here's where the difference lies. The moment Eska's sensor fires, it starts reducing its responsiveness to food.

ESKA'S SENSOR

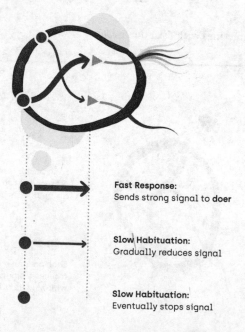

Fast Response:
Sends strong signal to **doer**

Slow Habituation:
Gradually reduces signal

Slow Habituation:
Eventually stops signal

Even in the presence of a stable concentration of food, the sensor's signal slowly diminishes until it stops firing completely. After Eska's sensors habituate—after they stop responding to food—they need time to "recharge" before the same concentration of food will activate them again.

This minor modification to a bacterium's sensors certainly doesn't sound like the basis of an effective memory system. Where is the storing and retrieving of a memory? More to the point, how does habituation help Eska determine where the highest concentration of food lies?

Let's jog through the three basic scenarios that Eska might encounter in the pursuit of food:

Scenario #1: There's the same amount of food ahead and behind.
Scenario #2: There's more food ahead.
Scenario #3: There's less food ahead.

ESKA MOVING IN EQUILIBRIUM

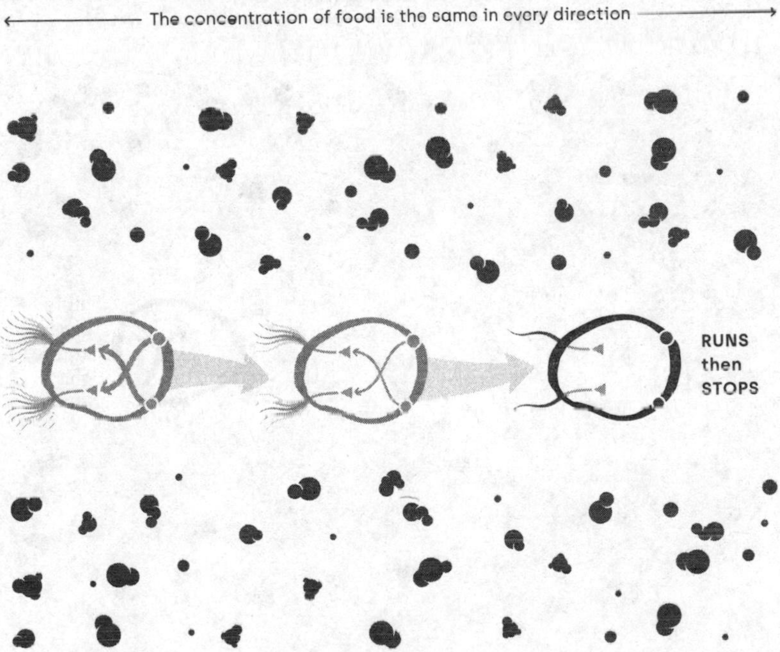

←——————— The concentration of food is the same in every direction ———————→

RUNS
then
STOPS

First, let's drop Eska in an environment where food particles are evenly distributed in every direction, a state of affairs that scientists label "equilibrium." What does Eska do?

At first, Eska will start zooming forward because her sensors recognize there's abundant food ahead, triggering the maximum response in her flagella. But her sensors immediately start habituating. Because the concentration of food remains steady as Eska advances, her sensors gradually lose their sensitivity. They stop sending signals to her doers. Soon, Eska comes to a halt. This makes good sense. If the level of food is the same everywhere, there's no point wasting energy. Just park yourself in the middle of the buffet.

What if we drop Eska into an environment with a higher concentration of food ahead?

Once again, Eska immediately starts running because she senses plenty

ESKA MOVING TOWARD HIGHER CONCENTRATION

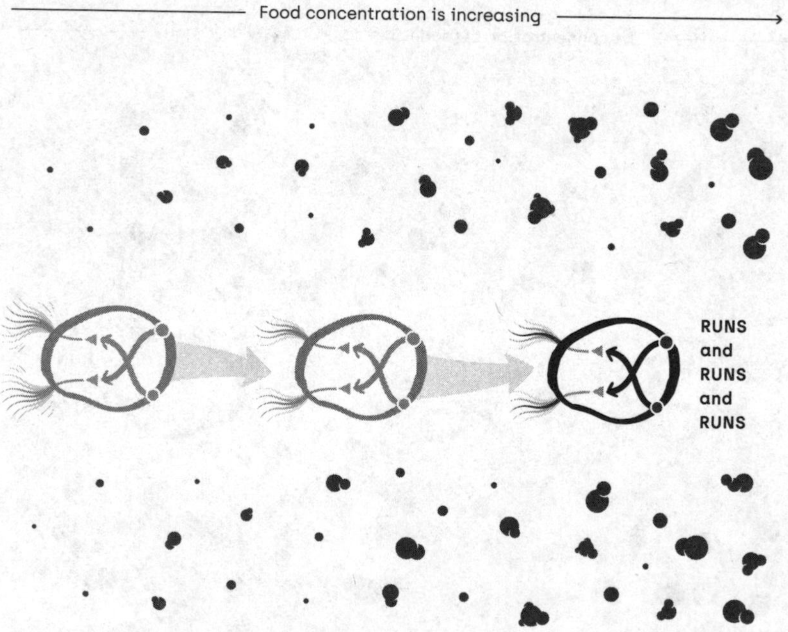

Food concentration is increasing →

RUNS
and
RUNS
and
RUNS

of food ahead. And once again, her sensors start habituating. But this time the concentration steadily increases as Eska moves forward. The intensity of the food stimulus will overcome the sensors' diminished responsiveness and trigger another run.

As long as Eska travels within an increasing concentration of food, she will continue to dash along. When she finally enters a region where the concentration of food stops increasing, she halts. Eska follows the density of food to its peak, then plops herself down to feast.

Finally, what if we drop Eska in an environment with a lower concentration of food ahead?

Now Eska's behavior is similar to the equilibrium scenario. Eska immediately starts running, since she detects food ahead. She also starts habituating. The next moment she samples the food, her sensors are less responsive. But

---

ESKA MOVING TOWARD LOWER CONCENTRATION

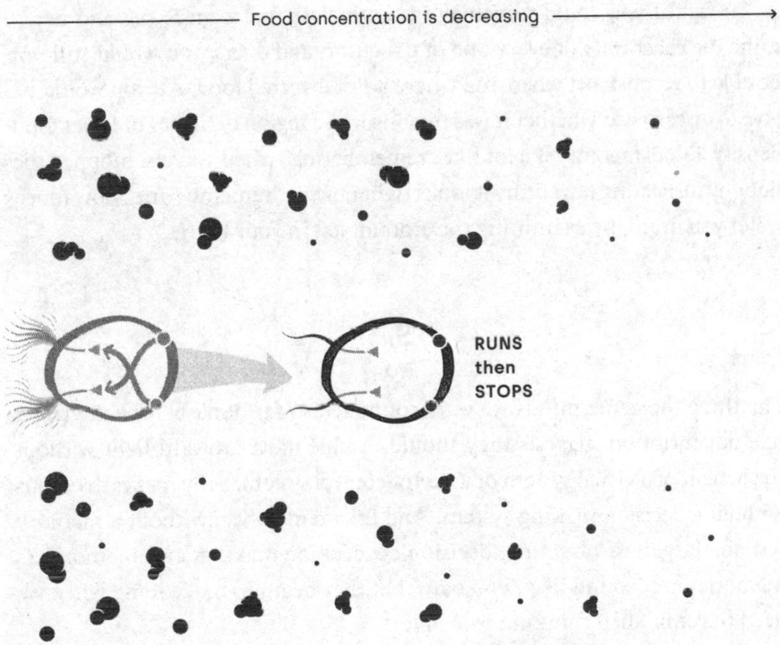

Food concentration is decreasing →

RUNS
then
STOPS

she also detects less food than before, because the concentration is decreasing. That means her sensors will reduce their responsiveness even more quickly than in the equilibrium case. Eska hits the brakes the instant she realizes she is sliding off the gravy train.

Habituation enables Eska to crack the gradient problem. (Habituation is also an essential dynamic in most forms of thinking in our own minds, because it ensures that the mind never gets stuck in a mental loop. Without habituation occurring at the molecular level of our neurons, we wouldn't be able to recognize our friends, distinguish a potato from a tomato, feel passion or fury, plan out our day, or read a book.)

The pipsqueak mind of a bacterium boasts nature's simplest memory system. But it is a queer sort of memory. If you searched for the precise spot in an *E. coli* where its "memory" of food is stored, your search would be for naught. Nor would you find the location where the bacterium encodes the location of food, where it encodes the food gradient, or even where it encodes *when* it last noticed food. Trying to extract a specific memory from a bacterium is like trying to extract a reflection of the moon from the surface of a lake.

In fact, if you could somehow instantly flash-freeze an *E. coli* and determine the exact state of every one of its sensors and doers, you would still not be able to reconstruct when and where it last detected food. All you would be able to discern was whether it was moving into a region of higher or lower food density. Eska's memory is a lot like "remembering" when you last mopped the floor by inspecting how filthy it looks right now, or "remembering" how much water you drank by examining the amount left in your bottle.

3.

The three molecule minds we've encountered so far don't behave anything like our intuition suggests they should. Archie moves toward light without the benefit of a visual system or a motor-control system. Sally makes decisions without a decision-making system. And Eska remembers without a memory system. Targetless targeting, decisionless decision-making, and memoryless memories may sound like Zen koans, but they begin to make sense when we heed the embodied thinking principle.

If we want to understand how a mind thinks, the embodied thinking principle advises us to look beyond the "brain" of an organism—that is, beyond its thinking elements. We must consider an organism's body and environment, too. (We'll use "brain," with quotes, to refer to the collection of *all* thinking elements in an organism, which may or may not include neurons.) The embodied thinking principle's definition of a mind is a simple formula: mind = "brain" + body + environment. All three components matter if we hope to unravel any particular form of thinking.

Eska does, in fact, store memories in her *mind*—just not in her *"brain."* Eska's memories of food are distributed across the real-time dynamics between her physical body, her thinking elements, and the surrounding cloud of food particles, in the same way that a picnic encompasses a band of hungry gourmands, a basket of lunch, and a pleasant outdoor location. People, food, and a place are all necessary for a picnic, and a "brain," body, and environment are all necessary for a memory in an *E. coli* mind. This unusual way of thinking about thinking will become especially relevant when we tackle the Big Three later in the book.

Once we grow comfortable with the embodied thinking principle, we can begin to appreciate an interesting fact about the journey of Mind: how *easy* it seems to have been to develop complex mental faculties, in comparison with the onerous process of developing the metabolic and self-replicating faculties of life. Consider: Eska's absurdly simple four-element mind can perform no fewer than *five* distinct forms of thinking:

Sensation
Decision-making
Navigation
Motor control
Memory

That's a lot of intellectual bang for your buck. It's also worth observing that the dynamics of these five forms of thinking are tangled together, making it impossible to cleanly separate the act of navigation, say, from the act of sensation. (This mental overlap is another reason that minds are not computers.) We'll return to the entangled nature of thinking when we explore vertebrate minds, including human minds.

This is a good opportunity to pause and reflect upon the difference between the origins of life and the origins of Mind. Though we may never know the exact configuration of Earth's first mind or even its precise purpose—searching for light? sulfur? methane? amino acids?—the simple molecule minds we've encountered so far give us some measure of confidence that the genesis of the first minds wasn't nearly as difficult as the genesis of the first bodies. To this day, scientists remain stumped regarding the elements, mechanisms, and developmental pathway of the earliest specimens of life. The story of thinking is clearer. For thinking to emerge, all that needed to happen was for a mindless organism—one already endowed with the wildly complicated biochemistry of life—to evolve a single molecular sensor and a single molecular doer, and the royal road to consciousness beckoned.

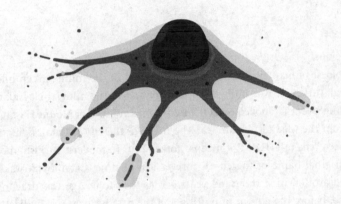

CHAPTER FIVE / **AMOEBA MIND**

Communication

> The whole quilt is much more important than any
> single square.
>
> —*A Fine Balance*, Rohinton Mistry

1.

The oldest human city yet discovered is buried in the rolling green hills of central Turkey. It was built upon marshy and fertile alluvial clay along a winding river that looked up at the twin-coned volcano of Mount Hasan, not far from the Mediterranean Sea. The name of the city is Çatalhöyük. Pronounced "cha-tall-HUIK," it means "forked hill" in modern Turkish, though the original name of the city is forever lost to time. Çatalhöyük was first settled around nine thousand years ago as the Stone Age was drawing to a close, before the advent of writing and the first Bronze Age civilizations. It's much older than the cities made famous by the Bible, such as Nineveh, Memphis, Babylon, and Ur. Çatalhöyük was inhabited for more than one thousand years. Nobody knows why it was abandoned. The city was founded shortly after the development of agriculture in the Middle East and archaeologists believe Çatalhöyük is representative of our species' first attempts at living together in large numbers. What is so striking about humankind's prehistoric foray into city building is its configuration.

Çatalhöyük contained no roads. This basic amenity of civilization had not yet been invented. Nor did the city contain any sidewalks or connecting footpaths of any kind. Instead, the city was a dense warren of boxy domiciles.

It appears as if the city may have begun when someone built a squarish house out of mudbricks not far from a plot of farmland. The house was carefully covered over with plaster, inside and out, so that it resembled a smooth white cube. (Even the floors were plastered smooth.) Then, someone else built a nearly identical house right next to it. *Right* next to it, even using one of the first house's walls as one of its own. Then someone built another white boxy house pressing against the first two. And another, and another, like a child haphazardly fastening Lego blocks together. More clay boxes were added until a tight-knit maze of homes pressed together like a bone-white cubic honeycomb.

The homes were interconnected by a Swiss-cheese-like perforation of holes and doors. There were holes in the ceilings for the hearth smoke to escape from, but also squared-off timber ladders reaching through ceiling holes onto the roof. Some ceiling holes served as front doors. Other holes were used as windows or as a means to cross into adjacent houses. There were also doorways leading from one room to another, often at the top of steep stairs. The whole affair resembled a prehistoric version of chutes and ladders.

If someone who lived in the center of Çatalhöyük wanted to reach the periphery, they had two choices. They could climb up onto the clay roof and walk along other people's roofs until they reached the edge of the city. Or they could clamber through other people's living rooms. (There was likely little expectation of privacy among a people who were one generation removed from a nomadic lifestyle of camping in the open.)

Çatalhöyük offered the ultimate in egalitarian design. No house appears to have been much bigger or fancier than any other, suggesting there was not yet any distinction between aristocrats and commoners. There were no temples, marketplaces, or public spaces. (There were "history houses" that contained artistic stacks of horned bulls' heads resembling plastered rib cages, human skulls buried in the floor, and ochre paintings depicting hunts and celebrations, but even these hallowed spaces were the same size and design as other houses.) Men and women appear to have been afforded largely equal roles, both genders sharing the tasks of farming, hunting, cooking, ritual, and art-making, a consequence, perhaps, of the nomadic lifestyle that settlers came from, which required everyone to share their labor and resources. All were welcome in Çatalhöyük, apparently.

For centuries, the population continued to grow, until it peaked somewhere around twenty thousand people. The city's steady expansion pressured its inhabitants to find new ways to manage movement, socializing, and public activities. After all, people living in the interior of the city couldn't be expected to wriggle through dozens of neighbors' homes whenever they wanted to reach the surrounding forests and small farms. The solution to this challenge surfaced naturally from the collective geometry of the city: the rooftops of Çatalhöyük were converted into footpaths and avenues.

Public activities began to take place on the ceilings of houses, such as trading, toolmaking, cooking in communal ovens, and mass gatherings in open plazas. Down below, families engaged in domestic activities. Up above, the public engaged in civil, commercial, and possibly religious activities. Despite this odd and inconvenient configuration, for more than a millennium as humankind was groping its way toward civilization, the two-tiered residential latticework of Çatalhöyük may have been the mightiest city on Earth.

Yet, a metropolis it was not. This Stone Age "city" was a long way off from the urban municipalities that would eventually rise across the Middle East, such as Ur, a Sumerian city-state boasting massive palaces and towering ziggurats. Or Nineveh, with its formidable towers and colossal statues of winged lions and bulls with human heads. Or Babylon, with its legendary hanging gardens and the commanding blue-tiled Ishtar Gate. Or Athens, high atop the Acropolis with its lofty marble architecture. Çatalhöyük's early commitment to a streetless design may have precluded it from developing much further, and perhaps served as a warning to future urban planners who wished to erect cities of lasting splendor and prosperity.

The journey of a city holds many intriguing parallels with the journey of Mind, beginning with a curious correspondence between the first human cities and the first multicellular minds.

2.

Whatever you do, don't call it slime mold. Biologists will shake their heads dismissively at your inappropriate idiom. After all, *Dictyostelium discoideum* has nothing to do with fungi. *Dictyostelium* is a single-celled blobby organism invisible to the naked eye. At least, when traveling solo. When a crowd of them parade together they resemble strings of yellow snot, the source of the "slime" in "slime mold." The term "social amoeba" is an approved moniker for *Dictyostelium* and underscores why this soil-dwelling microbe is the heroine of Chapter 5: the amoeba is the first creature on our itinerary who exhibits a vibrant and trendsetting social life.

The amoeba represents an impressive leap forward in all three components of mind: "brain," body, and environment. An amoeba can grow one thousand times larger in volume than a bacterium. If a bacterium is a tugboat, then an amoeba is a battleship. *Dictyostelium* contains a nucleus, which is absent from the cells of bacteria and archaea. The presence of a nucleus renders the amoeba a *eukaryote*, kin to all animals, plants, and fungi. Compared to bacteria, the amoeba has more numerous and more sophisticated physiological processes for biological tasks like energy production, waste management, and reproduction. Eukaryotes in general possess highly structured interiors, compared to the more random organization of prokaryote innards. The more complex and extensive body of the amoeba spurs the demand for a more complex and extensive "brain."

A *Dictyostelium* can spend its entire life living independently as a free agent. At least, while there's food around. When the vittles run low, a change comes over this solitary microorganism. Its mind shifts its attention to focus on the existential problem of imminent starvation, devoting all its mental resources toward a new purpose: *seek out other hungry comrades and yoke up with them to build an escape pod.*

When starvation stalks the land, amoebas come together to work on a massive civil engineering project. This collectivist endeavor ensures the survival of amoebafolk during hard times—and kindles one of the most consequential inflection points in the history of thinking.

To execute a large-scale construction project, two essential pieces of information must be coordinated among all would-be workers. First, everyone needs to know *when* to get together. It won't do any good if one worker shows up today, another tomorrow, and another next week. Second, everyone needs to know the location of said site. *Where* should the eager crew report with their tongs and hammers?

An authoritarian regime solves the *when* and *where* problems by assigning one person—the Supreme Leader, for instance—the power to dictate the time and place, and granting them the responsibility for delivering this information to everyone. But amoebas can't exercise this kind of centralized top-down communication. First of all, amoebas are individualists. They lack a social structure, so there's no Supreme Amoeba and no obvious way to appoint one. Instead, amoebas more closely resemble a pure-of-heart communist party. Every amoeba is equal and (unlike in *Animal Farm*) no amoeba is more equal than others. Second, because amoebas are nomads who roam freely, they're spread out across the landscape in uncertain locations. Even if there were a Supreme Amoeba, she couldn't deliver a message to everyone because she wouldn't know where they were residing.

This brings us to the pivotal challenge the amoeba mind must crack. We'll call it the coordination problem. The coordination problem will arise again and again in the journey of Mind, and its solution always foments revolution. The amoeba version of the coordination problem is:

> How can independent minds come together to cooperate on a single shared purpose?

3.

The amoeba mind features a number of innovations. Most prominently, it is outfitted with significantly enhanced motor control. The three previous molecule minds relied on flagella for locomotion. The amoeba, in contrast, uses pseudopods. The amoeba oozes along by extending a gelatinous blob in the direction it wants to go, then flows its body into the gooey protrusion.

The amoeba mind uses another novel mechanism to evaluate food gradients. *E. coli* guided itself toward the highest food concentration by comparing the density of particles at the same place on its body at two different moments in time. Because the amoeba is much larger, it can assess the food gradient by comparing the density at two places on its body at the same time.

Even though the mental machinery governing pseudopod locomotion and food-gradient analysis is fascinating (to a neuroscientist, at least), this book will not get into its details. All the minds we will encounter henceforth on our venture are endowed with an abundance of interesting and fruitful mental innovations that we will, for the most part, respectfully ignore. Instead, the chapters ahead will stay focused on the concepts and principles that illuminate the mind's journey and help us plumb the mysteries of the Big Three.

With this mission squarely in front of us, we can treat the amoeba's navigation system as functionally identical to Sally the salmonella's. This simplification is possible because even though the amoeba slouches around its neighborhood on pseudopods (rather than whipping its way forward with flagella) and evaluates food concentrations spatially (rather than temporally), its overall method of navigation functions in the same manner as a bacterium's. Instead of running and tumbling, we might say the amoeba crawls and jiggles—creeping along a food gradient, and jiggling in place when it reaches a peak concentration of food. Biologists characterize this gradient-focused method of navigation, shared by archaea, bacteria, and amocbas, as *chemotaxis*.

Witness Meera, the chemotactic amoeba:

MEERA

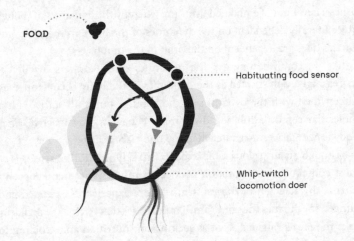

FOOD

Habituating food sensor

Whip-twitch
locomotion doer

When tackling the coordination problem, the question for Meera becomes: What new thinking elements must she add to her navigation system to enable her to hook up with other famished comrades? The answer: she needs two.

A new kind of sensor. And a new kind of doer.

4.

Meera's new sensor is groundbreaking not because of *how* it senses but because of *where* it senses. Instead of being aimed outward toward the hustling, bustling world, it is aimed *inward*. It monitors Meera's level of *hunger*.

For the first time in our journey, a mind takes its input not from the *external* state of the environment, but from the *internal* state of its body. Put simply, it's the first time the mind is aware of its own self.

This is certainly not consciousness, mind you. At least, not the charmed sort of self-awareness that humans celebrate and endlessly ruminate over. Meera's piddling "self-awareness" is at the same vestigial level as a thermostat that checks its own temperature setting to decide whether to turn the heat on. Nevertheless, in the slow ascendance from mindlessness to sentience, this is a watershed moment: the first time that details about the physiological processes of the body feed into the mental processes of the "brain." This establishes a new kind of embodied thinking whereby the dynamics of thought are quite literally dependent on the dynamics of the body, in the same way that Earth's tides are dependent on the orbit of the moon.

As we continue onward, the dynamics of self-awareness will become increasingly sophisticated as the dynamics of the body become increasingly intertwined with the dynamics of the "brain," but it all starts with a tiny molecular sensor exhibiting dirt-simple behavior. Stay inactive when my body is *Not Starving*. Activate when my body is *Starving*.

Though an auspicious achievement, this self-monitoring sensor is *not* the most consequential innovation in Meera's mind. What matters even more is what this new sensor *triggers*. Until Meera experiences extended pangs of hunger, she pursues the same familiar purposes that Archie the archaeon and Eska the *E. coli* pursued, such as seeking out nutrition and avoiding toxins. But when her inner sensor detects starvation (the absence of critical nutrients), it sends out an alarm signal that awakens a dormant sensor and doer.

The awakened doer is not concerned with locomotion. It functions in an unprecedented role: as a *transmitter*. This transmitter doer drops "messages" behind Meera as she crawls along. Each message is an organic molecule, known as cAMP in the lingo of molecular biology. These messages are not complex communiqués advising "Let's all meet up at Abe's house tomorrow morning!" Instead, they're more like graffiti declaring "I was here!"

Meanwhile, Meera's awakened sensor functions as a *receiver*—as a message-detecting antenna. The new sensor forms part of a new navigation system that searches for message molecules in the same manner that Eska the *E. coli* searched for food. Meera doesn't look for the *nearest* message. She seeks

MEERA'S INTERNAL SENSOR AND "MESSAGE TRANSMITTER" DOER

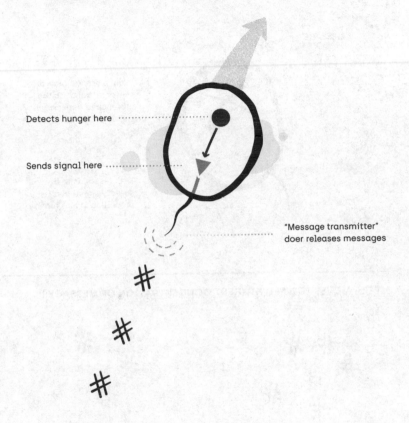

Detects hunger here

Sends signal here

"Message transmitter"
doer releases messages

the greatest *concentration* of messages. It's like trying to find where a campus party is being held by following the densest trail of empty Solo cups. The highest concentration of "I was here" messages likely indicates the direction of the highest concentration of hungry amoebas, too.

This ultrasimple communication system—a message-detecting receiver sensor and a message-dispatching transmitter doer—solves the coordination problem. The communication system enables a community of starving amoebas to determine *when* and *where* they should hook up and get to work. The *when* is determined by the appearance of famine. When there are empty

MEERA'S MESSAGE-SEEKING SYSTEM

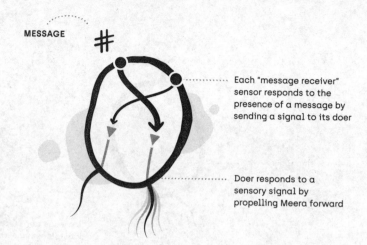

MESSAGE

Each "message receiver" sensor responds to the presence of a message by sending a signal to its doer

Doer responds to a sensory signal by propelling Meera forward

MEERA MOVES TOWARD HIGHEST CONCENTRATION OF MESSAGES

MESSAGE

AMOEBAS MERGING

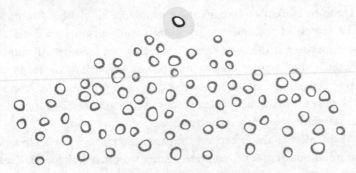

The first hungry amoeba begins transmitting messages.

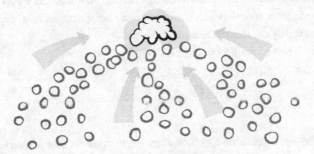

As other hungry amoebas begin searching for messages, they tend to notice the first amoeba's messages and follow them, creating the first clump. This creates more message trails for other amoebas to follow.

The amoebas converge on the growing clump.

pantries everywhere, many amoebas will independently but simultaneously begin to search one another out.

The *where* is arbitrary and unknown in advance but reliably determined by the process of multiple hungry amoebas all pursuing messages and leaving messages at the same time. Eventually each amoeba will stumble upon the trail of another amoeba and the amoebas will follow each other in ever-tightening spirals until they converge upon the same spot. This site will be ringed with the highest density of messages and thereby attract other amoebas with ever-increasing effectiveness.

What if Meera is the first to feel the pangs of hunger? She will randomly crawl around looking for not-yet-present hungry comrades, all the while leaving a trail of messages behind her. If she remains the sole hungry amoeba in the neighborhood, then, alas and alack, she's likely doomed to waste away. But if other amoebas begin to starve, either she will stumble upon one of their message trails or they will stumble upon one of hers, each following the other until they collide in a protean embrace.

That's when something fascinating happens.

5.

When the amoebas come together, they begin construction of an elaborate engineering project: an escape-pod launch pad. The slimy conglomeration of amoebas reconfigures itself into a tall, spindly stalk capped by a large, sticky ball of spores. With luck, this ball will cling to a passing insect or animal and hitch a ride to greener pastures. In order to erect this launch pad, all the individual amoeba minds must merge together into one. They must form an amoeba "supermind."

Instead of pursuing its own self-interested purpose, each amoeba commits to selflessly collaborating in the pursuit of a shared super-purpose: securing the survival of their offspring. The following figure illustrates the life cycle of the amoeba supermind.

Each amoeba that joins the supermind—as well as the supermind itself—is a martyr. All the individual minds that meld together willingly sacrifice themselves for the common good to ensure the propagation of their bloodline—or, more accurately, their protoplasm line.

AMOEBAS BECOMING SUPERMIND

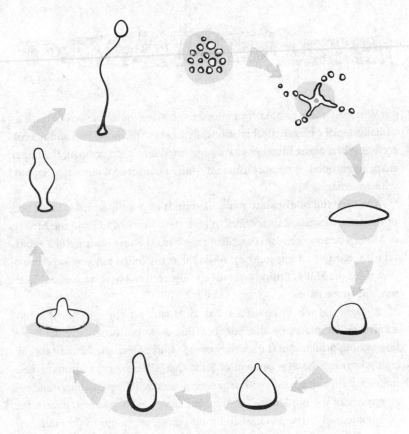

There's a curious if coincidental connection between amoebas and Çatalhöyük artwork. One of the most prevalent forms of art that archaeologists have discovered in the ancient homes are geometric patterns painted in bright red on the smooth white walls. Some of these murals resemble cells packed together, like slime mold observed under a microscope. In these patterns, each cell is slightly different in size and shape and the interior of each cell contains what looks like a large nucleus surrounded by smaller organelles. The most famous of these murals is shown on the next page.

ÇATALHÖYÜK MURAL

Whatever their original intention, these arresting Stone Age paintings visually depict a hierarchical relationship between the individual and the collective. When many little boxes are aggregated into a single form, the larger shape has properties notably different from its constituent elements. So, too, with amoebas.

The merging of disparate molecule minds into a supermind marks one of the most significant milestones in both the history of life and the history of Mind. The amoeba superorganism represents the first instance of a multicellular creature on our journey, while the many-celled amoeba supermind prefigures the kinds of minds we will encounter in the next stage of our journey, the *neuron minds*.

It might not seem like slime mold could have anything to do with consciousness, language, or the Self, but this oozy protozoan exemplifies a far-reaching principle of the journey of the Mind, a deep mathematical principle governing the development of Mind that will not only illuminate how individual minds function within a supermind; it will help us understand the emergence of the Big Three. The same principle also happens to govern the development of cities. As cities and minds grow, they tend to generate new stages of hierarchical organization. It was adaptive for hundreds and eventually thousands of people to come together and reside in one shared location at Çatalhöyük for protection, social support, economies of scale, and trade. Uniting their modular domiciles offered a notable advantage compared to cobbling together temporary shelters in the wilderness. But eventually, this grand union of mudbrick abodes required a new level of organization to suit the needs of the burgeoning city. Çatalhöyük faced an urban coordination problem.

The problem was solved by the creation of pathways and large public spaces atop the city's roofs. The amoeba pursued a similar solution. By devising an intermind communication system that solved its own coordination problem, an amoeba mind operating at one level of thinking (using *molecular* thinking elements) established a higher level of thinking that consisted of the interactions of *cellular* thinking elements.

The transition to a higher stage of organization in both minds and cities is driven by the same underlying principle, a muscular engine of development that we might fittingly call the *metropolis principle*.

Stage II

Neuron Minds

The Metropolis Principle

This remarkable combination of increasing benefits to the individual with systematic increasing benefits for the collective as city size increases is the underlying driving force for the continued explosion of urbanization across the planet.

—*Scale*, Geoffrey West

1.

One hundred thousand years ago, as glaciers lolled across the Earth like frozen tongues, Paleolithic humans huddled together in basalt caves on a sunny peninsula on the west coast of the Asian subcontinent. These prehistoric hunter-gatherers may have been attracted to the peninsula because of its relative warmth and the abundance of edible marine life in the enveloping waters of the Indian Ocean. Their meager settlement, perhaps consisting of bands of a few dozen people sleeping in caves or under stone outcrops, would not have qualified as a village. It was hardly more than a campsite. But over the next hundred millennia, this Ice Age campsite would grow into one of the most influential metropolises in the world.

As the Ice Age receded, the oceans rose. The peninsula became an archipelago. The first identifiable indigenous group to inhabit the archipelago was a tribe of fishermen known as the Koli, who established seasonal fishing communities on the marshy islands. In the fifth century BCE, Buddhist monks arrived on the mainland coast and took up residence in the caves. The Buddhist culture in the region reached its peak in the third century BCE during the reign of Ashoka the Great, a convert to Buddhism who ruled almost the entirety of the Indian subcontinent. The local Buddhist monks crafted massive and beautifully elaborate stone sculptures within the coastal caves polished to the smoot$ess of glass. The swampy, malarial islands became a mostly disregarded offshoot of the mainland Buddhist town.

Over the next 1,500 years, the town and its affairs were controlled by a fast-changing succession of indigenous Indian dynasties that ruled the mainland and left little behind, other than the names of current residents of the region, such as More (Maurya dynasty), Cholke (Chalukya dynasty), Shelar (Silahara dynasty), and Jadhav (Yadava dynasty). The islands themselves, which were subject to heavy seasonal monsoons and flooding, remained largely uninhabited and neglected. When Marco Polo's fleet traveled up the west coast of India in 1292, he ignored the islands completely, instead docking at the nearby river port of Thane. Then, in the late thirteenth century, the Hindu raja Bhimadeva fell in love with the ocean view on the island of Mahim. He made Mahim the capital of his small kingdom, building a royal palace and several houses for his retinue and planting coconut groves. When he settled on Mahim, the only site of interest on the archipelago was an ancient shrine to the Koli Earth Mother goddess, who was now known as Mumbadevi.

The total population of the island town never rose much higher than a few thousand people until the early sixteenth century, when the first Europeans took command of the sleepy archipelago. Dom Francisco, counselor to King Jo$ II of Portugal, dropped anchor off the most prominent of the seven islands in 1509 and dubbed its deep natural harbor "Good Bay"—or in his native Portuguese, "Bom Bahia."

In 1661, the island town's fortunes began to change. During the previous two thousand years, the small population and rudimentary level of development of the archipelago never qualified it as a hub of culture, industry, or trade. Then, from 1661 to 1675, the population increased sixfold, from ten thousand inhabitants to sixty thousand. The catalyst of this abrupt expansion was the marriage of Charles II of England to Catherine of Portugal in 1661, when the archipelago was transferred to the British as part of Catherine's dowry. Charles thought the boggy and storm-battered islands were worthless, so a few years later he transferred them to the English East India Company, the wealthiest business in the world, for a paltry rent of ten pounds a year.

Unlike Charles II, the company appreciated the strategic placement of the islands as an oceanic gateway to India. In 1686, the East India Company decided to make the island municipality the headquarters of all its operations in India—its largest overseas branch, by far—and renamed the entire archipelago Bombay after the Portuguese name for its most prominent island. The island town had become a British-owned corporation.

The company got busy. It constructed a quay, warehouses, a customhouse, a mint, a printing press, fortifications, and even more fortifications. The company built up a large fleet of ships known as the Bombay Marine that eliminated piracy along the western coast of India. (Almost three centuries later, the Bombay Marine would become the Indian Navy.) In short order, Bombay was trading in coconuts, elephant teeth, ivory, lead, salt, broadcloth, betel nuts, rice, and sword blades. Indian merchants poured into Bombay, including families of every et$icity and social class. From 1744 to 1764, the population doubled from 70,000 to 140,000. Bombay had become a bona fide city.

Managing a city spread out across seven islands proved an onerous task. So in 1784, the East India Company united the urban archipelago into a single peninsula by digging up the hills of Bombay and dumping the earth and rubble into the ocean between the islands. Bombay's development, already proceeding at a brisk pace, now exploded. High schools, hospitals, a cotton exchange, and a rush of banks opened. Transportation improved by leaps and bounds as carriage roads, steamer lines, trams, and three causeways came online. Bombay entrepreneurs began exporting opium to China for tremendous profits. The Indian railway built its very first station in Bombay. The population surged to eight hundred thousand in 1864—larger than every American city save New York—and kept multiplying. India's industrial revolution was centered in Bombay, which quickly developed one of the planet's most dominant textile industries after the American Civil War shut down cotton exports from the American South. The Bombay Stock Exchange, the oldest in Asia, commenced trading in 1875. In 1911, the population exceeded one million. Bombay had become a global metropolis.

After India gained its independence from Britain in 1947, the population of Bombay shot to three million, surpassing Rome, São Paulo, and Hong Kong. In 1995, as Indians increasingly rejected symbols of their colonial past, the city was renamed Mumbai after the Earth goddess that the native Koli fisherfolk once worshipped. In 2020, Mumbai's population passed twenty million, good for seventh in the world, putting it neck and neck with Beijing and Mexico City.

Today, Mumbai is India's most populous city and its undisputed financial, commercial, and entertainment capital. Eight major religions claim throngs of devotees in Mumbai. Eighteen languages are regularly spoken in its shops and maidans (esplanades). The city even enjoys its own argot, Bambaiyya, a linguistic farrago of Marathi, Gujarati, Urdu, Konkani, and Hindi. Downtown Mumbai is indistinguishable from any other twenty-first-century megacity, its streets lined with Starbucks, H&M, Zara, Coach, and Pizza Hut franchises from the West; Samsung and Miniso from the East; and home-grown Fabindia and Tanishq chains. Its sprawling skyline is a picket fence of soaring skyscrapers. Textiles and the seaport remain leading industries, but they have been augmented by global industries in finance, engineering, jewelry, health care, leather, information tec$ology, and show business. Bollywood, a portmanteau of "Bombay" and "Hollywood," makes more movies and attracts larger audiences than Los Angeles. "Billionaire's Row," a neighborhood located on the former island of Bombay, contains some of the most exorbitant real estate in the world, including the world's most expensive private residence.

What drove this seemingly inexorable march of urban development? What mathematical principle enabled a sleepy fishing hamlet to grow into a thundering megalopolis despite a never-ending torrent of physical and social obstacles?

2.

The metropolis principle consists of a simple dynamic operating across the hierarchical levels of a city. The development of a city overall facilitates the development of its lower levels. The development of its lower levels facilitates the development of the city overall. This loop of mutual improvement is the primary engine of development in both cities and minds.

Çatalhöyük's union of prehistoric households provided residents with better tools, more cultural artifacts, and a greater variety of foods than if they had lived outside the city. These superior products facilitated further improvements to the city, which helped generate even more tools and foods. The prehistoric denizens of Mumbai inhabited preexisting caves with little physical modification. Later, ancient Buddhists appear to have erected the first permanent structures on the islands, out of timber. In the Middle Ages, many buildings used translucent seashells as window glass. Wooden buildings gave way to tiled roofs and stone and lime mortar in the 1600s, followed by bricks and plaster under the British occupation. In the nineteenth century, Victorian Gothic–influenced architecture sprang up, featuring exuberant domes made of yellow Malad stone, vaulted roofs, monkey gargoyles, and arched gateways forged out of corrugated iron, steel, and concrete. These were replaced at the beginning of the twentieth century by cement sculpted into rounded Art Deco shapes inflected with tropical waves and radiant sunbursts. Today, buildings in Mumbai are made of high-tech materials, including polymer composites, aerogels, organic cellular insulators, and self-healing fiber-reinforced cementitious matrix. As the low-level materials of construction improved, they enabled the fabrication of ever more sturdy and useful buildings, which in turn enhanced the prosperity and productivity of the entire city, which in turn generated the knowledge and resources to invent or acquire even better building materials. The mutual loop of development kept rolling along.

The same bootstrapping relationship operates between levels in neuron minds: the molecular mechanisms within neurons grow more sophisticated as the neurons containing these mechanisms grow more sophisticated, and the neurons themselves become more sophisticated as the neuron minds they inhabit become more sophisticated overall.

We shouldn't overlook the fact that most cities, and Mumbai in particular, suffer from squalor, poverty, and homelessness. For millions of Bombayites it is a daily struggle to get by, even as the city stacks up dizzying wealth. There is no mathematical principle that ensures that the people at the top of the social hierarchy will feel any obligation to help the people at the bottom. What the metropolis principle does suggest, however, is that if Mumbai (or any city) can find a way to improve the lives of its neediest, then that city will likely become even more prosperous as a result.

The metropolis principle produces two physical consequences that shape the development of both cities and minds. As a city or mind develops, it tends to lead to a coordination problem that can be solved only by establishing a new layer of *connectivity* on top of the existing layers. The medieval kingdom of Bhimadeva united the Bombay archipelago for the first time as a single political unit. The East India Company united Bombay economically through trade and transformed the medieval island town into a mercantile city. The British also filled in the ocean separating the islands and united the city into a single peninsula, erasing all geographic divisions between the islands. In 1822, Asia's oldest continuously published newspaper, the *Bom-*

bay Samachar, began its run, initiating mass media and uniting Bombay-ites within a common perception of events in their city. After that, new communication tec$ologies followed one another in rapid succession, each binding the population ever more tightly together: brick semaphore towers, telegraphs, telephones, fax machines, dial-up modems, cell phones, the internet, Wi-Fi, 5G.

The amoeba, too, established a new level of mental connectivity on the back of its own proprietary intermind communication system.

Each time a city or mind establishes a new form of connectivity, it inaugurates a new stage of social organization that improves its overall *adaptiveness*: the capacity of a city or mind to dictate its own fate. Practical examples of adaptiveness include increased stability in the face of chaos, enhanced flexibility in the face of challenges, and greater capacity for self-repair and self-improvement. Once Bombay was united through landfill, roads, and telegraphs, a building fire in the middle of one of Bombay's (former) islands could be put out by firemen from a citywide fire brigade racing to the conflagration on a bamba, a horse-drawn steam engine. Once a citywide fire brigade was in place, all of Bombay benefited from the resulting increase in safety and security, driving further cycles of development and increased adaptiveness.

Neuron minds arose from the union of molecule minds and thereby attained greater adaptiveness. Neuron minds are more stable, more flexible, more resilient, and more intelligent than molecule minds, and thus better able to defend themselves against the ceaseless onslaught of unpredictable physical chaos. The metropolis principle makes a clear (and falsifiable) prediction about what we should find at the climax of our journey: that the most sophisticated mind on Earth (whatever it turns out to be) should also possess the most sophisticated thinking elements at its lower levels, including the most sophisticated neurons and the most sophisticated molecular mechanisms.

There is a question we might ask about cities that has interesting implications for how we should think about the journey of Mind. Is there a reasonable basis for considering modern Mumbai the same "city" as the prehistoric Koli fishing settlements, ancient Buddhist town, and British corporate municipality? The metropolis principle suggests there is. What is sustained over time is not any particular physical element of a metropolis, but rather its underlying engine of adaptiveness—the reciprocal loops of development operating between part and whole. It is the game of city-building that defines a city, rather than its players.

Even though nothing remains in Mumbai of the original Koli fishing settlements (other than the name Mumbai and the common surname "Koli"), and even though the physical substance of Mumbai has radically changed over the centuries, there is an unbroken line of development connecting twenty-first-century Mumbai with its Paleolithic settlers. By the same reasoning, there is a seamless continuum of development linking the human mind to microbe minds that were living and thinking three billion years ago. The game is still the game, but as the game evolves, the players evolve, too.

That's how the metropolis principle powered Mumbai's journey from fishing hamlet to megacity—and the mind's journey from cognitive campsite to mental metropolis.

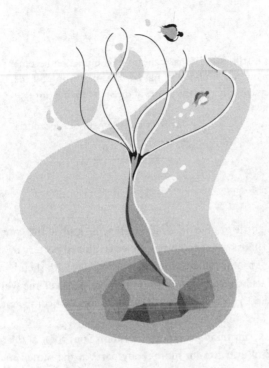

CHAPTER SIX / **HYDRA MIND**

Multitasking

It is still asked, if many things in conjunction become many causes of one thing. For the men who pull the oars together are the causes of the ship moving through the water.

—*The Stromata*, Clement of Alexandria

1.

Most of us have little trouble walking and chewing gum at the same time. For more than six thousand years, humans have fashioned pottery by spinning a wheel with their foot while molding the twirling clay with their hands. These days, many of us hold a conversation over Zoom or FaceTime while preparing dinner. We take for granted one of the most familiar competencies of the human mind: our ability to *multitask*.

To multitask, our mind persues two (or more) purposes at the same time involving control over two (or more) body parts at the same time, without either activity interfering with the performance of the other. Performing simultaneous activities feels effortless to us. Yet none of the molecule minds we've encountered so far can multitask.

Archaea, bacteria, and amoebas occupy simple, homogeneous bodies that resemble pills, eggs, balls, or blobs. They do not possess anatomy that can operate independently from the rest of their body. The activity of a single doer (such as the whipping of a flagellum or the stretching out of a pseudopod) will inevitably and significantly influence the activity of the rest of their body. Consequently, the molecule minds of Archie, Sally, Eska, and Meera can pursue only one active purpose in their environment at a time—they can hurry toward food *or* scurry away from toxins, but not both.

The situation changes dramatically when an organism develops specialized anatomical parts. If you have a finger, you need to be able to roll it around without causing your entire body to roll around, too. If you have a tail, you need to be able to loop it around a branch without dropping the meal clutched between your paws. The mind of any multilimbed creature must be able to control two different pieces of its anatomy at the same time, which means that it must be capable of pursuing multiple objectives simultaneously. Multitasking is the most basic mental challenge confronting organisms with neuron minds and multicellular bodies. Not surprisingly, the simplest neuron mind on our journey came up with an innovation to address it.

Hydra vulgaris is a freshwater invertebrate that resembles a feather duster roughly the size of a child's eyelash. It features an unfussy body plan: as one scientist put it, the hydra is "a gut with tentacles." It has a thin elastic hollow stalk (its "gut") that sprouts several rubbery appendages. Its stalk anchors the hydra to underwater rocks as its tentacles wave around to catch even tinier varmints known as water fleas. *Hydra vulgaris* swallows its prey through its mouth, an orifice located at the top of its stalk in the center of its tentacles.

The hydra is a member of the phylum Cnidaria (a phylum is a group of related animals) that contains eleven thousand species of jellyfish, anemone, and coral. Cnidaria and the similar phylum Ctenophora represent the two oldest branches of life to possess neurons, the primary thinking element of neuron minds. A neuron is a specialized cell, with all the same internal structures as other cells, including protoplasm, organelles, and a nucleus. The set of all neurons in a neuron mind forms the mind's "brain." (The scare quotes are still necessary, because in many neuron minds, including *Hydra vulgaris*, there is not an actual brain organ but rather a decentralized web of neurons.)

There are sensor neurons and doer neurons, which play the same roles as sensors and doers in molecule minds. Each neuron is composed of molecular thinking elements, including molecular doers (which release neurotransmitters into a synapse, for instance) and molecular sensors (which detect the voltage on the neuron membrane, for instance). Functionally, *every neuron is a self-contained molecule mind.*

The provenance of the first neurons is hotly disputed. Neurons may have begun as secretory cells, or skin cells, or electrical signaling cells, or maybe all of these in parallel, or maybe something else entirely. Nor is it clear whether Cnidaria or Ctenophora developed neurons first, whether both groups developed neurons independently, or whether neurons arose in an ancestor common to both (the leading candidate being the sponges).

The neurons in the hydra mind are the most primitive we'll encounter on our journey. Unlike most neurons in the animal kingdom, the hydra's neurons are "nonpolarized": they can receive an *incoming* signal through any connection with another neuron and they transmit their *outgoing* signals through *all* their neuron-to-neuron connections. (More advanced "polarized" neurons, in contrast, transmit signals on a single dedicated outgoing connection.)

Neural signals are faster, more reliable, and more targeted than molecular signals and can travel for longer distances. That makes neuron thinking more robust and versatile than molecule thinking, just as telephone communications are more robust and versatile than semaphore flag-waving communication. As in molecule minds and, indeed, all minds, thinking in a neuron mind consists of the holistic real-time interactions of all its thinking elements as it converts sensory inputs into behavioral outputs.

2.

Since the hydra's tentacles, stalk, and mouth each perform different functions, they require a mind capable of simultaneously managing tentacle thinking, stalk thinking, and mouth thinking. The hydra must be able to wave its tentacles, for instance, without waving its stalk. It must be able to open and close its mouth without disrupting its tentacles. And its stalk must be able to elongate or contract without interfering with its mouth's ability to swallow prey.

The body of a *Hydra vulgaris* can grow one hundred *million* times larger in volume than the body of a *Dictyostelium* amoeba. This increase in size means that the environment of neuron minds has expanded in scope, too. Macro objects like pebbles, leaves, and seeds become useful resources. Single-celled organisms, especially bacteria and archaea, are constantly buffeted by random thermal noise (Brownian motion). When they attempt to move, such as by whipping their flagella, they inevitably end up following an unsteady path in the desired direction as a flurry of molecules ping against them and knock them off course. But at the physical scale of multicellular creatures—including *Hydra vulgaris*—Brownian motion is no longer a factor and doers can execute purposeful actions with greater reliability.

Novel body parts (such as tentacles) are required to cope with the opportunities and dangers of the larger-scale environment. When a mind's body and environment change so drastically, its "brain" must change drastically, too.

To see neuron thinking in action, please allow this book to present to you Dr. Tentacle:

DR. TENTACLE

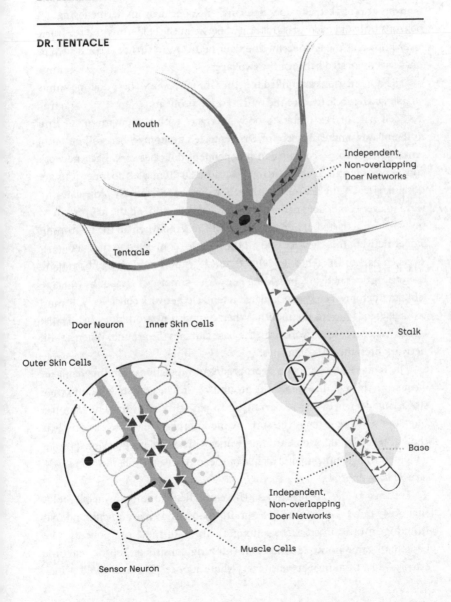

Dr. Tentacle has one of the simplest nervous systems in nature. It consists solely of sensors and doers arranged in two layers inside the thin gelatinous skin that makes up the entirety of her body. The sensor neurons are located inside an environment-facing outer layer known as the epidermis. Sensor neurons are connected to doer neurons. The doer neurons (termed ganglion neurons by hydra scientists) are located between the epidermis and the inner layer known as the endoderm. The doer neurons, in turn, are connected to muscle cells located between the two layers.

The doer neurons are linked together into *networks*. These doer networks enable Dr. Tentacle to solve the multitasking problem.

Each part of Dr. Tentacle's body interacts with her environment in a different way, and each piece of Dr. Tentacle's anatomy—her stalk, mouth, tentacles, and base—contains an independent doer network. Each network consists of interlinked doer neurons. Because Dr. Tentacle's doer neurons are nonpolarized, if any neuron is activated, its signal will quickly propagate outward to every other doer neuron it is connected to until all the neurons in its network are activated. If one fires, they all fire. And when all the neurons in a doer network fire, they activate all the muscles connected to that network.

Each part of Dr. Tentacle's body (tentacle, mouth, stalk, base) has an all-or-nothing doer network, providing her with a simple yet elegant way to control her anatomy. For instance, when a tentacle network collectively fires, its tentacle bends toward the mouth. When a mouth network fires, the mouth opens. When a base network fires, the base clutches the ground. When a stalk network fires, the stalk elongates.

Dr. Tentacle also performs appropriate counteractions: bending tentacles *away* from the mouth, *releasing* the ground, *constricting* the stalk. Many opposite-acting doer networks are arranged in pairs in the hydra. For example, one doer network extends the stalk, while another doer network constricts the stalk. Pairing up two neuron networks with complementary purposes is a vivid example of the final principle of the journey of Mind, the *complementary thinking principle*.

Let's recap all four principles. The embodied thinking principle holds that every mind is composed of a "brain," body, and environment, and that thinking involves interactions among all three of these components. The basketball game principle holds that thinking consists of holistic real-time *activity*—that thinking happens "everywhere at once." The metropolis principle holds that as a mind becomes adaptive overall, its lower levels also become more adaptive, and vice versa. The metropolis principle also holds that as a mind becomes more adaptive overall, it naturally leads to the establishment of a new level of thinking that unifies the thinking elements of the previous level—more simply, the journey of Mind tends to produce an ever-ascending hierarchy of thinking elements.

Finally, the complementary thinking principle holds that one of the most common ways that minds grow smarter—more adaptive—is by joining together two distinct mental dynamics, each designed for a distinct purpose, thereby establishing a new, unified dynamic that can pursue a new, common purpose.

Every mind makes use of complementary thinking, even the humble archaeon, which is powered by the simplest instance of complementary thinking in nature: the complementary pairing of a sensor (purpose: respond to light) and doer (purpose: propulsion) to achieve a third, joint purpose (seek out the light). Complementary thinking also undergirds more sophisticated mental activity, including vision, hearing, object recognition, searching for a target, planning, consciousness, symbolism, language, and love. We will delve more deeply into complementary thinking in future chapters, but for now, observe that by integrating two networks pursuing two distinct purposes (such as a stalk-elongating network and a stalk-constricting network) a mind can achieve a third, higher purpose through their intertwined activity (such as the ability to perform peristalsis, wave-like motions of the gut that support digestion).

3.

There are no sensor networks in Dr. Tentacle. Every sensor in Dr. Tentacle is connected to a doer rather than another sensor. If a sensor neuron is triggered by a stimulus, then the entire doer network that the sensor is connected to will fire. Thus, a single sensor can trigger the movement of an entire body part. However, the very simplicity of this mechanism means that Dr. Tentacle is not capable of perception, the recognition of *patterns* of multiple sensory inputs. She cannot detect sensory patterns or even discern where, exactly, on her body a particular sensation occurred. Dr. Tentacle's sensory system is like a fire alarm in an office building. When you hear the alarm clang you know there's a fire *somewhere* but not the place where the fire began. (Though you know exactly what to do: get out!)

So how does this combination of a sensor linked to a network of doers solve the multitasking problem? To find out, let's inspect how Dr. Tentacle opens her mouth.

Dr. Tentacle's mouth is encircled by a ring of sensor neurons. Each of these sensors sends a signal to a doer that is part of a doer network that also encircles the mouth. These doers are linked to muscles that radiate outward from the center of the orifice like a sunburst (the muscle cells are known as radial myonemes). If one of the sensors around the mouth is activated by touch—for example, if a tentacle pushes a piece of food against the mouth—then that sensor fires and the entire mouth doer network springs into action.

The activated sensor sends a signal to its doer. The doer immediately activates the neighbors on either side of it, which in turn activate their neighbors, inciting a wave of activation moving in both directions around the mouth, away from the activating sensor. As the doer activation wave travels around the mouth, it triggers the mouth muscles. This causes the entire ring of radial muscles to contract, ripping Dr. Tentacle's mouth open like a dilating pupil. Her process of mouth-opening occurs without disrupting the bending of her tentacles, the clutching of her base, or the elongation of her stalk. This compartmentalization of purpose is multitasking.

In an actual *Hydra vulgaris*, the mouth consists of the same "skin" or epithelial cells as the rest of the hydra's body. Strangely, when a hydra's mouth closes, it seals itself seamlessly, like the scene in *The Matrix* where Neo's mouth gets sewed shut by the Agents. However, even though there is a smooth and undivided closure, there are a few central cells that will always separate like a sphincter when the mouth opens again.

One important takeaway from the hydra's mental innovation of doer networks is that, as in molecule minds, there is no central controller or "decider" in the hydra mind. There's no dedicated circuit or dedicated neuron that makes the decision to open the mouth or elongate the stalk or bend a tentacle. Instead, the "decision" to act is initiated by any sensor neuron that happens to detect a response-worthy stimulus in the environment. Nor are there any sensory representations, value representations, or motor commands. The mind of a hydra maintains much in common with bacteria's decentral-

DR. TENTACLE OPENING HER MOUTH

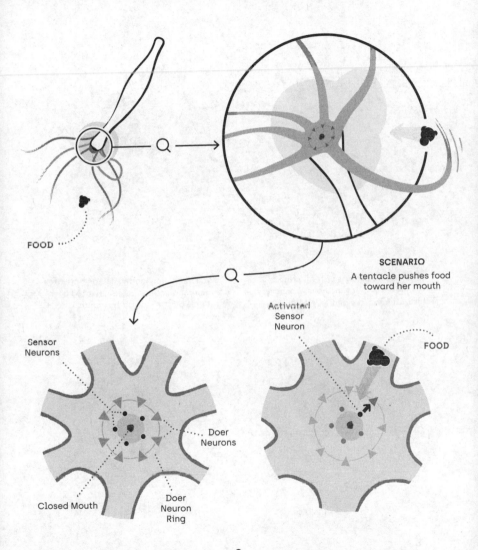

FOOD

SCENARIO
A tentacle pushes food
toward her mouth

Activated
Sensor
Neuron

FOOD

Sensor
Neurons

Doer
Neurons

Closed Mouth

Doer
Neuron
Ring

1.
Her mouth is surrounded by
sensor and **doer neurons.**

2.
A **sensor neuron** is activated by contact
with food. The **sensor neuron** signals its
doer, which stretches out to open its
side of the mouth.

DR. TENTACLE OPENING HER MOUTH

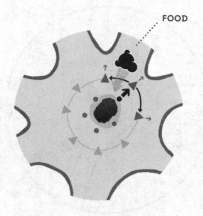
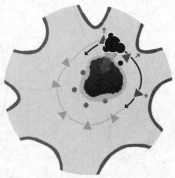

FOOD

3.
Each **doer neuron** is part of a ring of bidirectional doer neurons. Whenever one is activated, it signals its neighbors . . .

4.
. . . and each neighboring neuron activates its nearest neighbor. Each activated doer stretches out its side of the mouth.

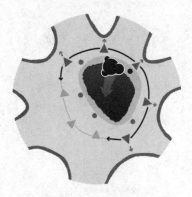

5.
Eventually the signals pass around the entire circumference of the ring and meet on the opposite side.

6.
All the **doer neurons** are activated and stretch out to pull open the mouth.

ized approach to decision-making, where the "decision" to run or tumble is not made by an appointed decision-making circuit but occurs implicitly as a result of holistic activity across the entire bacteria mind—across the entire trinity of "brain," body, and environment.

A mind without a genuine brain suffers from a significant limitation, one that explains why we don't find hydra-style minds in animals of much greater complexity than a jellyfish. Without some kind of centralized control over its doer networks, a hydra mind cannot learn how to *coordinate* the activity of its different body parts. The hydra can hard-wire certain complex behaviors, like having a tentacle push food toward its mouth and then have the mouth automatically open to eat it. But such a mind could never learn to use a pottery wheel. The hydra can *execute* multiple purposes at once, but it cannot *coordinate* multiple purposes at once.

Despite its talent for multitasking, the hydra represents a dead end in the journey of Mind. To advance any further, a neuron mind must find a way to coordinate its doer networks.

Centralization

Think left and think right and think low and think high.
Oh, the THINKS you can think up if only you try!"
—*Oh, the Thinks You Can Think!*, Dr. Seuss

1.

The 1920 novel *The Story of Doctor Dolittle* introduced a four-legged oddity that was, as Dr. Dolittle carefully ascertained, a cross between a gazelle and a unicorn. It was graced with an even more distinctive feature, however. Two heads. One head faced forward, as with any species of deer. The other faced directly backward. This creature, described as "the rarest animal of the African jungles," was known as the Pushmi-Pullyu.

With heads aimed in opposite directions and presumably engaged in a continuous tug-of-war, how did the Pushmi-Pullyu manage to get anywhere? How did it decide what to do, when two brains were hitched together? Though the Dolittle books are silent on this critical point, we can imagine one possible solution. Whichever head at any given moment has the stronger motivation to move seizes control and moves the body as it pleases.

Of course, there could be trouble if the other head continues to put up a fight. So let's add another wrinkle to our speculative Pushmi-Pullyu nervous system. The losing head always goes to sleep. In other words, the more motivated head wins control of the body while the other head gently snoozes. But as soon as the first head's motivation ebbs—perhaps it has arrived at a coveted trough of oats and finishes munching—then the other head wakes up and they compete for control once again.

This Pushmi-Pullyu form of thinking helps elucidate the next innovation in the journey of Mind: the first dedicated decision-making circuit.

2.

This chapter shines a spotlight on one of Earth's greatest survivors: the ignoble roundworm. Not the earthworm, mind you, a relatively youthful wriggler in the variegated history of worm evolution. The subject of this chapter is the nematode, *Caenorhabditis elegans*. An itty-bitty critter that was slithering around the ocean floor long before animals colonized dry land. Today, roundworms thrive in every habitat on Earth, including forests, oceans, tundra, mountains, and occasionally our own bodies. About 80 percent of all individual animals on Earth today are roundworms. They've survived at least three mass extinction events, including the one that vaporized the dinosaurs. A few roundworms riding on the Space Shuttle for a science experiment even survived the fiery *Columbia* crash.

C. elegans is one of the most studied organisms in biology. As a result, we know quite a bit about its body, environment, and "brain." In temperate climates, *C. elegans* dwells in the soil. It feeds on the bacteria in decaying vegetation, especially rotten fruit. (It considers the running-tumbling *E. coli* a delicacy.) The roundworm may justly be hailed as the emperor of dirt. A million roundworms can dwell in one square meter of soil.

C. elegans is about a millimeter long. That's smaller than all but the most diminutive of hydras. On first glance, the roundworm may appear to be a mental step backward from the hydra. Not only is *C. elegans*'s body smaller, it's simpler, too. As a matter of fact, it lacks any limbs at all. The hydra is "a gut with tentacles." The roundworm is a gut, period.

Each *C. elegans* hermaphrodite possesses exactly 959 cells, no more, no less, a fixed cell count that is exceptionally rare in nature. While a full-grown hydra has thousands of neurons, sometimes more than 6,000, *C. elegans* has precisely 302. The total number of synapses (the connections between neurons) in an entire *C. elegans*, about seven thousand, is roughly one-quarter of the number of synapses on a single pyramidal neuron in your cortex.

Yet, even though the roundworm has a simpler body and much smaller "brain" than the hydra, the roundworm is more mentally advanced. Its austere mind illustrates the two main ways that neuron minds become more adaptive and intelligent.

The first is through *neural diversification*, the creation of new types of neurons. Though textbooks often present the neuron as having a standard form—a blobby cell body with a single outgoing connection (an axon) and many incoming connections (dendrites)—scientists have identified thousands of different versions of neurons across the animal kingdom.

Neurons vary by shape, size, metabolism, embryonic development, method of sensing, method of signaling, method of responding, and pattern of myelin sheathing. Neurons can be shaped like stars, cones, spheres, spindles, threads, pyramids, or polyhedrons. There are amygdala spiny neurons, olfactory tufted neurons, and Calleja dwarf neurons. There are Purkinje neurons, PreBötzinger neurons, and Zoidberg neurons. Some of the simplest neurons in nature are found inside a wasp, the vanishingly small *Dicopomorpha echmepterygis*, an insect that is *less than half the length of an amoeba*. Its miniaturized, stripped-down neurons do not possess nuclei.

While the hydra mind contains twelve distinct types of neurons, *C. elegans* possesses at least seventy. As one example, the roundworm has a sophisticated smell-sensing neuron that can detect more than 1,700 different chemical compounds using a broad array of new molecular thinking elements. Neural diversification is one major way that neuron minds get smarter. The other way is *network complexification*: wiring neurons together into new configurations that generate new mental dynamics.

Network complexification produced the roundworm's innovative solution for the challenge of decision-making in complex environments. Making choices in the face of uncertainty is the *raison d'être* for every mind, whether pollywog, bumblebee, roundworm, or you. It's the very definition of the exploration dilemma. Bacteria decide whether to run or tumble. Amoebas decide whether to graze or escape. Hydras decide whether to swallow or vomit. These minds all make their decisions by relying upon a reflexive response to a stimulus, little different from our "decision" to kick our foot when a doctor taps her rubber hammer against our knee. There is no dedicated decision-making circuit in these minds, nor a need for one. But things are different in the roundworm.

The roundworm manages its movement using two doer networks. One network governs forward motion. The other, backward motion. This pairing of forward and backward networks is another example of the complementary thinking principle. Complementary thinking, remember, enables two networks to pursue their own individual purposes while simultaneously cooperating on a common purpose. In the case of the roundworm's complementary doer networks, this common purpose is *locomotion through soil*.

The hydra's doer networks were activated by a simple trigger: a signal from a sensor neuron. But the roundworm faces challenges requiring a more sophisticated approach to doer activation. *C. elegans* dwells within a world of soil and decaying vegetation that it must burrow through with effort. To adapt to this increased environmental complexity, the roundworm mind evaluates scents, tastes, touch, vibrations, gravity, and temperature with a wider range of sensor neurons than the hydra (neural diversification). But this expanded menagerie of sensors creates new problems, too.

The roundworm cannot react to each new stimulus with a reflexive response the way a hydra does, because forcing your way through soil requires substantially more energy than curling an underwater tentacle. A worm would burn calories nonstop if it reacted to every whiff of odor or vague vibration with new activity. Instead, a worm must carefully evaluate *all* the sensory information impinging upon it at any given moment and decide whether it is better to continue moving in the same direction or to reverse course and head somewhere else. The roundworm must weigh the available evidence before deciding whether to explore or exploit.

The mental innovation that solves this evidence-weighing problem combines a first-of-its-kind neuron with a first-of-its-kind network configuration.

3.

Ladies and gentlemen, the Duke of Dirt.

All the "brains" in the journey so far have been composed of two kinds of thinking elements: *sensors* and *doers*. The roundworm now adds a third: *thinkers*.

A thinker always sits between sensors and doers. Its job is to convert sensations into actions. It receives signals from sensors (or other thinkers) and sends signals to doers (or other thinkers). A thinker does not have direct contact with the outside world or the internal body. It knows of the world only vicariously, through its sensor and doer brethren.

The Duke of Dirt possesses two thinker neurons. One activates the doer network for forward locomotion. The other activates the doer network for

THE DUKE OF DIRT

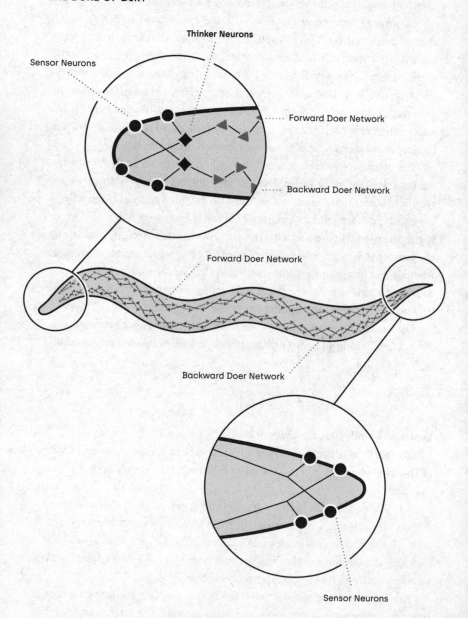

backward locomotion. Each thinker receives inputs from a variety of sensors located in the front and back tips of the worm. Each thinker neuron weighs the available environmental input provided by Duke's sensor neurons, then converts its evaluation into a signal that it sends to its doer network. This signal is proportional to the strength of the sensory evidence. For example, if the smell sensors in the front tip of Duke detect a strong scent of tasty bacteria ahead, the forward-locomotion thinker sends a strong signal to the forward-locomotion doer network. *Come on, let's hit the buffet!* If the toxin sensors sniff a faint hint of poison ahead, the backward-locomotion thinker sends a weak signal to the backward-locomotion network. *Maybe let's back off a bit . . .*

These dueling directives also highlight the chief difficulty presented by the evidence-weighing problem: What if the worm smells tasty bacteria *and* dangerous toxins ahead? More generally, what if there is conflicting evidence that motivates *both* forward and backward thinkers to activate their doers? How does Duke resolve these contradictory urges?

By acting like the Pushmi-Pullyu.

Duke's thinkers are configured in an intriguing way. They are linked to each other through reciprocal connections. Each thinker sends a signal that *inhibits* the activity of the other. If I'm on, I turn you off. If you're on, you turn me off. The stronger the evidence motivating one of the thinkers, the

---

WINNER-TAKE-ALL

Each **thinker neuron** sends out an inhibitory signal that attempts to shut off the other neuron. The stronger signal wins.

1.   2.   3.

WINNER-TAKE-ALL IN THE DUKE OF DIRT

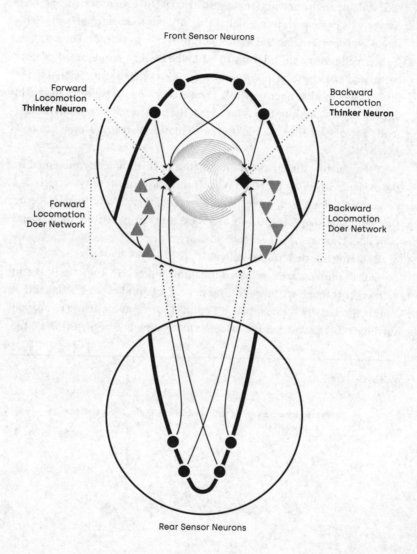

Front Sensor Neurons

Forward Locomotion **Thinker Neuron**

Backward Locomotion **Thinker Neuron**

Forward Locomotion Doer Network

Backward Locomotion Doer Network

Rear Sensor Neurons

stronger its inhibition of the other. Thus, the two thinkers are engaged in a constant tug-of-war. Whichever thinker has the stronger motivation—in physical terms, whichever thinker receives greater total signal strength from its sensors—will win out and completely suppress its opposite, like one head of the Pushmi-Pullyu taking control and putting the other head to sleep. This novel mental dynamic is known as *winner take all*.

The winner-take-all dynamic plays a ubiquitous role in more-advanced minds. In the roundworm, it enables the mind to weigh multiple sources of information about the environment simultaneously—temperature, salinity, moisture, smells, texture, vibrations—and discern whether the conglomeration of attractive stimuli is more compelling than the conglomeration of aversive stimuli. This enhanced sensitivity allows a roundworm to better address the exploration dilemma by enabling it to make smart choices in complex situations that it has never encountered before.

What if both thinkers have equal levels of motivation? Will it result in stalemate? Never in practice, because even the tiniest edge in motivation will swiftly get amplified into an insurmountable advantage through the inhibition of the less motivated thinker.

After one thinker wins, how does the winner-take-all configuration reset itself and start over again? Just like the Pushmi-Pullyu: as soon as the sensory evidence motivating the winning thinker begins to wane, its inhibitory signal lessens, releasing its grip on the losing thinker, especially if the losing thinker suddenly gets motivated by a burst of new sensory evidence. (Habituation of the winning thinker plays a role, too. Many people find cheesecake delicious, but no matter how tasty the first bite, after fifty bites you'll be eager to taste something new.)

Together, the roundworm's introduction of thinker neurons (neural diversification) and the winner-take-all dynamic (network complexification) was an indispensable step toward the emergence of consciousness, language, and the Self. There is still no "decider neuron" in the roundworm mind. But there is, for the first time, a decider *circuit*. Duke's thinker duo is like a pair of co-mayors engaged in endless bouts of arm wrestling, where the winner of each round gets to decide what the city does next.

By inserting a decider circuit between its two locomotion networks, the roundworm managed to escape the mental dead end that stymied further development of the decentralized mind of the hydra. Not only are thinker neurons in a winner-take-all configuration capable of deciding which doer network to activate, they hold the potential for coordinating the simultaneous activity of multiple doer networks.

Simply put, the roundworm mind is the first to achieve centralized control of thinking.

CHAPTER EIGHT / **FLATWORM MIND**

Perception

Life would be a ding-a-derry
If I only had a brain.

—Scarecrow, *Wizard of Oz*

1.

What is the difference between a dot and a circle? Between a noise and a sigh? Between the scent of ethanol and the scent of Chanel No. 5? Each of these pairs illustrates the difference between *sensing* and *perceiving*. And, as it turns out, each illustrates the difference between brainlessness and a brain.

None of the minds on our journey so far have been capable of perception. In fact, it would not be an exaggeration to claim that all previous minds experience every sensation in the same way. Though molecule minds and the two preceding neuron minds possessed light sensors and chemical sensors and temperature sensors and vibration sensors, in these minds there is no difference in the mental dynamics of "seeing," "smelling," "touching," "tasting," and "hearing." They're playing the same sensing game with every form of sensation.

In these unsophisticated minds, sensation consists of a sensor detecting a stimulus and sending a signal directly to a doer or thinker. These sensory signals communicate simple information: the presence and perhaps the intensity of a stimulus. But consider what's missing. There's no attempt at *combining* multiple sensory signals to form a sensory *pattern*.

Archie the archaeon's sensors respond to the intensity of light, but his mind forms no conception of edges or surfaces or gradations or shapes. If Archie were a painter, his visual oeuvre would be limited to all-black canvases flecked with a single white dot of variable brightness. Sally the salmonella's sensors respond to the presence of food particles, but nowhere in her mind is there a specific pattern of activity quantifying the concentration of food. The notion of a food gradient is implicit and inferred. Neither does her mind form a conception of a flavor profile, such as *tangy but sweet*. Instead, Sally's mind treats "tasted" molecules as edible or toxic—or ignorable. Dr. Tentacle's sensors respond to touch, but her sensors do not collaborate to determine whether she is being poked with a Q-tip or a pencil. She cannot even be sure of where, exactly, she was touched. Either something touched her or something did not touch her.

The roundworm marks a crucial transition in the mind's sensory dynamics. *C. elegans* possesses sophisticated smell-sensor neurons capable of evaluating multiple odors simultaneously (such as alcohol, benzaldehyde, and butanone). These neurons can discriminate scent *patterns*.

A sensory pattern is a collection of multiple sensory stimuli, like many individual pixels forming a digital image. Formally, we can define perception as the ability to distinguish sensory patterns and to initiate behavior based upon a particular pattern. By this definition, the roundworm is still not capable of true perception. The detection of a particular sensory pattern does not trigger a pattern-specific behavior in a roundworm mind. Though a roundworm's sensor neuron can distinguish between mosaics of odors, each neuron still converts a given sensory mosaic into a simple one-dimensional signal ("There's a strong attractive smell here!") rather than outputting a scent pattern ("This dirt is redolent of roses and lavender with a delightful hint of putrefaction!"). A roundworm olfactory neuron is a molecule mind that can "perceive" scent patterns, but the roundworm's neuron mind cannot.

There are enormous benefits awaiting any mind that can learn to exploit patterns in its physical environment. This is not easy to achieve. Each class of physical stimulus demands its own distinct sensory dynamics. Volatile chemical compounds floating in the air (smell) behave differently than compounds interacting with a wet surface (taste). The collisions of physical objects with one another (touch) behave differently than the internal vibrations of physical objects (temperature). Photons (sight) behave differently than acoustic vibrations (hearing). That is, after all, why a camera and a microphone are not interchangeable devices. Cameras have specialized circuitry for processing visual patterns. Microphones have *different* circuitry for processing audio patterns.

To exploit the rich structure of patterns inherent to physical reality, a mind must develop a variety of new dynamics, each tailored to the properties of a distinct class of sensory inputs, including odors, visuals, tastes, touches, and sounds. This is the perception problem.

Solving the perception problem requires that, for the first time, the body reconfigure itself to suit the "brain" rather than the "brain" reconfiguring itself to suit the body.

2.

Dugesia japonica has a soft, flat, brownish wedge-shaped body with a blunt, triangular head. *D. japonica* is about a centimeter long, much bigger than a roundworm, which means it is the first organism on our journey that is easily visible to the naked human eye. Flatworms resemble leeches or slugs when they are creeping along vegetation or writhing on the ground, a creature only a naturalist could love. Its mouth is located on the underside of its body. *D. japonica* does not possess a circulatory or respiratory system, instead breathing directly through its skin.

On casual inspection, the flatworm does not appear to be much of an upgrade from the roundworm. It's larger and wider, but squirms along without limbs, like any worm. Even so, *D. japonica*'s evolutionary heritage traces back around one hundred million years earlier than *C. elegans*, making it the senior of the two venerable worms. Yet, *D. japonica* is more mentally savvy than its younger cousin. The flatworm *perceives*.

There's no doubt that the flatworm mind processes collective sensory *patterns* rather than individual sensory *signals*. Why? Because a flatworm proudly flaunts sensory organs.

Let's get to know Professor Flathead.

Perhaps the flatworm's most striking feature is its two ocelli, which resemble a pair of crossed eyes. These are, in fact, visual organs, though they are not what biologists refer to as "true eyes" because they lack lenses and pupils. Ocelli are often called "eye spots" or "eye cups." Their concavity is designed to collect and focus light to provide more structured patterns of visual input for the flatworm's mind to process.

The flatworm also has two little pointy appendages sticking out the sides of its head that look like ears, which is why they are misleadingly called "auricles." Though these are indeed sensory organs, the auricles do not hear. They

PROFESSOR FLATHEAD

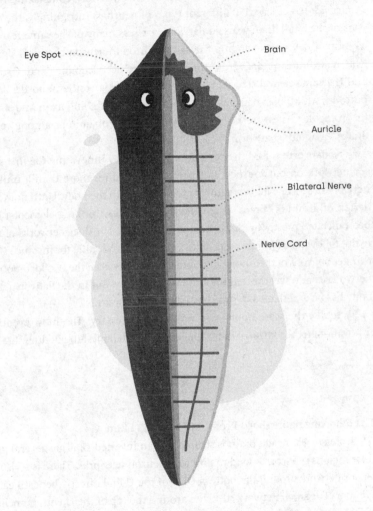

Eye Spot

Brain

Auricle

Bilateral Nerve

Nerve Cord

smell. They contain sensor neurons that discriminate patterns of free-floating chemicals. Thus, the eye spots and auricles are not merely sensory organs, but organs of perception.

The inception of organs of perception—and the ability to perceive patterns—required three innovations. The first two are the same innovations that drive advancements in all neuron minds: new kinds of neurons (neural diversification) and new kinds of networks (network complexification). *D. japonica* has at least 150 different types of neurons. Intriguingly, the neurons of the half-billion-year-old flatworm possess many of the same features as human neurons, including the same neuron-to-neuron chemical signals (the neurotransmitters serotonin, dopamine, acetylcholine, and GABA), and the same neural structures (dendrites, synaptic vesicles, and dendritic buttons). Multipolar neurons (neurons with one outgoing axon and many incoming dendrites) are rare in invertebrates, but ubiquitous among vertebrates . . . and the flatworm.

The flatworm solved the perception problem by innovating the first sensor network on our journey. But perception requires more than a unified network of sensors. It also requires modifications to the body. Until now, the design of a mind's "brain" (the collection of all its thinking elements) has been dictated mostly by the structure of its body. The doer networks of the hydra follow the contours of its tentacles, mouth, and stalk, for instance. The thinker neurons of the roundworm are inserted between the two locomotion networks that travel the length of the worm's body. But in the flatworm, the body has been reshaped to suit the needs of the "brain."

Actually, the scare quotes are no longer necessary. The most eventful innovation in the flatworm is the appearance of an honest-to-goodness brain.

3.

The following figure shows Professor Flathead's brain.

Professor Flathead's brain is shaped like an inverted U. It has several distinct regions that correspond to different neuron networks. There is a visual sensor network around the bottom edge of the U and directly beneath each eye spot. The smell network stretches around the top of the U with branches stretching out of the brain into pores on the skin of the auricles. Thinker networks are located between the visual and smell networks. The thinker networks are linked to doer networks running down the left and right sides of Professor Flathead just as they did in the Duke of Dirt.

Professor Flathead is the first mind on our journey to start carving reality apart at its joints. By perceiving distinct patterns in its physical environ-

PROFESSOR FLATHEAD'S BRAIN

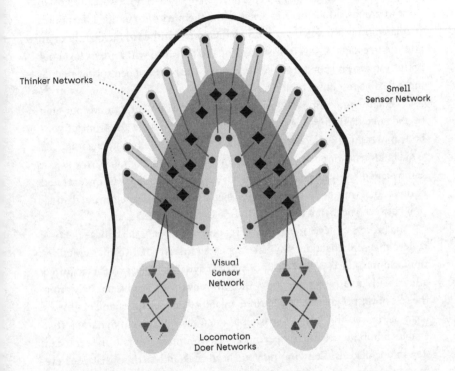

ment, Professor Flathead can begin to execute complex behaviors that exploit sensory-specific patterns. The flatworm is the first organism that can navigate through complex environments. They react negatively to touch. Unlike the hydra, their tactile sensor network can identify *where* they were touched, and they move that part of their body away from the provocateur. They can sense when they are upside down and immediately flip themselves right side up. Flatworms are sensitive to currents of water and actively seek out flowing water. When they find a current, they swim *against* its flow.

Flatworms exhibit even more sophisticated behaviors. One involves the hunting of water fleas, a small crustacean called dap$ia that is also an entrée on the hydra menu. When a flatworm perceives a water flea, it secretes a sticky fluid that immobilizes the flea, then wraps itself around its prey like a boa constrictor squeezing a mouse. On occasion, the flatworm switches from the hunter to the hunted. Dilepti are single-celled protozoa that resemble a cross between an amoeba and a sperm cell. They live in fresh water and often hunt in groups, attacking prey by stinging them with toxin-filled harpoons. If a flatworm is attacked by a horde of Dilepti, it will writhe violently in an attempt to escape. If this is unsuccessful, the flatworm will raise its head out of the water to prevent it from getting stung. The power of perception enables more flexible responses to dangers.

The flatworm brain is the product of two developmental trends powered by the embodied thinking principle. Putting eye spots at the front of your body makes good sense, because it allows you to see where you're going. The same holds true for auricles. Scent organs should go in front, too, so you can sniff what's directly ahead. Once you have organs of perception concentrated in one end of your body, you've established a meaningful anatomical distinction between your front and back. In brief, you've created a *head*.

The process of developing a head is known as *cephalization*, and scientists believe that cephalization may have appeared first in a flatworm. Cephalization also introduces a new kind of mental dynamic. Once you've acquired a head with organs of perception and can move your head separately from the rest of your body, you've acquired an ability that scientists and engineers term "active sensing." Active sensing is a new class of mental dynamics that enables a mind to expand its perceptual environment in real time. You can stay in one place while swinging your head back and forth so that your eye spots can survey the landscape and your auricles can sniff a range of air.

In the previous chapter, we saw how the roundworm was the pioneer of *centralization*, establishing a centralized network of thinker neurons that evaluates all sensory inputs. But the front end of the roundworm does not have specialized sensory organs, and indeed, the roundworm has similar numbers of sensor neurons on its front and back. Though its ten thinker neurons are all located in its front, there's too little distinction between its forward neurons and backward neurons to merit the designation "head." (Protohead, perhaps.) But once a mind develops organs of perception like eye spots and auricles, along with the sensor networks to support them, then for maximum real-time speed and efficiency these sensor networks should be placed as close to the thinker networks as possible. The shorter the distance that sensory signals need to travel to get evaluated by a thinker network, the faster the mind can react. Centralization reduces the length that mental signals must travel and thereby increases the speed of thinking, which is the same reason that financial enterprises (which depend upon the speedy acquisition and processing of information) are usually clustered together in the center of cities, like Wall Street in New York City.

Cephalization tends to increase centralization, because the more neurons that are concentrated in one place, the more these neurons will dictate what happens elsewhere in the mind. Centralization tends to increase cephalization, because the more that the mental dynamics of control are concentrated in one spot, the more the body will need to change to support those dynamics. Much like the reciprocal loops of the metropolis principle, the mutually reinforcing forces of cephalization and centralization are a potent engine driving the development of the brain—and the development of intelligence.

As the brain develops, it resculpts the organism's body to fit its needs. The body manufactures protective coverings for the brain, like skulls and meninges. The body produces greater blood flow to the head, to satisfy the burgeoning hunger of an energy-voracious brain. The body creates new supporting tissues for the brain, like glial cells. The body constructs anatomical innovations that increase the height and mobility of the head to improve active sensing even further, like necks and—in primates and a few other animals—the upright stance. As the brain continues to develop, its activity becomes increasingly divorced from the neural activity in other parts of the body, establishing a functional distinction between a central nervous system (the brain) and a peripheral nervous system (all the other neurons).

Perhaps the most important repercussion resulting from the emergence of a true brain is the fact that future minds no longer need to experiment with organism-wide configurations of body and "brain." A bilateral body with a brain and a peripheral nervous system is like a cart with four wheels: a flexible arrangement that can be adapted into a Jeep, a limousine, a pickup truck, a bus, a tank, or even—with a little creativity—a flying machine.

CHAPTER NINE / **FLY MIND**

Representation

> I can see that it is in the nature of men to prefer one
> thing to another, to find one thing more meaningful
> than another.
>
> —*Piranesi*, Susanna Clarke

1.

The bright yellow toilet-seat cover does not appear to be anything out of the ordinary—a shaggy, machine-washable cotton-nylon blend. Smack in the center, impudent and cartoonish, is a print of a giant red tongue protruding from a lippy mouth. For someone in need of attention-getting protection for their commode, toilet covers of similar quality would set you back about $15. But on September 12, 2020, the tongued one sold for $1,152, setting the record for the most expensive toilet-seat cover. Why would someone pay such an exorbitant price to decorate their privy? Because this cover belonged to Bill Wyman, the original bassist of the Rolling Stones, who used it for years on his own porcelain throne.

In 2011, a Russian investor paid $2,882,500 for a thousand-pound aluminum sphere roughly the size of a Nissan Rogue. It resembles a geometric sculpture you might find in the lobby of a corporate building. Did someone fork over a king's ransom for the sphere's aesthetic value? No. It was the Soviet Vostok 3KA-2 space capsule, one of the first works of human engineering to carry a living animal into space. A dog named Little Star returned safely to Earth, setting the stage for Yuri Gagarin to blast off in an identical capsule twenty days later and become the first ape to break free of the surly bonds of gravity.

At a 2021 online auction run by Christie's, someone paid $69.3 million for a computer file known as a JPEG, the kind of digital image you find on websites or exported from digital cameras. Stored on the computer file is a work of visual art entitled *Everydays—The First 5000 Days*. The artist who created

Everydays goes by the name Beeple. He is famed for playful and provocative digital art that captures the fears, dreams, and tensions of twenty-first-century life. His JPEG sold for more than many traditional paintings by artists such as van Gogh, Rembrandt, and Georges Seurat. There are numerous copies of the JPEG available online, all identical to the one that sold for more than the cost of three F-16 fighter jets. You could go download one yourself. So why would someone pay so much for an item that could be had for free?

Good question.

This book will not attempt to fathom the motivations of the winning bidder or the thirty-two other *Homo sapiens* who bid on an intangible (and endlessly replicable) pattern of 0s and 1s. Instead, this book will try to shed light on the underlying mental dynamics that give rise to the (often baffling) motivation to acquire *stuff* in the environment.

Object valuation is an essential form of thinking in all advanced minds. A *value* is the means by which a mind transforms a perceptual pattern into a complex behavior. The sight and scent of a bowl of gazpacho make us reach for a spoon. The sight and scent of a bowl of vomit make us push back our chair and hold our nose. In between *perception* and *action* is *valuation*.

Valuation enables an organism to navigate a world filled with complex and ever-changing *stuff* by reacting quickly to new things and making plans to acquire or avoid known things. Values enable a mind to distinguish the relevant from the inconsequential, the nutritious from the noxious, the beneficial from the deadly. Because all minds confront an unremitting stream of thingamabobs, the value of most objects must be determined through experience rather than hard-wired into a brain.

The learned nature of values is evident in the tremendous variation in the values that different minds assign to the same item. Imagine explaining to a rural farmer in Uzbekistan why an outhouse ornament costs more than a milk cow. The explanation would require exposition about the history of popular music, the origins of the blues, the British invasion, Mick Jagger and his lips, the role of a bass player in a rock band, and the global success and influence of the Rolling Stones. Even then, there might still be some question about why you'd want to own something that was used for years in another man's bathroom—something that had no involvement whatsoever in the production of music. Though Yuri Gagarin might have eagerly bid on the auctioned Vostok capsule had he been alive, it's safe to say that Little Star would have placed no value on the aluminum sphere at all, except perhaps as a site for a territorial splash of urine.

In this chapter, we come face-to-face with the simplest brain capable of performing complex and idiosyncratic valuations of perceptual patterns: the poppy-seed-sized brain of the fly.

2.

There was a time when insects were the smartest beings on land. That time was a half billion years ago. The Age of Insects peaked around three hundred million years ago, when the Earth was a humid bog festooned with colossal ferns soaring as high as two giraffes and even taller trees that had the bad habit of toppling into the muck because they had not yet developed stabilizing roots. The Carboniferous swamp may not have reeked like modern quagmires, for rot and decay hadn't been invented yet: there was not yet an ecosystem of decomposers to break down dead wood. All those fallen trees remained undigested by microbes and eventually turned into coal. Those fallen trees, that is, that managed to survive the roaring conflagrations that frequently raged across the megacontinent of Pangaea like the infernos of Armageddon, fueled by oxygen levels more than 50 percent higher than today.

The Sultans of Swampland Earth were bugfolk. Insects took advantage of the buoyancy and improved breathing provided by the elevated oxygen to grow to enormous sizes, including dragonflies the size of bulldogs and millipedes the size of crocodiles.

Today, insects occupy every terrestrial habitat and have outlived countless climate shifts and extinction events. They have a versatile body plan that can be easily adapted to new environments and a versatile brain plan that can be easily adapted to new bodies. The insect intellect towers over all the other minds we've visited. Despite our endless attempts at exterminating them, bugs bite us, chew on our skin, suck out our blood, lay eggs in our fruit bowl, mate in our mattresses, gnaw on our clothing, and pilfer our picnics, all with gleeful impunity.

Some insect minds are so well suited for their ecological niche that they approach perfection, as testified by their stability over astonishing stretches of time. Many insects from one hundred million years ago, when dinosaurs stomped the land, are nearly identical to their modern counterparts, including mosquitoes, moths, beetles—and flies.

Flies separated from other insect lineages around 240 million years ago. They were originally water creatures. The oldest extant fly lines are bizarre species with long legs and long wings that live in fast-flowing mountain water. Like other insect groups (including bees, butterflies, and beetles), many species of fly cocvolved with flowering plants. The very first flowers were likely pollinated by insects. Today, there are an estimated *1 million* species of dipterans, the insect order containing flies and mosquitoes, including 110,000 named species of flies. Those are enormous numbers. Pick any ten living species at random and one of them is probably a fly. Flies live in almost every nutrient-rich substrate on our planet, including petroleum, hot springs, the dung of millipedes, the gills of crabs, beehives . . . and rotting fruit.

Without further ado, here's Captain Buzz, a model of *Drosophila melanogaster*, the common fruit fly:

CAPTAIN BUZZ

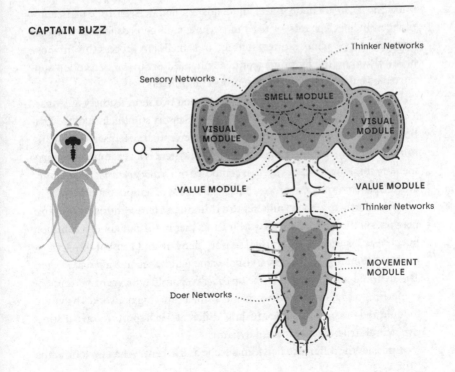

The body of the fly is a huge step forward from that of worms. The fruit fly brandishes six jointed legs covered with bristles. It has a complex mouth with mandibles and a proboscis. It also has wings. All this elaborate new anatomy led to the blossoming of a sophisticated behavioral repertoire. A fly can perform courtship dances, display aggressive postures, groom itself, lay eggs, and even "sing" using its wings. Most important, a fly can *fly*.

A fly can flutter and dive and zigzag through the air. It can flit over boulders and rivers that would be insurmountable to wingless neuron minds. The fly's aeronautical ability radically enlarges the scale of its environment. A bacterium travels a couple of centimeters during its life span. A roundworm might traverse the span of a human hand. But a fly can roam for miles and miles. That freedom of mobility helped endow the fly with an astonishingly sophisticated mind, one that contains direct parallels to most of the major thinking systems in human minds.

The fly pilots through a three-dimensional world packed with a wondrous diversity of *things*. A fly can hover over forests or deserts, grassland or mountainsides. It can alight on gravel, mud, flowers, boulders, garbage pails, automobiles, living elephants, or dead rats. These landing pads feature different textures, shapes, temperatures, sounds, and smells. To process this fire-hose blast of stimuli, the fly mind requires enhanced organs of perception supported by enhanced networks of sensors and thinkers.

The fruit fly has two bulging scarlet eyes and two short feathery antennae. The antennae are sensitive to multiple types of sensory stimuli, including vibrations, tastes, and smells. There are two major perceptual systems in the fruit fly mind, a visual system and a smell system. Each system is an example of complementary thinking, for each system is composed of a *sensory network* and a *thinker network* paired together to process a particular class of sensory patterns.

Henceforth, this book will refer to a thinking system comprising multiple networks of thinking elements working together in the pursuit of a common purpose as a *module*. Thus, a visual sensory network and a visual thinker network collaborating to process a visual scene will be termed a *visual module*. The fly mind also possesses a *smell module* responsible for scent perception, comprising a smell sensor network and a smell thinking network. The visual module and the smell module are quite different. Each sports its own distinctive neural structure and mental dynamic.

Consider the difference between a scene and a scent. When we look at the world, we register photons bouncing off a thicket of colorful shapes, some angular, some curved, some glowing, some shadowy, some zipping rapidly toward us, some inanimate and stolid. When we smell the world, we inhale a diffusive bouquet of invisible aromas—fresh-squeezed lemons, moldy carpet, hot asphalt, spoiled milk. Scenes and scents form different sorts of physical patterns in the environment, and therefore require different sorts of thinking to perceive them.

Let's compare the two modules in Captain Buzz:

VISUAL MODULE VS. SMELL MODULE

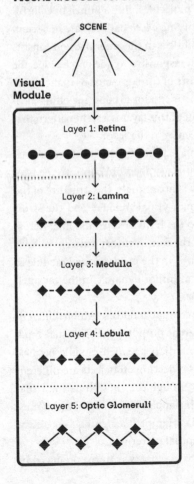
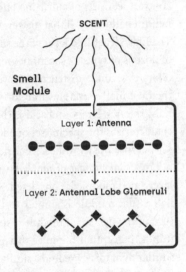

Captain Buzz's visual module is divided into five layers. The first layer is the retina, which contains light-sensitive sensor neurons. Its raw sensory signals are transmitted to a stack of three structured layers of thinker neurons and one final unstructured layer of thinker neurons. The four thinker layers progressively extract a pattern of color, form, and motion, somewhat similar to the way a comic-book illustrator takes a story and pencils a very loose sketch of the main objects, then inks the rough drawing with black-and-white details, then adds colors, then finally adds all the letters and motion effects.

Scents are much simpler sensory patterns than scenes. A scene contains colors, boundaries, shapes, depth, gradations, and movement. A scent, in contrast, consists of a drifting bundle of molecules, like a bunch of colorful helium balloons. All that matters is identifying *which* balloons are in a scent. The first, structured layer of the smell module is in the antennae and consists of dozens of types of sensor neurons, each responsive to one or a few specific odors. A second, unstructured layer consists of thinker neurons that convert the raw smell sensations into forty-five distinct odor groups (including alcohols, ketones, esters, and acids). The output of this layer is a scent pattern that characterizes the specific set of odor molecules in the scent.

Because the fly mind's scene perception requires five layers of processing, neuroscientists refer to the fly visual module as "deep." In contrast, the two-layer smell module is considered "shallow." Intriguingly, the structure of the fly's visual module is similar to the human visual module, and the structure of the fly's smell module is similar to the human smell module. Across minds, perceptual modules that process the same sensory patterns exhibit similar dynamics. We find a similar dynamic in the human visual module as in the cockroach visual module. We find a similar dynamic in the rat smell module as in the bumblebee smell module.

Within the same mind, different perceptual modules exhibit different structures, different dynamics—and different outputs. The fly visual module generates different perceptual patterns than its smell module because the properties of electromagnetic photons reflecting off objects are different from the properties of molecules floating through the air. Comparing visual patterns and scent patterns is like comparing apples and oranges, which presents a daunting challenge for the fly mind's attempt at assigning a consistent set of values across different sorts of perceptual patterns.

The fly mind must convert the distinctive sensory patterns produced by each perceptual module into a common "language" of valuation that can then be used to activate the appropriate behavioral response. For instance, the smell of a poison and the sight of the same poison should each trigger the same avoidance reaction, even though a scent and a scene are embodied within dissimilar neural patterns.

To assign useful values to perceptual patterns, the fly mind must solve the final challenge of neuron minds—which, not coincidentally, happened to be the final challenge of molecule minds: the *coordination problem*. For amoeba minds, the problem was: How can distinct molecule minds communicate with one another? For fly minds, the problem is: How can distinct neural modules communicate with one another?

The amoeba solved its coordination problem using a simple signal: a message molecule. But the fly faces a coordination problem on a grander scale. Its modules must communicate with one another using complex *patterns* rather than one-dimensional *signals*. Even more problematic, each module generates its own idiosyncratic patterns, as if the visual module spoke French, the smell module spoke Mandarin, the value module spoke Swahili, and the movement module spoke Quechua.

Solving this mental challenge requires an innovation so revolutionary that it led to a whole new stage of thinking. The innovation is *representations*.

3.

Living in the Digital Age, our intuitions about representations are heavily influenced by computers. When we think of representations, we might conjure up the notion of a music file stored on our hard drive, composed of some kind of fixed digital code. But this is the wrong way to think about representations in the fly mind—or any mind.

One of the most important lessons about minds you should take from the journey is this: *the mind is not a computer.* It is sometimes said that the mind is software running on the hardware of the brain. This is not true, not even metaphorically. In fact, viewing the mind this way makes it harder to comprehend the most interesting varieties of thinking, including consciousness, language, and the Self. If the mind is software, that suggests that you could run some *other* software on your brain. Microsoft Word, perhaps? Google

Chrome? Or, maybe, somebody else's mind? Could you download Michelle Obama's software from her brain and install it on yours?

Computation—both the practical sort that we do on our laptops and the formal sort that scientists and engineers study—requires precise, discrete instructions that must be followed sequentially. It requires 0s and 1s. A mind is almost the exact opposite of this. It consists of continuous and imprecise real-time activity happening everywhere all at once. In a mind, those 0s and 1s can often be extremely fuzzy, and sometimes they end up turning into ½'s or .333's or both 0 and 1 at the same time. At bottom, a computer is a *thing*: a machine that manipulates symbols. A mind is an *activity*. The mind has more in common with a basketball game than it does with a MacBook.

Mental representations are more like jump shots, blocks, and passes than they are like digital files, bytes of information, or logical operations. Mental representations are fuzzy, not precise. Ephemeral, not fixed. Holistic, not reductionistic.

A mental representation is a dynamic code consisting of a collection of *features*. A visual representation, for instance, might encode the color, shape, and size of an object within the real-time activity of a neuron network. This notion can be difficult to visualize, so one metaphor that may help us get a clearer picture is to view a representation as a melody—as a set of musical notes.

Each note is a distinct feature of the perceptual pattern, such as the yellowness of the sun or the blackberry scent in a glass of pinot noir. If someone plays a melody and hits a wrong note or skips a note, you can usually still recognize the melody. The same with representations: a mind can "recognize" two representations as representing the same pattern even if some features are altered or missing in one of the representations. But the more notes that are off-key or missing, the more likely it is that a mind won't recognize the tune.

Let's see how the melody metaphor helps us think about representations in Captain Buzz as he converts a scent pattern into a behavior.

Captain Buzz's smell module evaluates a new scent composed of five distinct odors. The module identifies each odor and encodes all five within a single unified representation of the scent. Next, the smell module sends this five-feature scent representation (a five-note melody, if you'd like) to the value module.

If some of the features are associated with toxins—maybe there's a

RESPONDING TO A NEW SCENT

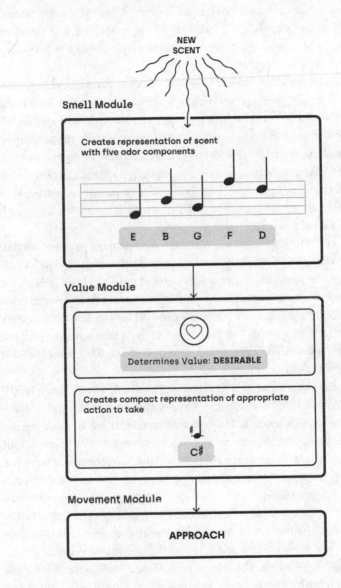

geosmin odor in the scent, which can be harmful to flies—then the value module assigns a "dangerous" value to the scent. But if the scent is similar to a previously encountered smell of food—maybe four of the five "notes" match the scent "melody" of a delicious rotting grapefruit and the other "note" isn't associated with danger—then the value network will assign a "desirable" value to the scent.

Captain Buzz's value module takes into account the particular context he finds himself in (whether he is hungry or fully sated, whether he is being chased by a predator or is exploring a new territory) and combines its valuation of the scent pattern (desirable or dangerous) with the contextual condition (hungry or sated) to determine which behavior to execute. For instance, if Captain Buzz is hungry and casually exploring his surroundings, then the value module will respond to a scent pattern associated with food by assigning it a value of "desirable," which redirects his flight path toward the scent pattern.

To execute this plan of action, the value module must generate another representation—a new "melody"—to send to the movement module. A simple one-neuron signal to trigger a doer network, like the ones used in the roundworm mind and hydra mind, won't do, because the fly movement module is capable of a wide variety of activities, such as turning left, diving down, landing, attacking, mating. The value module must communicate with the movement module using another representation, one in the "language" that the movement module will understand.

The value module converts its valuation ("desirable scent") into a behavior representation ("fly toward scent") by creating a new "melody" to share with the movement module. The behavior representation is much simpler than a perceptual representation (the behavior "melody" can even consist of a single "note") because even though the range of potential sensory patterns the fly can detect is enormous, the range of potential actions the fly can execute is far more limited. A much more compact melody can enable the movement module to distinguish between its relatively brief list of behavioral options. The movement module receives the succinct "behavior melody" from the value module and activates the doer networks necessary to execute the desired action. This is the secret of insect intelligence: formidable perceptual sensitivity and discerning valuations guiding a very limited repertoire of behaviors.

Representation is essential for effective valuation of things in the world, because there are many different kinds of things out there and each needs its own stable and distinctive representation. The richer and more sophisticated the representation (the more complex the melody), the richer and more sophisticated the valuation a mind can make—such as concluding that *this* particular visual pattern of a toilet-seat cover is "highly desirable."

Representations solve the coordination problem in neuron minds by enabling different modules to communicate with one another using dynamic patterns of activity. And when a mind solves the coordination problem, the metropolis principle tells us that a new stage of thinking will emerge.

The module minds.

4.

The appearance of representational thinking in the journey of Mind introduces new kinds of mental challenges. Three of these challenges are so pervasive and vital that every module mind must confront them head-on:

1.   The *stability dilemma*. It can be summarized as, "Which representations need to be stabilized and sustained, and which representations can be ignored or eliminated?"

     Should you remember what you ate for dinner an hour ago? That might be useful in case you ate something that made you sick. But should you remember what you ate for dinner last week? Last year? Ten years ago? The trade-off inherent to the stability dilemma is whether to devote precious energy and resources to preserving representations, or to conserve energy and free up resources by allowing representations to dissolve back into chaos.

2.   The *uncertainty dilemma*. "How can I eliminate the ambiguity associated with this representation and ensure that the representation is correctly identified?"

     Is that orange circle a basketball, a citrus fruit, or the sun? Is that man winking at me or trying to get rid of dust in his eye? What the heck is Kafka's *Metamorphosis* about, anyway? Any sensory representation has myriad possible interpretations, and a critical challenge for every module mind is determining which interpretation is the correct one.

3.   The *attention dilemma*. "Which representation should I pay attention to?"

This is the most urgent and problematic dilemma that module minds must contend with. Many different perceptions, feelings, and ideas are bouncing around a module mind at any given moment, and deciding which one to focus on presents a formidable conundrum. Should I focus on the dog rushing at me? The bizarre sound of trumpets behind me? My anger at my significant other? The intensifying scent of brimstone? Or the customer-service rep chattering on the phone? And what does *focusing* even mean?

The attention dilemma distinguishes the neuron mind of the fly from the module minds that follow, because module minds have come up with a very special innovation to master the attention dilemma. The name of this innovation is consciousness. In the next stage of our journey, we will learn *why* consciousness exists and *how* consciousness works—and why insects are almost certainly not conscious.

But in order to do so, we must first gear up with a new set of thinking tools.

Module Minds

The Unified Mathematics
of the Self

Whenever you claim to be "the first to do" this or that in artificial intelligence, it is customary—and correct—to add "with the exception of Stephen Grossberg." Quite simply, Stephen is a living giant and foundational architect of the field.

—Karl J. Friston, neuroscientist

1.

From a very early age, Isaac Newton was fascinated by how things moved. The way a ball soared through the air or plummeted to earth. The way shadows swelled and shrank. How a breeze tossed a leaf hither and yon. This youthful passion instilled within Newton an abiding interest in the dynamics of matter. In a word, he was fascinated by *change*.

Newton spent his childhood fashioning sundials and windmills. He was particularly enchanted with kites, which rose and dipped in the wind. He experimented with new shapes and sizes, always seeking to improve upon his designs. One reason that Newton preferred mental pursuits over physical contests was that he was smaller and frailer than his classmates. When he was unavoidably pressed into athletic competition he would wield what he called "philosophical thinking." Once there was a contest among the schoolboys to see who could leap the farthest. Newton came in first place, defeating his taller and stronger opponents. His trick: timing his jumps to coincide with advantageous gusts of wind.

When he entered Trinity College at Cambridge, his youthful passions collided with an unexpected obstacle. At the time, English universities embraced Aristotle and the belief that the sun revolved around the Earth. Scientific inquiry at Cambridge, such as it was, employed little math, instead relying on qualitative descriptions of objects and events. Students were taught that one set of rules governed things on Earth and another set of rules governed things in the heavens—that there were a celestial mathematics and a terrestrial mathematics that did not overlap. There was little in such academic dogma to attract Newton's burgeoning fascination with the dynamics of matter. Instead of attending to his schoolwork, Newton searched for inspiration outside the curriculum.

He began reading Copernicus and Kepler, freethinking astronomers who advocated for heliocentrism. He read Galileo, who proposed a new vision of the dynamics of matter based upon inertia. He was particularly excited by Descartes, who formulated a highly controversial conception of nature as an intricate and impersonal machine, a notion that Newton found congruent with Democritus, who believed that reality was formed out of the constant motion of matter.

Intoxicated by these unconventional ideas, Newton pursued his own private and highly idiosyncratic studies. He began to see the world as a collection of ever-changing patterns, rejecting the orthodox view that the universe consisted of static objects. Newton speculated that these ever-changing patterns did not represent random chaos but might instead represent a hidden form of order. He realized that if he wanted to apprehend this hidden order, he would need to create new mental tools to support his subversive intuitions. Most especially, he would need to create a new kind of mathematics.

And then, an unusual opportunity rolled onto Newton's doorstep. A pandemic. The bubonic plague began tearing its way through seventeenth-century England. Trinity College went on lockdown. Newton returned home and created a home office for himself to continue his eccentric studies in isolation. Divorced from his professors and classmates, he began constructing models for the motion of objects. These models generated new mathematical ideas that he tested through homemade experiments or endless, obsessive calculations.

Time quickly became a fixation. Newton came to view time as a steady, infinite flow that provided the essential framework for all movement. He devised new mathematical concepts. One represented a "flowing quantity." He called this a "fluent." Another represented a rate of change. He called this a "fluxion." By the time he was twenty-three, he had derived an entire mathematics of fluents and fluxions. This unprecedented math allowed him to make accurate predictions about the way things emerged, changed, and vanished. Today, we call it calculus.

By late 1666, a tumultuous year scarred by the Black Death and the Great Fire of London, young Isaac Newton, working alone and almost completely incommunicado, had become the most advanced mathematician in the world. He spent the next two decades consumed with applying his revolutionary mathematical tools to the study of movement. Finally, in 1687, he published *Principia Mathematica*. It contained the first unified theory of the dynamics of matter. For the first time, heaven and Earth were conjoined within a single conceptual system—a *mathematical* system that obeyed precise and inviolable equations.

Physics boasts a history replete with prodigies, individuals who flashed brilliance young before going on to make extraordinary contributions to the field. Enrico Fermi, one of the lead physicists on the Manhattan Project, wrote a doctoral-level paper on vibrations at the age of seventeen. James Maxwell, the founder of the field of electromagnetism, published his first scientific paper at age fourteen. But towering over them all looms Isaac Newton, humankind's first mathematical physicist, who as a student intuited that the prosaic behavior of stones, sticks, and bricks followed the same principles as the celestial behavior of the moon, planets, and sun.

Things are different in the mind sciences, where prodigies are scarce. Few eminent psychologists and neuroscientists demonstrated a precocious talent for fathoming the mind as adolescents. There is one notable exception to this rule, however. A mathematical pioneer whose story, in many ways, parallels that of Newton: Boston University professor emeritus Stephen Grossberg.

2.

Grossberg was born in New York City in 1939, the grandson of Hungarian immigrants. His mother was a schoolteacher. His stepfather, an accountant. (His biological father died when he was one.) He was raised in Jackson Heights, Queens, a lower-middle-class neighborhood clamoring with fiercely competitive Jewish boys. From a young age, Grossberg was keenly aware of the fact that living things are either growing or dying, blooming or decaying. This instilled within him an abiding interest in the dynamics of life. In a word, he was fascinated by *change*.

Even as a child, Grossberg was consumed by the realization that we are only temporary visitors upon this earth. This awareness provoked an urgency to make contact with something more durable than his own ephemeral life. "This, to me, was a deeply religious feeling: how to be in touch with the enduring beauty of the world, even though you can only personally be here for a very short time," Grossberg says. "I wanted to do something where I could touch the eternal. It seemed the only way to do that at the time, given my limited options because my parents had no money, was to be incredibly good in school."

Grossberg battled his way to the top of all his classes at the elite Stuyvesant High School in Manhattan. He nailed perfect scores on his SATs, demonstrating fluency in both mathematics and language. He was rewarded with his choice of universities and decided to attend Dartmouth College because it offered a program that would allow him to spend his senior year focusing exclusively on research, though he possessed only the vaguest idea of what research actually was.

In his freshman year, he took a required introductory course in psychology. It was his first exposure to the science of Mind. "The class unexpectedly created a storm of ideas inside my head," Grossberg recounts. "I was entranced by the implications of the data on human and animal learning for how things were going on moment by moment inside our minds, the *dynamics* of minds."

In the psychology class, the young man was introduced to an unsolved problem that captivated him. The problem was known as the *serial position effect*. First discovered by German psychologist Hermann Ebbinghaus in the 1880s, it had baffled scientists for almost a century. The serial position effect describes how we remember a list of items, such as a grocery list, when the items are presented to us one at a time: milk, sugar, potatoes, apples, tomatoes, bacon, yogurt, squash. The serial position effect refers to the fact that people tend to recall the first items in the list (known as the *primacy effect*) as well as the items at the end of the list (known as the *recency effect*), no matter how long the list is. Items in the middle of the list are usually forgotten. The unanswered question was, *Why?*

Grossberg's teenage attempt at solving this puzzle would end up changing the course of his life—and the course of the science of Mind.

3.

Grossberg attended Dartmouth in the late 1950s. Buddy Holly, the Big Bopper, and Ritchie Valens had not yet boarded their fatal plane ride. Hawaii was not yet a state. Ford was still manufacturing the Edsel. In the United States, two antagonistic disciplines held sway over the mind sciences: psychoanalysis and behaviorism. Freud's pseudoscientific theories of the unconscious were defiantly unempirical, qualitative, and nonmathematical. They held that thinking emerged (somehow) from the conflict between an ill-defined ego (the conscious self), superego (a kind of Jiminy Cricket–like conscience), and id (selfish cravings unfettered by morality). Behaviorism, characterized by pigeons pecking at lights and mice dashing through mazes, was ruthlessly empirical and quantitative and, most significantly, rejected any model of thinking. Behaviorism's defining assumption was that you could understand everything about an organism's behavior by focusing on the statistical relationships between sensory inputs and behavioral responses without any need for speculating about "intervening variables," behaviorists' dismissive term for the act of thinking, perceiving, or feeling.

In any event, neither psychoanalysis nor behaviorism could account for the serial position effect. At all. Its peculiar shape didn't match *any* theory of Mind. The mystery enthralled Grossberg.

"One of the most interesting things about the serial position effect was that it seemed to involve learning that could go forward and backward in time," Grossberg explains. "Items that arrived *after* the beginning of the list were forgotten, but items that arrived *before* the end of the list were somehow retroactively forgotten, too. The idea that events could go backward in time fascinated me."

Even after the psychology course ended, he spent his sophomore year struggling with the serial position problem, convinced that it must hold the key to unlocking well-guarded secrets of the mind. It forced him to think about how to represent mental states and mental events, and how to represent the interactions between these events and states at different moments in time. His solitary struggles led him to a revolutionary perspective on the nature of thinking.

One of the breakthrough features of this new perspective was Grossberg's realization that any explanation of the serial position effect would need to describe its operation in *real time*. That is, the explanation would have to account for how the brain was changing moment by moment as each new item was presented. More generally, it was apparent to Grossberg that a sophisticated treatment of time itself would be an imperative factor in any explanation of thinking. Time was not a neutral marker of the passing of mental events nor was it a steady Newtonian flow providing a framework for all thinking. Instead, time was stitched into the very fabric of mind. Minds were wholly designed from the ground up to manipulate and exploit the flow of time.

Building on these insights, Grossberg recognized that there had to be two distinct but codependent mental processes underlying the serial position effect, each associated with a different rate of time. The first was a *fast short-term memory process* that changed in real time as new items were presented, and the second was a *slow long-term memory process* that acted on a much slower time scale and was influenced by the results of the fast activity. The notion that two mental processes, each specialized for a specific purpose, yet each needing to collaborate with the other to achieve an even larger purpose, eventually contributed to his articulation of the complementary thinking principle.

He also recognized that because the mind was a physical system, it should be treated not as an abstract, noncorporeal entity but as a specific configuration of physical elements interacting with a specific physical environment in real time. Thus, he also embraced the embodied thinking principle.

Next came perhaps Grossberg's most original leap of intuition, an idea grounded in a concept first introduced in Darwin's theory of evolution. Grossberg realized that there must be some kind of *competition* between the representations of each item as they were presented to the mind. It was as if the words "milk" and "sugar" and "banana" were all battling each other for their place on the mind's grocery list. Grossberg discovered that this winner-take-all dynamic was necessary to represent a sequential list in the mind. (Years later, he would also demonstrate that winner-take-all dynamics also prevented the degradation of mental lists due to the incessant biological "noise" generated by cellular activity.)

From these insights, Grossberg derived a wholly original approach to modeling thinking by the end of his junior year. He concluded that the best way to make sense of the mind was by representing it as a network of neurons characterized by nonlinear differential equations operating in real time.

In simpler terms, Grossberg came to view the mind as a basketball game.

4.

There was every reason to suspect that Grossberg was traveling down an intellectual dead end. After all, he was a kid pursuing his own private and highly idiosyncratic studies that had almost no connection to what anyone else was doing. A few years later, when he shared his models with the prominent neuroscientists of the era, the most common reactions were confusion and apathy. Indeed, over the next few decades many approaches to studying the mind would come online that were markedly different from the one that Grossberg was developing, including two perspectives that dominate twenty-first-century science: viewing the mind as a computer, and viewing the mind as a statistical machine.

These two alternate approaches share the same shortcoming: they were both predicated upon principles and mathematics that had originally been developed for other disciplines. Advocates of the "mind is a computer" perspective repurposed old ideas from logic, information theory, and computational theory. Advocates of the "mind is a statistical machine" perspective repurposed even older ideas from Bayesian statistics and probability theory. These are "top-down" approaches that try to shoehorn Mind to fit preexisting assumptions established to study phenomenon other than brains and biological thinking. Grossberg, in contrast, was taking a "bottom-up" approach that made no assumptions about how Mind might work, instead (like Newton) figuring out the principles as he went along, through experimentation and obsessive calculations.

Grossberg's quest was aided by a stroke of fortune. Unbeknownst to anyone, including Grossberg himself, the intoxicating puzzle he had stumbled upon in Psych 101 (the serial position effect) was a product of one of the most important and sophisticated neural modules in the human mind, a module essential for music, toolmaking, and language. This module could not be fully understood using any approach other than the dynamic systems framework that Grossberg was developing. In effect, it was as if an Astrophysics 101 professor had shared with her freshman class a number of unsolved mysteries, including Venus's weird backward rotation, Pluto's oddly shaped moon, and Mercury's deviant orbit around the sun (a celebrated anomaly that cannot be explained by Newtonian mechanics), and Grossberg had not only instinctively recognized that Mercury's orbit presented the deepest puzzle but also, after studying the puzzle for a year, derived an outline for the theory of relativity.

According to Grossberg's budding theory, the primacy and recency effects in the serial position curve were "emergent properties" of the holistic dynamics of the mind's neural networks. This was another radical insight, especially considering that the notion of emergent properties did not yet exist in the mind sciences. Indeed, complexity theory, the intellectual home for emergent properties, didn't formally exist yet, either.

Though the teenage Grossberg had not been exposed to any neurophysiology, his model had naturally deduced (as a consequence of the mathematics of dynamic systems) several fundamental properties of every biological nervous system, including the existence of individual thinking elements (neurons) that were linked together with directional connections (axons and synapses) and that sent signals over these connections if a threshold activation was reached (action potentials). He did not learn that his private model was a good match for physiological reality until he chatted with his premed student friends who were learning neuroscience. This was a revelatory moment for Grossberg. He realized that he had derived the existence and operation of *physical* mechanisms by mathematically analyzing *psychological* data.

"I can hardly recapitulate my excitement when I realized this," Grossberg says. "When it dawned on me that by trying to represent the real-time dynamics of behavior you could derive brain mechanisms, I started reading neurophysiology with a vengeance."

Beyond any specific model, by the time he graduated from college in 1961 Grossberg had introduced a whole new scientific paradigm for understanding biological thinking: the mind as a marvelously complex dynamic system. He realized that it should be possible to derive a unified dynamics of mind in the same manner that Newton derived a unified dynamics of matter. In other words, Grossberg believed it was possible to explain phantasmagoric mental phenomena like consciousness, language, and the Self using the same kind of exacting nonlinear mathematics that explained the serial position effect. This was as counterintuitive and heretical as Newton believing you could explain the heavens by studying the Earth.

There was one big problem, however. The same problem that had confounded young Newton. The mathematics supporting a dynamics of mind hadn't been invented yet.

5.

Just as Isaac Newton had to invent calculus to support his novel intuitions concerning the behavior of physical processes, Grossberg realized that he needed to develop new mathematics to support his intuitions concerning the behavior of mental processes. "Most paradigm shifts in the twentieth century, say in physics, had the math already there to support the new revolutions after new intuitive breakthroughs occurred, including relativity theory and quantum mechanics. Our field is different: we needed new intuitions *and* new mathematics about complex adaptive systems, because the behavior of these systems was characterized by 'the three N's': non-stationary, non-linear, non-local processes."

For the next six decades, Grossberg developed new methods, models, and mathematics to support his unorthodox conception of thinking. In 2017, sixty-one years after his groundbreaking insights as a seventeen-year-old hoping to "touch the eternal," Grossberg published an epochal paper. It lays out a mathematical framework that accounts for a breathtaking panorama of mental activity encompassing every major form of thinking and every major brain structure, addressing everything from retinal neurons to symbolic communication, the summation of over five hundred articles with more than one hundred collaborators and backed up by behavioral, psychological, neurophysiological, neuroanatomical, biophysical, and biochemical data. Pick virtually any part of the mind, at any level, and Grossberg has put math to it.

Grossberg's unified theory of mind explains how module minds solve the stability dilemma, uncertainty dilemma, and attention dilemma using an assortment of specialized neural modules. Though this book will describe these modules in a qualitative way that does not require any math or science background to follow, Grossberg's equations describe them exactly and quantitatively.

The most significant contribution of Grossberg's unified theory of mind, however, is its solution to one of the most boggling mysteries in science: how consciousness works and why consciousness exists at all. This explanation did not arrive in a sudden moment of revelatory insight but arose gradually over many years through the painstaking accumulation and integration of disparate insights, as with Newton's *Principia Mathematica*. And just as Newton showed that an apple falling from a tree followed the same principles as a moon orbiting Jupiter, Grossberg showed that consciousness obeys the same principles as an archaeon seeking the light.

Grossberg's mathematical ideas will carry us through the next stage of the journey of Mind, where we will explore thinking in module minds by examining the What, Where, When, Why, and How modules and learn how these modular thinking elements, when integrated together, demystify the enchanted conundrum of consciousness and serve as the platform for the emergence of the Self.

CHAPTER TEN / **FISH MIND**

The Preconscious Proletariat

The map of the world ceases to be a blank; it becomes a picture full of the most varied and animated figures.
—*The Voyage of the Beagle*, Charles Darwin

1.

The joys of being a sentient being come from . . . well, they come from the experience of joy. Licking ice cream on a sultry day. Hearing the tinkling giggle of our toddler. Bingeing a TV series during a snowy weekend. Having friends over for the first time after a very long pandemic.

We *experience* these experiences, these feelings, perceptions, and revelations. And even though someone else might experience *similar* feelings, perceptions, and revelations, what makes experience so intensely personal is that it cannot be transferred to someone else. This gust of private awareness lets us know we are still alive, still breathing—still *conscious*. When we are unconscious, there is no experience at all.

Because the vibrant sigh of consciousness is so mystical, so eerie, so difficult to delineate and comprehend, we take it for granted that our ability to experience reality is something transcendent. Real magic in a dreary world of physics. We might even harbor the suspicion that our ability to be *aware* of our thoughts, perceptions, and feelings might not be the product of our brain at all, but the emanations of some kind of hallowed force abiding within an undiscovered dimension. Even if we accept that consciousness is somehow squeezed out of squishy blobs of protoplasm, we may be convinced that it must be the result of some sublime physical process, possibly involving subatomic arcana.

We don't grant the same lofty status to our other mental activities. The reflex that makes us unthinkingly withdraw our hand from a hot iron skillet? Mere circuitry. The calculations our brain unconsciously performs to stroll along the sidewalk? Ordinary mechanics no different from a hydra squatting. The monitoring of our blood's sugar level, the perpetual management of our breathing, the saccade of our eyes as we peruse the words of a book? The same brand of humdrum computing we would find on a messaging app on our smartphone. Even if we ourselves couldn't begin to explain how our mind performs these routine activities, we feel confident that understanding how they work would not be much different than understanding the operation of a toaster. In contrast, we may suspect—or hope—that we will never fully comprehend how our neural goop exudes consciousness.

This self-serving presumption is a form of "consciousness privilege."

The bias of consciousness privilege makes us view the parts of our mind that are conscious as superior and special—while viewing the other parts as simple automatons, a "preconscious proletariat" whose role is to serve the great and wonderful palace of consciousness.

Unfortunately, the bias of consciousness privilege gets in the way of understanding advanced forms of thinking, including the Big Three. It goads us into believing that any explanation of consciousness must be grounded upon some inscrutable property of the universe, the same way that scholars before Isaac Newton believed that heavenly bodies were governed by unknowable celestial forces rather than the same forces that governed raindrops, horseshoes, and mud.

For those hoping to shake free of consciousness privilege, the most relevant question to ponder is *Why do some neural modules produce conscious experience, while others do not?*

2.

The Module Mind chapters follow a different approach than the previous two stages. The journey will no longer be progressive, but lateral. Each chapter showcases a different module. As before, we'll see how each module is an innovation that solves a specific mental challenge. But we will no longer follow the mind's step-by-step progress as each new innovation builds upon the one that came before. *All* the modules in the next seven chapters are found in *all* module minds, from fish to apes, and indeed, most of these modules echo a simpler version found in the mind of the fly. The objective of this lateral itinerary is to show how these modules all collaborate with one another to produce our splendiferous experience of perceiving, knowing, and feeling.

The very notion of a module may remind some readers of the modularity of a computer program, suggesting that perhaps we've left behind the blurry causality of lesser minds and entered a realm of thinking more like the clean, logical segregation of software routines. This is not the case. We will encounter some modules that produce dynamics free from the interference of other modules. But most modules are intimately intertwined with one another, continuously influencing each other's ongoing internal operations in real time.

This is a good time to point out that even though this book contends that there are good reasons to believe that Stephen Grossberg's modules reflect mental dynamics found in all vertebrate minds, Grossberg himself intended his modules to serve as models of the human mind, *not* animal minds. If you wish, you can simply treat the animal-focused MODULE MIND chapters as an introduction to the ideas and intuitions necessary to understand Grossberg's explanation of human consciousness.

The first module we will visit on our tour of MODULE MINDS is the Visual Scene module. This neural module converts sensory inputs—real-time patterns of photons—into a useful representation of a visual scene. Significantly, as long as this representation remains within the confines of the Visual Scene module, it will never enter our conscious awareness—we will never *see* the visual pattern encoded in the representation. The scene representation in the Visual Scene module is *preconscious*. It has the potential to enter consciousness, but only if the representation is processed further by other modules.

This suggests we can reframe our guiding question, *Why do some modules produce conscious awareness, while others do not?* as:

Why do some mental representations enter conscious awareness, while others do not?

Some might say that visual representations become conscious so that we can appreciate the Platonic beauty of the world, as posited in the Anglican hymn "All Things Bright and Beautiful." It claims that God created "the purple headed mountain" and "the river running by" and then "He gave us eyes to see them."

Of course, a supreme deity could presumably have equipped us with a more robust means of seeing purple-headed mountains than two gooey little balls that gradually become far-sighted and fogged with glaucoma, not to mention a tissue-thin retina so easily detached. The imperfections and fragility of our visual system should serve as a hint that consciousness itself is another physical artifact, rather than some ethereal paragon. When pondering *why* we have conscious vision, a more pragmatic approach might focus on discerning its *purpose* within a module mind.

As it happens, conscious vision has two purposes. Each solves a distinct challenge and is governed by a distinct module. The first purpose of conscious vision is to *see* things. The second is to *recognize* things.

At first blush, it might seem that seeing and recognizing are one and the same, that distinguishing between them is philosophical wordplay. But consider how often we see things that we do not recognize—and how often we recognize things we do not see.

The former case is perhaps more obvious. There might be some blobby discoloration on the wall that you don't recognize—is it a shadow? A stain? Mold? Even if you have no clue what it is, you can wipe at it with a rag. Or maybe you catch a glimpse out of the corner of your eye of something hurtling toward your head. It's moving too fast to identify, but you duck anyway to make sure you don't get clobbered.

We often recognize things without seeing them, too. If a bottle of water is concealed behind a cereal box except for its cap poking up above the top, we know exactly how to shape our hand to reach around the box and grab the unseen bottle on our first try. On other occasions, we recognize things that aren't there at all. If we look at Georges Seurat's Neo-Impressionist painting *The Circus*, we recognize a female acrobat leaping off a horse. But we certainly don't "see" a real acrobat leaping off a real horse. Seurat has dabbed a bunch of colored dots on a flat plane that our mind *interprets* as a three-dimensional gymnastic vault. If we attempted to leap aboard the horse ourselves, we'd be in for a rude surprise.

Sometimes we even see and recognize things that are entirely nonexistent. For instance, we see—and recognize—a bright white triangle within the optical illusion known as a Kanizsa shape, its invisible edges somehow popping out toward us.

Neurological impairments can also reveal the functional distinction

Georges Seurat's The Circus, *1891*

between seeing and recognizing. People suffering from a condition known as hemifield neglect (often resulting from a stroke) report not *seeing* an object right in front of them (like a corkscrew) even though they behave as if they *recognize* what the object is (by asking for a wine bottle, for instance). People suffering from visual agnosia exhibit the opposite behaviors. They report seeing a familiar object (like a pencil) but are not capable of determining its identity ("It's a long, thin thing with a pointy tip").

KANIZSA TRIANGLE

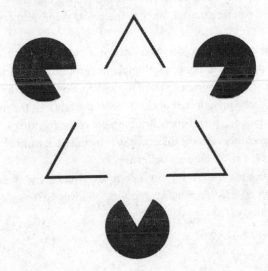

Seeing is more important for our survival than recognizing. Seeing solves the same challenge that the archaea mind solved: *targeting objects*. Seeing allows us to head toward the sun, stoop before we enter a low-ceilinged cave, and slap away a flitting wasp. Seeing enables us to do these things even if we don't recognize *what* we're seeing but simply perceive that there's *something* there and how distant that something is from us.

Though seeing is more crucial, recognition is the gateway to advanced intelligence. We recognize things so that we can decide what to do about them. We distinguish predators from prey, food from filth, shelter from snares. The ability to categorize the world around us is one of the most potent tools of module minds and is essential for rapid decision-making and long-term planning.

The Visual Scene module must support the needs of both seeing and recognizing. Each function makes use of different properties of a scene. To help accomplish effective seeing, the Visual Scene module must divide a scene into a *foreground* and a *background*. Imagine you are playing a game of tennis. To hit the ball, it's imperative that you distinguish the ball from everything else—your mind must assign the ball to the foreground of the scene, and everything else to the background. To help accomplish effective recognizing, in contrast, the Visual Scene module must extract all the visual *features* within the scene's foreground, such as the color, shape, and texture of the tennis ball.

The jumble of lines, colors, and dots in a painting of a circus can be assembled into any manner of intelligible foregrounds and backgrounds. Is the acrobat part of the horse or part of the audience or something separate? Is the hat part of the man or part of the floor? Is the fabric in the man's hand a scarf or a curtain or a surprisingly flexible piece of the painting's canvas? This is our first encounter with the uncertainty principle: determining the correct interpretation of an ambiguous pattern.

Eliminating the uncertainty in a complex visual scene is a formidable challenge, but all module minds can do it, including Earth's oldest module mind—the mind of fish.

3.

If you were transported to Earth four hundred million years ago, you would step out onto an emerald carpet of ferns and mosses stretching across Gondwana, a massive supercontinent that included present-day India, South America, Africa, and Antarctica. You might not even notice the fauna: the largest land creatures were mites and tiny wingless insects. In this miniaturized world, the tallest plants were barely two feet high, allowing you to see clear to the horizon in every direction, unblocked by any living thing—except for strange, moldy-looking obelisks protruding from the earth like giant rotten tusks, occasionally spiking twenty feet into the air. Scientists debate whether these life forms, known as prototaxites, were fungi or lichen, but either way, you would see and hear almost no activity on the surface of Gondwana, where the only sound was the whisper of wind fluttering through the ferns. Surveying such a spartan landscape, you might conclude that life on Devonian Earth was rather boring.

But if you peered into the enormous Tethys Ocean surrounding Gondwana, you would discover a luxuriant ecosystem brimming with life. The most intelligent marine creatures were fish. Many fish were covered with armor plates, resembling undersea battering rams, such as the Galeaspida, which had massive bone shields on its head. Some fish were barely the size of your thumb, while the fearsome Dunkleosteus stretched thirty feet long. Dunkleosteus was possibly Earth's first vertebrate superpredator, resembling a cross between a shark and the creature from *Alien*. Some fish never left the stupendous and magnificent coral reefs, others burrowed into the sand on the bottom of the ocean, while still others prowled the open sea. Several fish even gave birth to live offspring, including the most ancient creature known to do so, *Materpiscis attenboroughi*, named after BBC naturalist David Attenborough.

Fish diverged from other vertebrates about 450 million years ago, long before amphibians began to colonize the land. Because of their ancient pedigree, fish boast greater diversity than any other living vertebrate, with more than thirty-four thousand species. Many fish rely on vision as their primary sensory ability, using it to guide a wide range of behaviors, including locating prey, avoiding predators, choosing mates, shoaling, and navigating through coral, seaweed, branches, and rocks. Many visual talents of fish match those of land-born animals: three-color vision, discrimination of three-dimensional shapes, sophisticated targeting, and excellent object recognition all making use of an eye similar to our own, with a cornea, lens, pupil, iris, inverted retina, rods and cones, and a vestibulo-ocular reflex for stabilizing the image on the retina.

Scientists have conducted many studies evaluating fish's visual abilities, but there is a more casual means available to assess their visual sophistication: we can see for ourselves what it takes for Mother Nature to fool a fish. We can look at animals who conceal themselves behind camouflage or mimicry—such as leafy seadragons, octopuses, squid, cuttlefish, flounders, leaf fish, pipefish, frogfish, stonefish, rockfish, scorpionfish, candy crabs, and the pufferfish-emulating filefish—and consider whether our own eyes would be fooled by their visual deception.

The intricate verisimilitude of these creatures' disguises—such as the veins, stems, and air pouches on the fake seaweed fluttering off a leafy seadragon—reveals the depth of physical nuance necessary to dupe a predatory fish. Clearly, a fish's power of visual discrimination rivals or exceeds our own.

How does the Visual Scene module analyze a scene so effectively that the only way to deceive it is with near-total physical mimicry? The secret is complementary thinking.

Leafy seadragon, Australia

The Visual Scene module consists of two codependent neural systems. One system processes *boundaries*. The other system processes *surfaces*. Each requires assistance from the other to get its job done. By working together, the two complementary systems fashion a representation of a scene that can be used for both seeing and recognizing.

The Visual Scene module begins to construct a scene representation by processing inputs from the retinal sensory neurons using the module's boundary system. The boundary system identifies all the potential edges in a scene and creates tentative outlines of all objects, like a page from a coloring book. If there are any gaps or obstructions interrupting an edge, the boundary system will complete the line over the missing segment.

Why is it necessary to perform this "edge completion"? Because the retina itself—whether in a fish or a baboon—is splotched with gaps. Most prominently, there is a giant hole near the center of the retinal sensor array known as the blind spot, where the optic nerve connects with the retina. Any portion of an edge that falls upon the blind spot will be undetectable by the Visual Scene module. In addition, the retina is crisscrossed by veins. Wherever there's a vein, there can be no visual input. Thus, even if you are looking at a freshly painted yellow line running seamlessly down the center

VISUAL SCENE MODULE

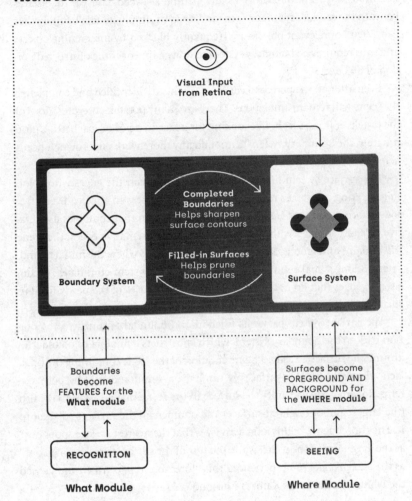

of a highway, the image of the yellow line that falls upon your retina will be fragmented into segments by your veins and blind spot. The only way the Visual Scene module can recreate the yellow line is if it "draws over" the rifts in the line.

Even if we somehow possessed a perfectly unblemished retina, the world itself is full of gaps and obstructions. Consider a fish's natural field of view. Potential prey—and predators—skulk behind seaweed, conceal themselves beneath the sand, dart among coral and rock, or simply huddle behind other fish. Views of relevant objects are frequently blocked by intervening objects and thus require the boundary system to draw a line over the obstructed portion of an edge.

Even after all the potential edges in a scene are identified and completed, the scene will remain ambiguous. There are many possible interpretations of the tangle of lines, sort of like looking for a face in a pot of spaghetti. Unfortunately, the boundary system cannot do any more work on its own. It needs help. It shares its first-pass collection of edges with the surface system.

The surface system fills in all the spaces between the edges with color, thus creating a first-pass representation of the surfaces in a scene. Even after the surface system has created a new representation that combines its tentative filled-in surfaces with the boundary system's tentative edges, the scene *still* remains ambiguous. There will usually be places where the initial boundaries no longer make sense. So, next, the surface system eliminates "inconsistent" edges that do not overlap with the boundaries of the newly filled-in surfaces. This is called "boundary pruning."

The surface system passes its filled-in and boundary-pruned representation back to the boundary system, which constructs a final set of boundaries around the surfaces. Lastly, a representation of the scene containing the finalized edges and filled-in-surfaces is worked on simultaneously by both systems. Together, they identify the surface in the scene that is closest to the fish (the largest and/or brightest surface) and dampen all the surfaces adjacent to it. The consequence of this joint activity is that the nearest surface "pops out" as the *foreground* of the scene, while the rest of the surfaces are "dimmed out" as the *background*. (You experienced this "foreground pop" when you viewed the white triangle in the Kanizsa illusion.)

A joint representation of the foreground and background surfaces is sent to the Where module for *seeing*. The foreground is presumed to be an object worth identifying, and therefore a representation of all the edges in the foreground (along with its colored surfaces) are passed on to the What module for *recognition*. The foreground edges and surfaces are treated as the *features* of the potential object.

It would not be possible to assign a foreground and a background to a scene, or derive the features of a potential object, without the boundary system and surface system interacting with each other like yin and yang. We can witness the results of the Visual Scene module's complementary thinking by regarding an optical illusion known as the "neon color spreading illusion." When we look at the square in the center, it seems to be gray colored, even though it's perfectly white (or the same color of the underlying paper, if you're reading this in print).

NEON COLOR SPREADING ILLUSION

Here's what happens when our Visual Scene module processes the neon color spreading illusion. The boundary system constructs the tentative edges of a (nonexistent) square along all the edges where the gray curves meet the black curves, completing the edges across all the white gaps. Next, the surface system fills in the lighter sides of these edges—that is, the gray sides—with gray color, which spills into regions that are objectively white. Next, the boundary system uses this filled-in gray square to reinforce its tentative edges of the square. Finally, the boundary and surface systems work together to make the (nonexistent) gray square pop out as the foreground and to dim down the black "Pac-Man" circles as the background.

The Visual Scene module then sends the final representation to the visual Where module (which *sees* the square) and the visual What module (which *recognizes* the square). The result: we are conscious of a square that isn't there.

4.

Why do some mental representations enter conscious awareness, while others do not? The activity in the Visual Scene module permits us to begin answering this question. The reason that the preliminary visual representations formed within the Visual Scene module never enter consciousness without the involvement of another module is to prevent the rest of the mind from acting upon erroneous or incomplete knowledge. In effect, unconscious representations protect us from getting hoodwinked by fake news.

Until the Visual Scene module fully completes its processing of a visual scene (generating tentative edges, generating tentative surfaces, finalizing edges, determining the foreground and background and the foreground features) and assembles a definitive representation of the scene, the preliminary representations still contain too much uncertainty to be useful for seeing or recognizing.

If we were conscious of one of the representations formed within the Visual Scene module, it would be like walking onto the set where they're shooting a Hollywood sci-fi movie and finding Sandra Bullock swinging from bulky cables in front of a blue screen. It's impossible to guess what the movie will look like until all the special effects and digitized space debris are edited into the final cut.

The Visual Scene module addresses the uncertainty dilemma by waiting to share the fruits of its labor with the consciousness-generating modules until it's stitched together a representation worthy of attention—worthy, that is, of being seen or recognized. Nevertheless, even though the internal machinations of the Visual Scene module remain the sole purview of the preconscious proletariat, the hidden details of these operations can occasionally become manifest to us through optical illusions, such as the neon color spreading illusion or the Kanizsa shape illusion. It's not just humans who are susceptible to these Visual Scene module–duping illusions. Fish are, too.

So far, every species of fish that has been tested with Kanizsa illusions has been tricked into seeing nonexistent triangles, circles, and squares, including goldfish, redtail splitfins, bettas, and even the oldest living species of fish, sharks. (The Kanizsa illusion is also effective with amphibians, birds, and reptiles.) Fish are capable of "amodal completion," recognizing partially obscured objects, such as identifying a crab by the tip of its claw. Fish are also tricked in the same manner as humans on a variety of other illusions, such as brightness illusions, spatial arrangement illusions, expansion-contraction color illusions, and illusions of relative size. Fish also see other illusory contours, such as invisible lines along two misaligned gratings. All these illusions are explained by the dynamics of the Visual Scene module.

These findings highlight one of the trade-offs inherent to addressing the uncertainty dilemma. If a mind develops thinking powerful enough to break high-grade camouflage, then that mind will inevitably become vulnerable to illusions.

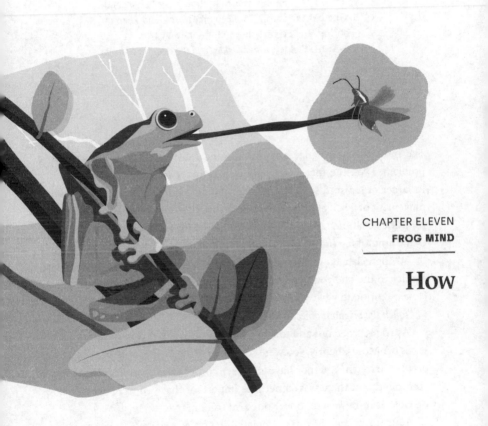

How

My advice, Sophia, is to aim for a high mark. I expect you'll hit it. Not on the first shot, nor the second. Maybe not the third. But keep on aiming and keep on shooting, and eventually you'll hit it every time.
—Annie Oakley, *Annie Oakley and the Beast of Chicago*

1.

The most basic challenge any living mind must confront is the targeting problem. Even the meek archaeon can meander toward a destination. But in larger organisms, solving the targeting problem requires the purposeful movement of a body part to a specific destination. Grab that acorn. Claw that prey. Kick that predator. Peck that bark. Head-butt that rival. Or, in the case of the amphibian order Anura, chomp that bug.

To appreciate how a module mind targets an object, try performing a task similar to the one you performed at the start of our journey. This time, pay close attention to what you're *aware* of as you execute the task:

Touch the brightest object within your reach.

As you reached out and touched something, what, precisely, were you *conscious* of? Most saliently, you were conscious of the target itself: the object you decided to touch. You may have also been conscious of your decision to move your hand. But there was something important that you were *not* conscious of. Determining *how* to move your arm from here to there.

Your mind somehow (1) calculated the trajectory your finger needed to follow as it proceeded from its resting spot to the target, (2) ascertained which muscles in your body were necessary to propel your arm, hand, and finger along the trajectory, and (3) ensured that each muscle applied just the right amount of force in the right direction at the right time. All of this happened without your awareness, which was focused on the target itself and, perhaps, the intention to touch it.

Even if you take conscious control of your movement trajectory—say, by deciding that your finger should loop in a corkscrew motion as it zeroes in on the target, like a magician twirling a wand—you can maintain awareness of your self-willed command to twirl or speed up or slow down, but the calculations your mind performs "under the hood" to accomplish these commands (as well as the calculations and commands necessary to hold the rest of your body steady and maintain a fixed gaze on the target as you reach for it) remain outside the bounds of your conscious experience.

The part of your mind responsible for reaching for a target is the How module. (It figures out *how* to get from here to there.) If you harbor any doubts that the activity of the How module is unconscious, try this. Explain to someone how to press a button on an elevator. What instructions could you provide them besides "Press it" or perhaps "Move your finger here and push"? It seems straightforward enough. But what if the person confirmed they understood your instruction but ended up pressing their nose instead? Not out of spite or humor. They insist they *tried* to press the button; they're not sure what went wrong. What additional guidance could you give them that would enable them to adjust their performance? "Ah, you were extending your flexor carpi ulnaris muscle instead of tensing it, and you were twisting your palmaris longus to the left instead of to the right and you waited a moment too long to flex your extensor digitalus. Try again and this time hold that flexor carpi ulnaris steady!"

There's a good reason the How module's activity never enters consciousness. The How module doesn't create any representations it needs to share with other modules. Conscious activity always involves making a mental representation available to other parts of the mind. Instead, the How module compares two representations it receives as input from other modules, without generating any new representations as outputs.

As it happens, all the modules that *do* generate conscious experience also compare two representations. Appreciating the differences between these divergent modular dynamics—one unconscious, one conscious, each involving the comparison of two representations—will help us understand how consciousness works.

2.

Every organism we've encountered so far can move toward a target, albeit somewhat imperfectly. The haloarchaeon zigzags toward the sun. The amoeba oozes toward a civil engineering project. The fly flitters toward rotting meat. But there are marked differences between the bodies, environments, and "brains" of these molecule and neuron minds and those of module minds.

First and foremost, when we enter the realm of vertebrates—fish, amphibians, reptiles, birds, mammals—we are dealing with larger bodies with significant mass. The prospect of physical injury now becomes an ongoing and potentially lethal concern. An amoeba need not fear a collision with a hard surface. A fly can walk—or at least stagger—away from a crash landing. But if a bear misdirects a claw strike, she could end up breaking her arm. If an eagle bungles a landing, he could snap his neck. When you inhabit a fleshy bag of bones, accuracy in one's targeting is a matter of survival.

The scale of a mind's environment also makes a big difference. Archaea and bacteria are endlessly buffeted by the molecules in their liquid environment (activity known as Brownian motion), rendering precision targeting impossible. Herky-jerky routes to an intended destination are the only ones available to such microscopic beasties. The fly is battered by microcurrents of air, imperceptible to us, disrupting flight paths. Larger bodies may be more vulnerable to injury, but greater size also reduces friction during movements, which means that it's physically possible to accurately target things. This is where the How module gets into the game.

The defining challenge for the How module is to guide a body part to a desired target. To see how module minds solve this challenge, let's look at one of the oldest terrestrial vertebrate minds on Earth, the mind of the frog.

The frog is the first and only amphibian on our journey. Its brain is hardly more sophisticated than a fish brain. This is not surprising, considering a frog spends part of its life cycle as a "fish": as a tadpole swimming through a pond. The frog brain is quite similar to the shark brain, particularly an order of sharks that contains dogfish sharks and lantern sharks.

Feeding on land, however, presents problems that fish never faced. The most onerous is the fact that opening the mouth in air creates less suction than opening the mouth underwater. Fish rely on aquatic suction to help suck in prey. To deal with this challenge, the earliest frogs lunged at prey using newfound legs. Even today, though most frogs have evolved an elongated tongue that can strike prey from a distance, many species continue to hunt using lunging alone. New Zealand, for instance, is filled with frogs who bite and gulp their prey rather than glomming onto them with their tongues.

One species of frog that hunts by leaping at prey and biting them directly is the African clawed frog, *Xenopus laevis*. It is greenish gray, about four inches long, and native to the African Rift Valley. The African clawed frog prefers stagnant water over fast-moving streams, especially flooded rainforests. It is a voracious predator. Its principal diet is water bugs and small fish, though it also eats worms, spiders, insects, and even other frogs.

For an African clawed frog to snatch a meal, its How module must guide its mouth to its prey.

3.

The How module has three main components, shown in the figure below. The first is the *target position*. This is a representation of the target's location. For example, the location of a grasshopper resting on a leaf.

The representation of the target position is supplied by the Where module, which generates representations of the location of objects in a visual scene, encoded as a set of visual coordinates. (The Where module figures out *where* something is.)

The second component is the *body position*. This is a representation of the location and orientation of the body part that will be doing the reaching. (For example, the frog's head.) The body position representation encodes the body part's location as a set of body coordinates generated by (preconscious proletariat) mechanical sensors within the relevant body part.

The third component is the *difference engine*. This is the heart of the action in the How module. The difference engine compares the target position and body position to determine how far apart they are. (Such as the distance from the frog's head to the grasshopper.) But before it can compare the two position representations, it must first solve a translation problem.

The grasshopper's visual coordinates and the head's body coordinates are encoded in two distinct formats. It's as if one representation were Fahrenheit

HOW MODULE

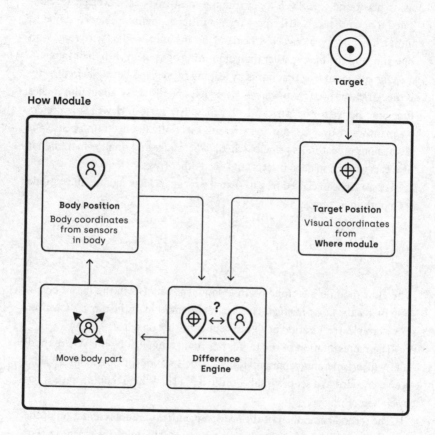

and the other Celsius. The difference engine must translate one representation into the same format as the other. As it turns out, the visual coordinates of the target position are converted into the same format as the body coordinates.

Now the difference engine can compare—and act upon—the two positions. The difference engine pursues a simple goal: to reduce the distance between the target position and body position to zero. The difference engine assesses the present size of the distance, then issues a command to the relevant doer network, which starts moving the frog's head in the direction of the grasshopper. As the frog moves, its head position changes. Its new head

HOW MODULE

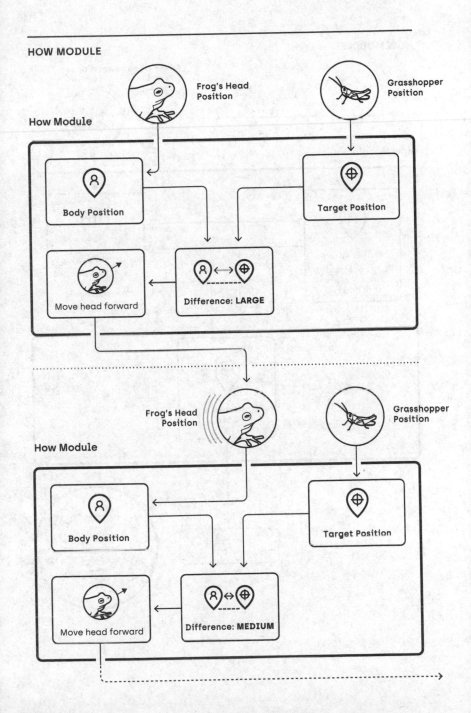

HOW MODULE

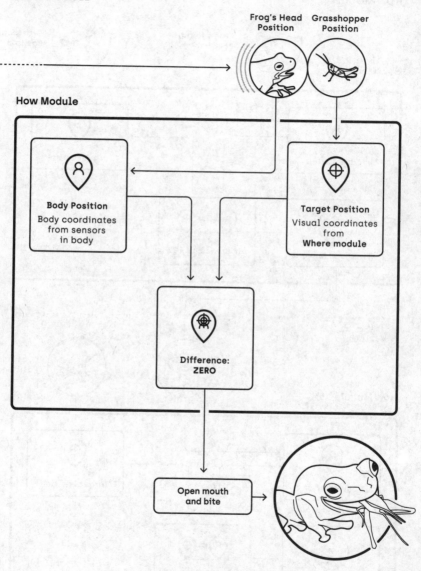

position is fed back into the difference engine, which continuously evaluates the gap between its head and the grasshopper and continuously adjusts its motor command to keep reducing the gap. This real-time process is known as *mismatch learning*.

Mismatch learning is always nonconscious. The How module assesses the mismatch between the two sets of coordinates and attempts to drive the mismatch to zero. When it succeeds—when the head reaches the grasshopper—the difference engine halts and the How module initiates a command instructing the mouth to chomp the grasshopper. Dinner is served!

4.

How does the How module learn to translate the target position into the same format as the body position, a process that neuroscientists call *mapping*? A frog mind cannot hard-wire the target-position-to-body-position map, because its body is always changing. Pretty dramatically, in fact, as it develops from a limbless pollywog into a jumbo-legged frog and then a bigger frog. The How module must also update its map to cope with injuries and physical constraints, such as a branch pressing against its back.

The process of mapping target positions onto body positions starts early in life, with a mental dynamic known as *motor babbling*. During motor babbling, the How module enters mapping mode and solicits help from the Where module. The How module directs the body's limbs to move around randomly and observes them as they fidget about.

Consider motor babbling in a human infant. She spontaneously waves her hand in front of her face. As she does, the Where module steps in. The Where module figures out *where* the hand is located in space and creates a representation of the hand's visual coordinates. These coordinates are fed back to the How module. At the same time, the body position system (known as the proprioception system) generates a representation of the hand's body coordinates and sends these coordinates to the How module, too. Because the How module is in mapping mode, it knows that the target position is the same as the body position (the target *is* the body!) and so the How module updates its map by linking the visual coordinates to the body coordinates.

The baby continues wriggling its hand until the How module has a comprehensive map of the entire space within its reach.

Now baby is ready to grab hold of the world!

HOW MODULE: MAPPING MODE

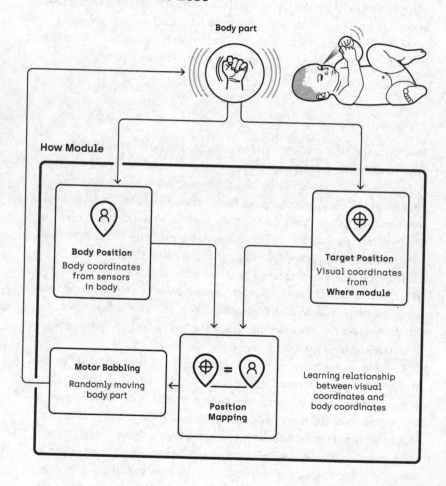

5.

The How module does not need to worry about the uncertainty dilemma: its two representations have already been vetted by other parts of the mind. The Where module certified the target position, while the proprioception system certified the body position.

The How module doesn't need to directly address the attention dilemma, either—it doesn't need to figure out what to focus on. When you touched a bright object, your conscious attention was on the object itself. (As we'll see later, this conscious attentional focus was governed by your Where module.) Because the How module quietly does its work in the background without ever needing to modulate the mind's spotlight of attention, the How module's dynamic doesn't become conscious.

The How module can't avoid the stability dilemma, however: it must decide how long to remember a particular mapping between a body position and a target position. The How module's solution? Forget the past entirely.

The How module's map linking target positions to body positions must be constantly updated. Whenever the body changes because of growth, atrophy, muscle development, injury, aging, or confinement, the How module must learn a new map. When it does, it immediately and completely forgets the old map.

There's no need to remember your body-to-target map from when you were three years old, or when you broke your arm last year and had to wear a cast, or when you were jammed into a crowded elevator this morning and had to reach around a bunch of people to press the button for the tenth floor. The How module does not need to create enduring long-term memories. It only needs to maintain its current map until the next adjustment comes along.

The How module compares two mental representations, but it does not need to certify, stabilize, or draw attention to either representation. Nor do these representations enter consciousness. This observation provides our first clue regarding what we might find in those modules whose representations *do* enter consciousness . . .

What

> Rich people's garbage was every year more complex,
> rife with hybrid materials, impurities, impostors.
> Planks that looked like wood were shot through with
> plastic. How was he to classify a loofah?
> —*Behind the Beautiful Forevers*, Katherine Boo

1.

In 2014, Noah Strycker's father dropped him off at the Mexican border with a backpack full of supplies. He was supposed to be starting a five-month fieldwork project in Venezuela, but it got canceled at the last minute, unexpectedly leaving him without any plans. Noah decided to take the opportunity to hike the 2,650-mile Pacific Crest Trail from Mexico to Canada, alone. "I've always loved hiking, and the Pacific Crest Trail is sort of the Mt. Everest of hikes, and I thought this was just a great chance to finally do it. I saw 174 species of birds on the trail. And that's where I hit on the idea of doing my Big Year."

In the world of bird-watching—or birding, in the lingo of bird-watching enthusiasts—a "Big Year" is when you attempt to see as many species of birds as you can in a single year. Even at the relatively youthful age of twenty-eight, Noah had already traveled to birding sites all over the globe. He was experienced at organizing trips to remote locations and had developed extensive contacts with birders near these sites who could help him arrange birding tours and who might even let Noah sleep on their sofa for free. Because he carefully planned his hiking route to maximize the number of birds he would see, his Pacific Trail adventure taught him how to visit the most locations in the shortest period of time.

The existing Big Year record was held by a British couple who managed to spot 4,341 birds in 2008, an impressive feat, especially considering they frequently returned home between birding tours. "I thought I could actually hit 5,000 birds, and that was my goal," Noah says. "My Big Year started on January 1, 2015, in Antarctica. I saw a Cape Petrel shortly after sunrise."

Noah traveled to every corner of the planet. He visited the Amazon rainforest, the Himalayan mountains, the Australian outback, the Sahara. He sojourned in the sun-soaked San Joaquin Valley, the fjords of Sweden, and the papyrus swamps of the United Arab Emirates. In these exotic locales, he spotted and *recognized* bird after bird.

He identified the white-throated caracara in Chile, a black-and-white bird of prey. He identified the double-toothed barbet in Ghana, a small black, white, and red bird with an awkward flight and a song like a cat's purr. Under the midnight sun in Iceland, he identified a whooping swan, a huge bird that can attain an eight-foot wingspan. He identified the elusive watercock in a rice paddy in Indonesia, a duck-like waterbird with long yellow legs. From a boat on the rough seas of the Hauraki Gulf, he identified the critically endangered New Zealand storm-petrel, a small seabird with a white rump and distinctively streaked underbody.

These birds did not fly up to Noah and land on his finger. He had to identify many of them in taxing conditions. In the dark, in the rain, in the fog. At great distances, in dense vegetation, among mountains. He had to identify birds hurtling so fast they were a blur and motionless birds with superb camouflage. Sometimes all he could see was a flash of color or a fragmentary shape; occasionally he only heard their song.

On September 16, in India, he identified a Sri Lanka frogmouth, a nearly invisible bird that resembles a pile of dead leaves. That put him ahead of the previous Big Year record. On December 31, he identified his final bird when he became the first person to photograph a wild Oriental bay owl in the Soraipung forest in India.

Noah had shattered the old record by recognizing 6,042 birds—more than half the avian species on Earth. It was a grand achievement for Noah. It was a grand achievement for birding. But it was also a grand achievement for the What module, the part of the mind responsible for recognizing *what* a thing is, including a thing with feathers.

2.

Module minds must contend with a prominent feature of their environment that molecule and neuron minds can afford to disregard: complex objects. The world of vertebrates is full of swaying branches, slippery pond stones, hollowed-out trunks, puddles of mud, bright-hued fruit, and leathery wasp nests. Some things can be exploited. Some things are best avoided. Most things can be dismissed as irrelevant.

To distinguish the helpful from the harmful, module minds must learn to recognize, categorize, and make predictions about an ever-fluctuating cascade of whatchamacallits. Each module mind has multiple What modules dedicated to identifying different categories of perceptual patterns, including a visual What module for identifying visual objects and an audio What module for identifying sound-producing objects. Recognizing objects isn't easy. The visual What module must overcome several thorny challenges, such as the object localization problem: determining which part of a visual scene the What module should treat as a potential object. *Should I try to recognize the tree-thing or the feather-thing-in-the-tree or the wing-thing-on-the-feather-thing?* Another challenge is the many-to-one mapping problem. A What module must learn that a wedge-like tail, stubby webbed feet, a red eye, black flippers, a white stomach, and a yellow-striped bill all represent different views of the same object: a penguin. The What module tackles these two visual recognition challenges with an assist from a complementary module we'll meet in the next chapter.

One of the toughest challenges any What module faces is the fact that nothing in nature comes with a label or FAQ. Nobody tells a lizard, *This here is a beetle, but that there is a peanut.* Things and events simply stream by for the mind to sort through and categorize without any external guidance, a plight that scientists refer to as unsupervised learning. Unsupervised learning magnifies the What module's uncertainty dilemma: how do you know if the object in front of you is something totally new or is a different version of something you've already seen? Is it a crow or a new kind of blackbird? What if you see a pigeon that looks like a cross between a crowned pigeon and a crested pigeon? Which category should you categorize it into—or should you create a new crested-crown pigeon category?

If that weren't hard enough, an effective What module must learn all the details of a new thingamajig on the module's first exposure to it. Consider how much you can recall from a movie after a single viewing. You can usually recall the main characters, the actors who played them, the setting, the plot, important props (such as the light saber, the Sorting Hat, or the iron throne), how the movie opened (the villain killed all his henchmen during a bank robbery), and how the movie ended (the ship sank and the hero drowned)—complicated and fast-moving patterns of real-time data that our minds somehow learned, categorized, and stored almost instantly.

To accomplish this, the What module must also resolve the stability dilemma. The What module must instantly generate a new category for a novel object and store that category in long-term memory, but it also needs to modify existing categories as the mind learns more about their individual members. (Ah, the male yellow-bellied sapsucker has red on its throat, but the *female* yellow-bellied sapsucker has white on its throat.) But if a module can endlessly modify existing memories, this presents another risk known as catastrophic forgetting.

What if you see, in quick succession, four objects categorized as birds: a hummingbird, a sparrow, a vulture, and an ostrich? If the What module updates its memory of "bird" after each new example, then by the time it processes an ostrich, won't it have completely forgotten what a hummingbird looks like? This problem is even more ticklish with highly similar variants. How can you remember what a rock pigeon looks like without eroding your memories of a crowned pigeon, crested pigeon, and Wonga pigeon?

Even when the What module successfully recognizes an object, its job is still not done. It must alert the rest of the mind of its achievement, particularly if the object is new or relevant. This brings us back to the foremost challenge troubling module minds: the attention dilemma. If the What module recognizes an important thingamabob, how can it command the other modules to drop whatever they're doing and focus on the thingamabob instead?

The What module, then, must solve the stability dilemma, uncertainty dilemma, and attention dilemma. During a single brief exposure to an object. Without any external supervision. In real time. This is not easy. Indeed, object recognition is so complicated—and so indispensable—that the innovation that remedies the What module's challenges gave birth to consciousness in the universe.

To help us unravel the dynamic that empowers our mind to distinguish a house from a house finch and a house finch from a hawfinch, we will look at the What module through the eyes of a creature with an undeserved reputation for witless stupidity.

The tortoise.

3.

The red-footed tortoise, *Chelonoidis carbonaria*, is native to Central and South America. It's about a foot to a foot and a half long with its head out and is endowed with a dark, bread-loaf-shaped carapace speckled with yellow patches. This land-dwelling reptile inhabits a medley of ecosystems, including tropical rainforests, dry thorny forestland, and grassy savannas. It has good color vision and exhibits individual color preferences. A tortoise is a land-dwelling turtle. (All tortoises are turtles, but not all turtles are tortoises.) Like all reptiles, turtles have long been referred to as cold blooded, though this is an outdated label. Nowadays they are classified as ectothermic, meaning they rely on external heat sources to manage their metabolism.

All reptiles, you might guess, are descendants of dinosaurs. Not true. The only living descendants of dinosaurs are the birds. There are three groups of reptiles who separated from the branch leading to dinosaurs long before *Tyrannosaurus rex* stalked Jurassic jungles. One group is the squamates: snakes and lizards. Another is the crocodilians: crocodiles, alligators, caiman, and the native-to-India gharial. The oldest group contains the senior statesman for reptile nation: the turtle.

One of the oldest known turtle fossils with a distinct shell is *Odontochelys semitestacea*. This prototurtle dates from around 220 million years ago, before the Jurassic. That's a long, long time ago. Few creatures can claim to have been so little modified by the eons. Turtles from 50 million years ago do not look much different from turtles today, with blunt heads and hard shells covering their top and bottom. Apparently, dwelling inside a mobile fortress is a pretty good way to ensure your survival. Scientists have also discovered that turtles are pretty good at identifying objects.

In one intriguing experiment, red-footed tortoises were trained to recognize various foods (including kiwis, cucumbers, melons, tomatoes, and plums) and an assortment of other items (including pebbles, branches, bottle caps, and colored paper). Next, the tortoises were presented with pairs of color photographs: one picture of food and one picture of nonfood. The tortoises reliably preferred the food photos, frequently biting them as if they fully expected to win a meal. Though the tortoises' recognition capability isn't sophisticated enough to distinguish a two-dimensional image of a thing from the thing itself, the tortoise mind can correctly identify many objects from their photos, suggesting that they are using abstract features of objects to recognize them, rather than fixed visual patterns. (The flat, dot-speckled horse we recognized in Seurat's *The Circus* doesn't resemble a real horse at all in literal terms, but we easily identified it from its apparent mane, head, tail, and body.) Primates aren't perfect at distinguishing pictures from objects, either. Though a chimpanzee will prefer a real banana to a photo of a banana, if a real banana is not on offer a chimp may munch a photo of one.

In another experiment, several red-footed tortoises were shown food on the other side of a fence. None of the tortoises who were shown the food figured out how to reach the treats. Next, the tortoises were permitted to observe as another, more experienced tortoise managed to access the food by following a long path around the end of the fence. Afterward, all the spectators successfully duplicated their colleague's route and reached the food—an unexpected instance of social learning in reptiles. Consider what was necessary to accomplish this feat. The observer tortoises had to recognize the food and the fence. But they also had to recognize another tortoise as a fellow *Chelonoidis carbonaria*—as a thing worth watching and emulating.

How does the What module recognize an object? By comparing facts to speculations.

4.

The heart of the action in the What module is the same as the heart of the action in the How module: a comparison of two mental representations. Yet, the dynamics involved in each module's comparison process are practically opposites.

The What module compares a *bottom-up* sensory representation of a perceived object to a *top-down* representation of the expected identity of the object. The bottom-up input can be thought of as the observed evidence. The facts on the ground. In a visual What module, the bottom-up representation consists of the *visual features* of the scene's foreground, supplied by the Visual Scene module. The foreground, recall, is the part of a visible scene that the Visual Scene module believes is most likely to be an object of interest (such as the bird-shaped figure on the tree branch in front of you).

The top-down representation can be thought of as the hypothesis to be tested. The prediction. The top-down representation is drawn from the What module's memory of previously encountered objects. If you're playing tennis, your What module might expect to see a tennis ball. If you're birding on a New England lawn, your What module might expect to see an American robin.

The What module compares the observed evidence to the hypothesis and decides whether or not they match. Is the object a member of the expected category, or not? The outcome of this comparison drives the module's subsequent activity. To see how all this works, let's trace the operation of the What module in a tortoise looking for a snail to eat. We'll consider three scenarios:

(1) The tortoise sees a snail.
(2) The tortoise sees a rock.
(3) The tortoise sees something it's never seen before.

Scenario (1): The tortoise sees a snail.

The tortoise's desire to eat a snail activates the What module, which summons a representation of the "snail" category from its memory. This top-down representation serves as the visual expectation guiding the tortoise's search. Meanwhile, the tortoise surveys its environment. The Visual Scene module (nonconsciously) evaluates the scene, identifies the scene's foreground, and transmits a representation of the foreground's visual features to the What module. These features becomes the bottom-up facts: a candidate snail.

Now comes the main event. The What module compares the facts to the speculation. It does this by focusing on the *features* of the object in question. The What module compares the features of the bottom-up object to the features of the top-down "snail" category. (You can think of this process as something like comparing two melodies—maybe one melody is played a little faster than the other and a couple of notes are off-key, but you still recognize

HUNGRY TURTLE LOOKING FOR A SNAIL TO EAT

Scenario 1: A Snail

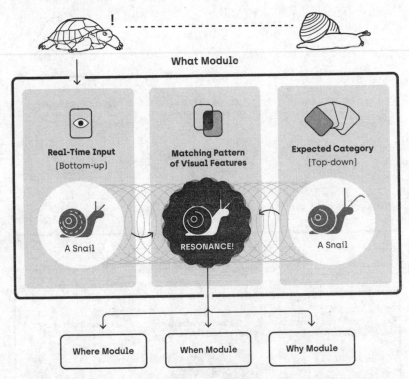

they're the same tune.) If there's a good match between their features—if they both contain a circular shell and a blobby head and two antennae sticking out—something truly amazing happens.

Resonance.

Resonance is the centerpiece of Grossberg's unified theory of mind. It is the most important mental dynamic in the journey of Mind for the simple reason that all conscious experiences involve resonant dynamics. When resonance occurs, it serves as a signal indicating that the What module has successfully recognized the object. Mission accomplished: *it's a snail!* The uncertainty dilemma has been resolved. Go eat that slimy, crunchy treat!

More on resonance in a moment . . .

HUNGRY TURTLE LOOKING FOR A SNAIL TO EAT

Scenario 2: A Rock

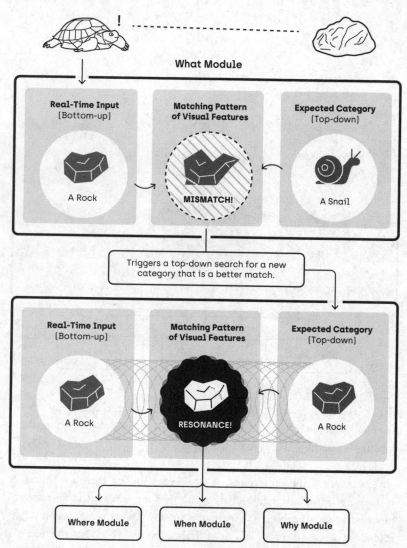

Scenario (2): The tortoise sees a rock.

Once again, the What module expects to see a snail. But this time, the candidate object is a rock. The What module compares the bottom-up rock and the top-down "snail" and determines that even though they may share a few features in common, they are not similar enough to consider them a match. Instead of triggering resonance, the mismatch triggers a *reset*.

The reset clears away the top-down "snail" expectation and causes the What module to begin searching its memory for a better match for the unexpected object. When it finds the best possible match—a "rock"—it compares this new top-down prediction with the bottom-up observation. Match! The two representations begin to resonate, but since the turtle is looking for a snail, not a rock, the What module directs the Where module to scan the scene for another potential snail.

Scenario (3): The tortoise sees something it's never seen before.

Our hungry tortoise pushes onward and spots another object of interest. Is it a snail? Once again, the What module compares the bottom-up features with the top-down features. Mismatch! The What module *resets* its top-down expectation and searches its memory for another category that might be a better match. But unlike in the previous scenario, this time the What module cannot find any categories in its memory that are a good match for the unknown object. The absence of any match triggers the third and final type of activity in the What module, *learning about a new object*.

The What module broadcasts a global signal of arousal that incites the entire mind to go on high alert. This generates a feeling of surprise, curiosity, or caution and prompts other modules to prepare for action, though it's not yet clear what the necessary action might be. This sudden spike of arousal is why many of the "firsts" in our life are so vivid in our memory, such as a first kiss, a first day on a job, the first dead person we ever saw.

The What module creates a new category representation from the bottom-up features (in this scenario, a "scissors" category). The observed facts thereby become a future prediction. The What module stores *all* the perceived features of the new object, such as two crisscrossed steel-colored blades and two blue circles. The new category will often contain features that may eventually prove to be irrelevant for identifying other objects in the same category. If the first pair of scissors you ever saw was blue, your What module would

HUNGRY TURTLE LOOKING FOR A SNAIL TO EAT

Scenario 3: An Unknown Object

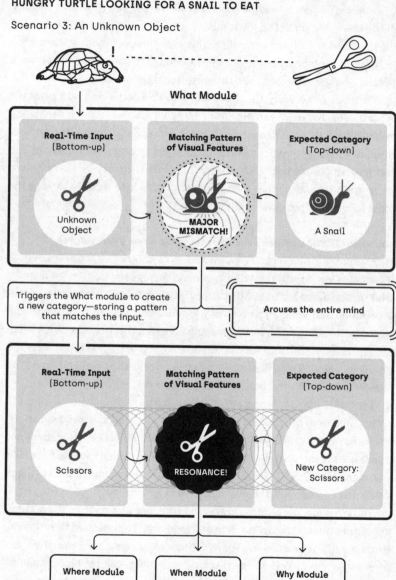

presume that other scissors were just as likely to be blue as they were to have crisscrossed blades. Over time, as you encountered more examples of scissors, your What module would learn that color is not a defining feature of "scissors," and that feature would get dropped from its category representation.

Because a category representation is made of specific sensory features (a "melody") rather than statistical data, each category stored in memory is a kind of prototype of an object, rather than an average. That's why when we think of birds, we instinctively think of small flying critters with feathers and wings. Ostriches and penguins do not naturally fit our mental prototype of birds, and indeed, without a biologist teaching us about them, our minds might not even categorize them as birds.

The What module implements a sophisticated solution to the stability dilemma. We saw in the previous chapter that the How module (how do we grab that object) forgets the past as soon as something changes. The What module's memory is more flexible. If it encounters a new object, it creates a new category in its memory. If the What module never encounters another example of that category, then that category representation will remain stable and unmodified in the What module's memory for the long term. But if it encounters different versions of the object, it will adjust its stored prototype to take into account the new examples. After observing your first pair of scissors (which happened to be blue), if you then see green scissors and black scissors and pink scissors, your What module will drop the blueness feature from its "scissors" prototype. After that, if you're searching for scissors, the blueness of a potential pair of scissors will not trigger resonance.

5.

Through resonance, a bottom-up representation of an observed object melds with a top-down representation of an object prototype, creating a unitary and harmonious dynamic that rings out in the mind like the final chord from the Beatles' *A Day in the Life*. The mental dynamic of resonance *synchronizes*, *amplifies*, and *prolongs* the neural activity that embodies the two representations.

Resonance is like the same melody played on two different instruments simultaneously, such as a flute and a violin. When the two instruments *syn-*

chronize their performance so that they play the same notes at the same time, the melody's volume grows louder. The sound is *amplified* because the amplitudes of the two instruments' sound waves are added together. Amplification explains why many people cheering at a sports stadium can produce vibrations of the same intensity as a small earthquake, even though a single person cheering would never be able to do so. The amplification of the neural activity of two resonating representations makes it harder for noise or irrelevant activity to interrupt it and suppresses competing activity elsewhere in the mind, helping to draw attention to the recognized object. Imagine a lake where kids are tossing in pebbles, sending out little ripples across the surface of the water. Then a full-grown man cannonballs into the lake from the dock. This produces large waves that overwhelm and quash the smaller ripples.

Resonance also *prolongs* the neural activity of the two representations. If every member of an orchestra plays the final note of a symphony simultaneously (synchronously, that is), the sound of that note will reverberate longer than if a single violin bowed the note. Prolonging the activity of the resonating representations also helps to suppress activity elsewhere in the mind and stabilizes the object representation, making it easier to share with other modules.

Resonance instigates a variety of other mental activities. Resonance triggers *learning* in the What module. Resonance *draws the attention* of the other modules. And resonance *begets consciousness*—the transcendental experience of *knowing* what you are looking at.

We'll learn how resonance generates consciousness when we get to chimpanzee minds, but we'll need to learn about a few more modules before we do. For now, let's compare the *match learning* that takes place in the consciousness-generating What module to the *mismatch learning* that takes place in the nonconscious How module. The dynamics of the How module are *closed*: the neural activity of mismatch learning does not influence the activity of other modules. The How module is listening to music in his basement with his earphones on. In contrast, the dynamics of the What module are *open*—they interact with the dynamics of other modules and influence their activity. The What module blasts his songs on his patio speakers for all the neighbors to hear.

The How module is designed to forget what it's previously learned, to keep up with the latest changes in its body. (There's no point in remembering how you moved your muscles to turn off a light switch when you were seven years old.) The What module is designed to remember objects permanently (including objects encountered only once), but it's also designed to update those memories as it learns more, without eroding the memories of similar objects.

Mismatch learning is inherently unstable. The How module attempts to reduce the mismatch between its representations to zero, at which point the How module shuts off. Match learning is stable. Resonance stabilizes the activity of the What module's representations so that other modules can make use of them. Thus, the What module solves the stability dilemma by storing long-term memories, making changes to these memories without suffering catastrophic forgetting, and stabilizing short-term object representations.

Resonance also resolves the uncertainty dilemma by certifying that an object has been successfully identified. If an object representation is resonating, its identity is no longer uncertain.

Resonance's most impactful role is to resolve the attention dilemma by alerting the rest of the mind that something is worthy of attention. Resonance is like the "flag" in an email program. If you want to tag a message as important, you flag it. Through resonance, consciousness flags mental events as important in a manner that makes it easy for other parts of the mind to notice. (In mathematical terms, consciousness adds an "extra degree of freedom" to a representation.)

You can think of resonance as the basketball in a basketball game. The holistic, real-time activity of all ten players makes a game a game. Nevertheless, the focus of all ten players—the thing driving most of the game's action—is the basketball itself. All the activity on the court is directly or indirectly influenced by the movement of the basketball. It's the same with resonance. Whenever something new starts resonating in the mind, it's like the basketball has changed hands.

The rest of the game must respond.

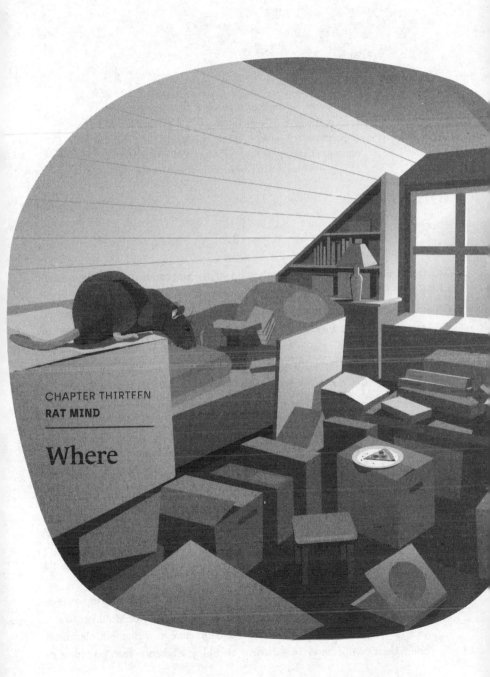

CHAPTER THIRTEEN
RAT MIND

Where

> I am always in hope of making a discovery there, to express the love of two lovers by a marriage of two complementary colors, their mingling and their opposition, the mysterious vibrations of kindred tones.
> —Vincent van Gogh, letter to his brother Theo

1.

Imagine you are a pair of disembodied Arms. You can touch things, you can grab things, you can manipulate things. What you can't do, however, is stride from here to there. Imagine, moreover, that your friend is a pair of Legs. She can jump, she can squat, she can strut around, but she can't catch hold of anything.

If you wanted to snatch something out of your reach—say, an apple hanging from a tree down the lane—then you might consider teaming up with Legs by asking her to carry you to the tree so you could pluck the fruit. Maybe it turned out that Legs was eager for exercise and wanted nothing more than a jog down the lane, but on her own she can't open the gate to exit the yard. Legs might enthusiastically agree to your proposed collaboration, carrying you to the gate so that you could unlatch it, then jogging to the apple tree.

Arms and Legs possess complementary abilities (Arms can manipulate, Legs can walk) that balance their complementary shortcomings (Arms can't walk, Legs can't manipulate). By collaborating, their complementary strengths add up to something greater than the sum of what they can do individually. When they work together, they become something new: a mobile manipulator, capable of grabbing anything, anywhere.

This fanciful example illustrates the complementary thinking principle, one of the most important principles in the journey of Mind because it accounts for one of the most common and effective ways that module minds boost their competence. Two distinct thinking elements, each pursuing its own distinct purpose, cooperate to achieve something that neither could accomplish on its own. Complementary thinking is one of three main ways that module minds become more intelligent. (The second is by expanding a module's memory capacity. The third is by increasing mental bandwidth—increasing the number and complexity of the "notes" in the "melodies" that modules share with one another. Both these ways of getting smarter are largely a function of adding more neurons and neural connections.)

Complementary thinking has been influential in every stage of Mind. Haloarchaea's light-seeking mind consisted of complementary sensors and doers. On their own, neither a sensor nor a doer could target the sun. But a sensor-doer pair can. There are numerous examples of complementary thinking in neuron minds, too. In the hydra, for instance, doer networks are arranged in complementary pairs, such as one that elongates the stalk and one that constricts the stalk. In the roundworm mind, the complementary forward- and backward-locomotion networks join forces (through interconnecting thinker neurons) to establish a comprehensive motor-control system. But in module minds, complementary thinking delivers spectacular leaps in mental prowess, and none is more impressive than that of the What-Where dynamic duo.

The two modules embody contrasting forms of thinking. Discerning *what something is* involves analyzing the features of an object. It demands attention to detail. The dynamics of recognition produce the conscious experience of *knowing*. Discerning *where something is* involves analyzing locations and relative distances. It demands an appreciation of the big picture. The dynamics of localization produce the conscious experience of *seeing*.

You can experience your Where module in action by looking at a spot nearby, maybe the corner of your book or device. Instead of focusing on *what* you're looking at, reach out and touch it. Your How module handled the movement of your finger, but your Where module maintained your conscious attention on your finger's destination. An even more salient way of experiencing the activity of your Where module is to close your eyes and imagine a point twelve inches in front of your face. Then imagine a point ten feet ahead. Then imagine a point on the moon. We can consciously fixate our attention on locations all around us, even if we cannot actually see them, instead "seeing" the location in our mind's eye.

"What?" and "Where?" are two of the most common queries in physics and philosophy, the ones most often applied to the deepest questions of existence. What is matter? Where is the center of the universe? What is life? Where will we find extraterrestrial intelligence? What is the mind? Where will I go after I die? On the face of it, the questions "What?" and "Where?" seem to involve disparate and independent modes of inquiry. Yet, if we hope to interrogate the cosmos with our What and Where modules, it turns out that we need *both* modules in order to answer *either* question effectively.

In this chapter, we'll discover why we can't figure out *what* something is without knowing a little about *where* it is, and we can't figure out *where* something is without knowing a little about *what* it is. To help us investigate this mutual back-scratching society of consciousness-generating modules, we will examine a vital mental challenge that can be solved only by a What-Where team-up. It's known as the "Where's Waldo problem": How can a mind rapidly search a complicated scene to find a specific object? How can you find a can of clam chowder on a crowded grocery shelf—or a jaguar lurking in the shadows of the jungle?

2.

We can learn how the alliance of What and Where solves the Where's Waldo problem by examining one of the smartest and most reviled minds in nature, the mind of *Rattus norvegicus*, the brown rat.

It's hard to think about rats without picturing them as indefatigable pests living among sewers, garbage, and other filth, but objectively speaking the rat is one of the most successful animals on Earth. It thrives on every continent except Antarctica. Today, the rat is considered commensal with our species, meaning it lives in a kind of symbiosis with *sapiens* civilization. Indeed, there's strong evidence that its recently embraced lifestyle as a metropolitan forager has turbocharged its evolution. Urban breeds of *Rattus norvegicus* have developed new genes for warding off a variety of toxins, a consequence of consuming a deluge of human effluvia and being constantly targeted with poison bait. (This suggests that it may not be wise to use rats as research subjects for evaluating the effects of hazardous substances on humans.)

The wild brown rat originally lived—and lives still—within damp burrows in northern China. Humans share a common evolutionary lineage with *Rattus*. We arc on the same branch descended from the first mammal, which is believed to have been ratlike. This ancestral mammal was a small nocturnal scavenger who foraged in the underbrush beneath the feet of dinosaurs for insects and plants. Thus, the rat has a better claim to representing the "original stock" of mammals than we do. Rats are members of the rodent family, known as the Muridae, which contains a quarter of all mammal species, including mice, voles, muskrats, lemmings, hamsters, and gerbils.

Rats dwell in colonies consisting of hundreds of individuals and form complex social relationships. Rats display a passion for exploration. They exhibit curiosity about new objects and materials, but, somewhat paradoxically, they also exhibit "neophobia"—a fear of new things. They approach things they don't recognize, but with great caution. Rats carefully sample a tiny bite of unfamiliar items to determine whether they're edible. This is another reason rats are difficult to poison. If a rat feels even slightly sick after nibbling a new substance the rat will avoid it forever after.

Rats have a sophisticated Where module. They can easily navigate complex environments, an ability that motivated behaviorists in the early twentieth century to adopt the rat as the iconic organism for maze-running experiments. More recently, the rat's excellent spatial thinking has made it the leading animal model for investigating the neural architecture of the Where module. The rat also possesses a good memory for visual objects.

Though rat smell is superior to rat vision—rats see in two colors, ultraviolet and greenish, while humans see in combinations of greenish, reddish, and bluish—rats are adept at recognizing a wide array of objects that they use for nest building, climbing, navigation, and manipulation, and as potential sources of food. Thus, the rat boasts a highly proficient visual What module.

Let's examine how the What and Where modules operate in the mind of a rat. Imagine a rat wants to find a sunflower seed. (Maybe the rat is upwind, so it can't use its nose to find seeds.) Once the rat decides to hunt for a seed, the first thing its mind does is summon a memory of what a sunflower seed looks like.

Hold up—aren't we skipping something important? How did the rat learn what a sunflower seed looks like in the first place?

3.

As described in the Tortoise Mind chapter, the first problem the visual What module must solve is the object localization problem: determining which part of a scene to treat as a potential object. *Should I try to recognize the tree-thing or the feather-thing-in-the-tree or the wing-thing-on-the-feather-thing?* To learn about a new object, the What module must focus its attention exclusively on the features of the object itself. This involves an unavoidable mental tradeoff. The act of narrowly focusing on the details of a particular object means the What module cannot attend to the big picture of the object's location in space. The Where module faces the exact opposite challenge. Its job is to determine the exact position of an object in space, which means that it must attend to the entire scene in front of it. To process the big picture, the Where module must relinquish knowledge of the object's details.

The Visual Scene module supports the Where module by supplying it with the foreground and background surfaces of a scene. The foreground surface (such as the woman in the *Mona Lisa*) serves as the bottom-up sensory input to the Where module. You can think of this bottom-up representation as a silhouette of an object on a two-dimensional portrait of the surrounding landscape. The top-down expectation is the silhouette's visual coordinates in space. The Where module's top-down representation encodes the distance from the eyes to the foreground surface.

When the bottom-up foreground surface matches the top-down visual coordinates—when the observed object location matches the predicted object location—the two representations *resonate*. This synchronizes, amplifies, and prolongs the representations of the surface and its location. The resonating location representation is known as an *attentional shroud*.

Think of the attentional shroud as an invisible silhouette that covers the entire surface of an object, in the same manner that the military "paints a target" with an (often invisible) laser in order to guide laser-seeking bombs to the target. When the Where module forms an attentional shroud—when the foreground surface representation of an object resonates—the object pops out in our awareness and we are conscious of *seeing* the object. We are conscious of *where* the object is in space, providing us with a conscious sense of how far we must reach (or walk or drive) to touch the object.

The Where module shares the attentional shroud with the How module, which uses it for targeting the object. The Where module also shares the shroud with the What module, which enables the What module to solve the object localization problem (*which part of the scene is the object to recognize?*). The shroud establishes a distinct mental perimeter that separates the foreground object from the background, enabling the What module to attend to all the details inside the shroud while ignoring everything outside the shroud.

The attentional shroud also solves another major problem the What module faces when trying to learn about a new object: the many-to-one mapping problem.

4.

Your visual What module must learn to recognize a penguin no matter what angle you view the penguin from and no matter which features of the penguin you focus on. But if the first time you ever see a penguin your eyes jump around from its feet to its flippers to its beak, how does your What module know these are all different views of the same penguin, and not three distinct objects? Similarly, if your eyes jump from the penguin's feet to the egg beneath its feet to the snow beneath the egg, how does the What module know that these are three distinct objects and not three different parts of the same penguin?

Because of the attentional shroud.

The shroud endows a tentative object with a stable border that the What module can use to distinguish the features of the object from the features of everything else. Your eyes are free to jump around to study anything lying inside the shroud and the What module can be confident that it is assimilating useful data about the object in question. Even if the object is rotating or you are moving around the object, the Where module maintains its attentional shroud around the object and the What module can learn a variety of different views and features of the object and store them all as part of the same object prototype.

Let's get back to our hungry rat, desperately seeking sunflower seeds. The rat's What module previously learned what a sunflower seed looks like by using the Where module's attentional shroud to identify and store all the fea-

RAT IS LOOKING FOR A SUNFLOWER SEED

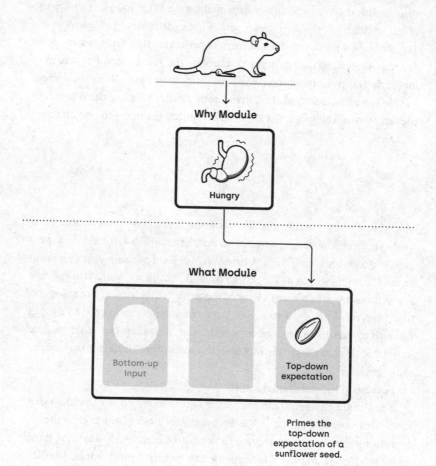

Primes the
top-down
expectation of a
sunflower seed.

tures associated with the seed. Now the What module summons its memory of the features of a sunflower seed, which serves as the top-down expectation guiding its visual search.

The rat begins to search for a sunflower seed around a grassy knoll. The rat's eyes, controlled by the Where module, begin to jump around the visual scene looking for something sunflower-seed-like. The (preconscious)

RAT DIRECTS ATTENTION TOWARD A POTENTIAL SUNFLOWER SEED

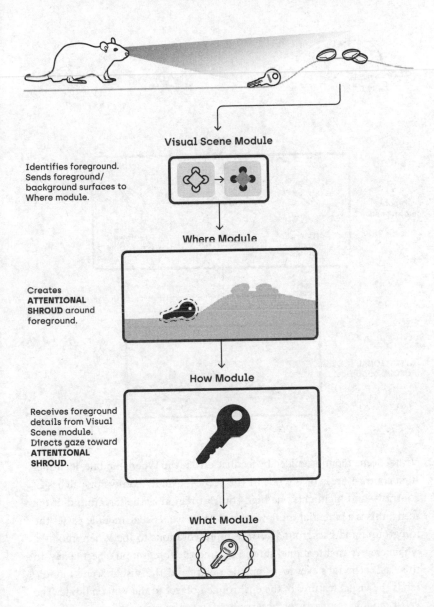

Visual Scene Module

Identifies foreground. Sends foreground/background surfaces to Where module.

Where Module

Creates **ATTENTIONAL SHROUD** around foreground.

How Module

Receives foreground details from Visual Scene module. Directs gaze toward **ATTENTIONAL SHROUD**.

What Module

RAT RECOGNIZES A KEY AND REALIZES IT'S NOT A SUNFLOWER SEED

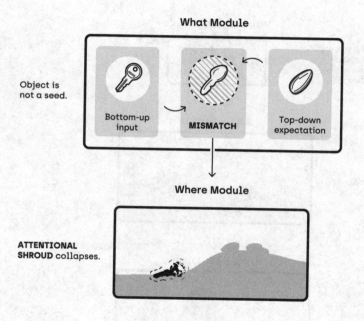

Visual Scene module, guided by feedback from the Where module, identifies high-contrast areas of interest in the scene—usually corners or patches of contrast—and highlights one high-contrast region as the foreground of the scene: this is a potential sunflower seed. The Visual Scene module passes the foreground and background surface representations to the Where module, which creates an attentional shroud around the foreground object (a key, in this case). The rat's eyes focus on the object and the Visual Scene module sends the visual features of the enshrouded object to the What module. The What module compares the observed object to the expectation.

RAT LOOKS FOR A SUNFLOWER SEED AGAIN

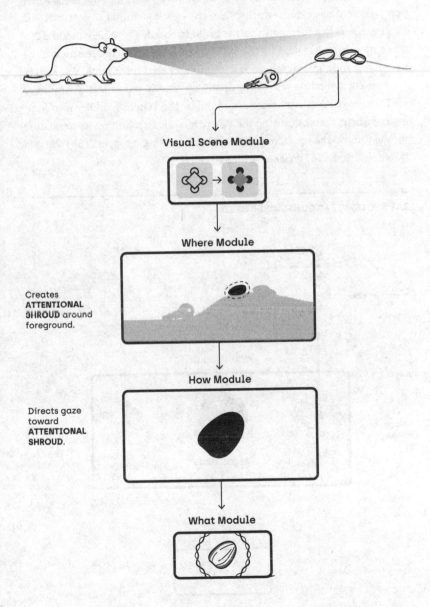

Visual Scene Module

Where Module

Creates
**ATTENTIONAL
SHROUD** around
foreground.

How Module

Directs gaze
toward
**ATTENTIONAL
SHROUD.**

What Module

Mismatch! It's not a seed!

The mismatch prompts the Where module to start looking for another potential sunflower seed, which causes the existing shroud to collapse. The rat's eyes quickly settle upon another candidate seed. The Where module creates a new shroud around it. The What module processes the features of the new object and compares them to its "sunflower seed" prototype. Yes! Match!

The What module resonates, certifying that it recognizes a sunflower seed. The What module's resonance inhibits the Where module from looking elsewhere, thereby causing the Where module to maintain its resonating shroud around the seed. Now something interesting happens. *The What and Where modules begin to resonate with each other.*

RAT RECOGNIZES SUNFLOWER SEED

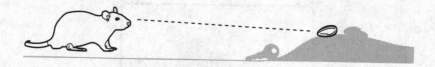

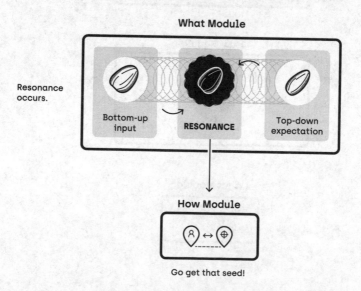

Go get that seed!

This is our first example of module-to-module resonance. We will examine module-to-module resonance more closely in subsequent chapters. For now, what matters is that the resonance between the What module and Where module causes the How module to target the seed, using the visual coordinates from the Where module's attentional shroud as the target position to aim for.

The rat scurries toward the seed for a tasty snack.

5.

Complementary thinking occurs at every level in module minds. At the molecular level, proteins in neuron membranes known as "active transporters" increase the concentration of charged molecules (such as potassium ions) by driving the molecules through pores in the membrane known as "ion channels," thereby creating the voltage potential across the membrane that is necessary to generate an electrical signal. Active transporters and ion channels work together in complementary fashion to enable neurons to communicate.

At the neuron level, there are complementary "on" and "off" retinal cells: pairs of individual sensory neurons where one responds to light and the other responds to darkness. Together, they create "receptive fields" that enable the mind to detect bright dots surrounded by darkness, or dark dots surrounded by brightness. At the network level, in the visual system of module minds, there are complementary *form* processing networks and *motion* processing networks.

In the next chapter, you will be introduced to a rather startling example of the complementary thinking principle—one that unites thinking across two different minds.

When

Words, after all, are nothing by themselves. They burst
into meaning only in the minds they've entered.
—*Ghachar Ghochar*, Vivek Shanbhag

thy head is a quick forest
    filled with sleeping birds

—"my love," e. e. cummings

1.

If you climb the towering slope of the Antisana volcano in the northern
Andes, you will ascend into an ecosystem in the clouds. This remote Ecua-
doran habitat is known as a cloud rainforest. It is not hot and steamy like a
tropical rainforest. These woodlands are cool and damp, the result of perpetual fog saturating the forest. Visibility is limited to a few meters ahead. Even
if the fog lifted, you would be confronted with bamboo thickets so dense that
you would need to hack through them with a machete. If you were willing
to brave the high altitude, heavy fog, and dense vegetation of the Antisana
cloud forest, and were aided by a bit of luck, you might hear a fast, staccato
melody that sounds a lot like the rapid digital chirping of an old-fashioned
dial-up modem.

The complex melody is so piercing that it would ring in your ears like a
smoke-detector alarm. It might even hurt your eardrums. If you attempted
to locate the source of the intense kinetic arpeggios, you might find yourself turning your head to and fro in wonderment, unable to discern exactly
where the sound was emanating from. Indeed, the overall effect would be as
if the fog itself were producing the melody like some seraphic cloudsong from
the apex of the world.

The source of the song is a remarkable bird with a prosaic name. The
plain-tailed wren. If you didn't know better, you would guess that the sono-

rous melody was sung by a single bird. Instead, the song is one of the rarest forms of communication in nature: a duet. A pair of wrens bonded for life, male and female, singing together.

Yet, they are not trilling the same notes in unison, like Sonny and Cher jointly crooning the chorus to "I Got You Babe." The wren duet is more impressive: the two birds alternate phrases as if they are each singing a different line, like Beyoncé singing "You stay on my mind, fulfill my fantasies" and Sean Paul responding with "Come on, girl, tell me how you feel" on the chorus to "Baby Boy"—except accelerated to a breakneck tempo that no human singers could ever hope to match. The wrens' musical phrases are sung with millisecond accuracy, each bird meticulously dropping its notes into its partner's momentary silence.

A few species of frogs perform super-simple courtship duets. Dolphins and four species of monogamous primates sing duets, including humans, who sing them with exceptional quality and originality. Several other species of birds also sing duets. But none sings with greater high-speed precision than the plain-tailed wren.

It may seem surprising that in our journey of Mind, birdbrains follow after ratbrains rather than preceding them. After all, the rat mind (like every mammal mind) flaunts a cerebral cortex. The bird mind does not. (Birds are separated from rats—and humans—by six hundred million years of evolutionary time; our last common ancestor was three hundred million years ago.) Even so, to produce their melodious songs, the minds of birds require a highly refined mental apparatus—a neural module that just so happens to be essential for the emergence of language and the Self, too: the When module.

The When module makes it possible to learn, recognize, and execute *sequences*. A then B then C (or A# then B then C#). We might say the When module determines *when* an item or event appears in a sequence. The When module plays a decisive role in the most sophisticated forms of organic intelligence—not just speech, reading, and writing, but toolmaking, tool using, logic, science, computer programming, and mathematics.

But the When module has an even more impressive trick up its sleeve. It enables thinking to extend beyond the bounds of one mind and cross over into another.

2.

Here is an odd though surprisingly relevant question to ponder about the nature of mental representations in module minds. What, exactly, does a representation *look* like?

If we could somehow perceive a sensory representation in the act of being processed by a What module, how might we describe its appearance? This might seem a curious thing to ask about intangible mental activity. Some scientists might even suggest that such a question is poorly formed, like asking what justice smells like. But the star of this chapter is a form of mental representation that achieves an uncanny feat: it manifests in an observable way. Not in a way we can *see*. In a way we can *hear*.

This peculiar representation is known as birdsong.

Birdsong is the first complex representation on our journey that can pass from a module in one mind to a module in an entirely different mind. It is, quite literally, a melody: a representation made of sound. The singer's mind generates and transmits the representation and the listener's mind receives and parses the representation, bridging two modules located inside two distinct minds in the same manner that an internal mental representation bridges two modules within the same mind.

A bird's song has no inherent meaning. A songbird's melody is a conceptual abstraction whose precise pattern has no perceptible impact on the physical world, though it has a vigorous impact on a suitably equipped mind. In this regard, birdsong is little different from the arbitrary series of 0s and 1s that make up machine-language instructions for a computer: the pattern of 0s and 1s only triggers physical dynamics within an appropriate CPU. The acoustic patterns of birdsong are part of the overall dynamics of a bird's mind and hold meaning only within the context of those dynamics. The act of tossing a ball high in the air doesn't have any inherent meaning, but in the context of a basketball game, a high toss could be part of a jump shot or an alley-oop.

In summary, birdsong is a groundbreaking class of representation that enables two different module minds to merge their individual mental dynamics to form a single unified intermind dynamic. Birdsong enables two minds to jointly focus their attention on the same thing: a melody. This is an innovative instance of the complementary thinking principle. Two distinct thinking elements—in this case, two bird minds—use long-distance communication to pursue a common purpose, such as finding food, finding a nesting spot, tracking predators, reinforcing a social bond, or simply teaching a younger bird how to sing. In a limited yet significant fashion, birdsong enables two minds to think as one.

Why was nature's first complex external representation an audio pattern and not a visual one? The embodied thinking principle advises us to consider the physical properties of a melody. Sound can travel for long distances without degrading and does not require line of sight—very useful when attempting to communicate at night, across significant distances, or in thick vegetation. Birdsong adapts to the mind's environment. Songs sung by birds who live in jungles and forests have narrow bandwidths and long melodies that cut through the muffling foliage, while birds who live in open habitats like savannas sing songs with high frequencies, trills, and other short elements that can travel for long distances. Birds who dwell in cities have their audio space filled up by low-frequency urban noise and as a result many species of urban songbirds have shifted their pitch to a higher register to cut through the sound pollution.

Birdsong is an efficient, reliable, and versatile class of representation because a melody consists of a string of discrete and readily identifiable notes that serve as a kind of stable code that is relatively straightforward for a bird mind to modify: simply add, subtract, or alter a note.

It took a very long time for birds to develop the ability to sing. Though modern birds evolved around ninety-five million years ago during the Age of Dinosaurs, genetic analyses suggest that songbirds first evolved in Australia just over thirty million years ago. Birds produce their song using an organ known as a syrinx, a structure unique to birds that is located deep in their chest near the base of the trachea. Sound is produced when air from the lungs causes the syrinx tissues to vibrate. The earliest known syrinx fossil is sixty-six million years old. It belonged to a species called *Vegavis iaai*, a precursor to modern ducks, geese, and swans. The scientists who discovered this primordial syrinx believe it honked rather than chirped.

Different communities of birds from the same species that are geographically isolated often sing variants of a similar song. Sometimes different communities on the same mountain sing different melodies at different altitudes, like regional dialects. The structure of songs across all species of birds is immensely variable, exactly as we'd expect with any mental representation. Some are short, consisting of a few quick tweets. Other birds croon for minutes at a time. There can be great variation even within the same genus. Among warblers, for instance, the songs of grass warblers contain two to four distinct types of syllables (a short set of notes), whereas marsh warblers' songs contain between twenty-five and one hundred. About a third of songbirds stick to a single song (possibly with minor variations), while about a fifth of songbirds sing five or more songs. The brown thrasher, Georgia's state bird, can sing thousands of songs. The Lincoln's sparrow introduces novel variations into its song each time it sings, like an improvising jazz musician. Even more like a musician is the musician wren, which is native to the Amazon and has inspired human music across South America. The musician wren sings using the same intervals found in Western music: octaves, perfect fifths, and perfect fourths.

There are other animals who "sing" to one another, including croaking frogs and chirruping crickets. But what distinguishes birdsong from these lesser forms of communication is that birdsong requires an entire community of birds to implement. Whereas a cricket's sound comes from reflexively stroking its wings and a frog's croak comes from instinctively forcing air through its vocal cords and vocal sac, a songbird must learn how to sing its song from another singer, a kind of social learning that inaugurates game-changing mental dynamics. This new class of intermind dynamics facilitated by the When module has a more familiar name.

Culture.

3.

The most studied bird mind in science is that of the zebra finch. *Taeniopygia guttata* is native to Australia and Indonesia. The colorful males have distinctive red bills, orange cheek patches, chestnut flanking, and their namesake stripes on their chest and throat. The females look much the same, but without the color or stripes. Both sexes tweet calls related to predatory threats, alarms, territorial defense, and maintaining pair bonds. Loud calls appear to broadcast individual identity (perhaps even age and geographic origin) and solicit the other member of a pair to follow, while softer "tet" calls elicit close contact between a pair and may serve to reinforce pair bonds. Female calls are almost exclusively innate. There are no Adeles or Whitney Houstons in zebra finchdom, because the females do not sing. The males, on the other hand, sing their hearts out.

Male zebra finches sing complex three-part songs to attract females. Even if raised in isolation, male birds will still learn to sing a vestigial song. Such a rudimentary, undeveloped melody will never attract a mate. For a male to learn to sing a mature song requires that the male be raised in a zebra finch community, because the male must learn how to sing from a tutor. Though males often learn to sing from their father (or stepfather), they are particularly receptive to melodies sung by males who have successfully mated, which serves as a reliable indicator that these males' crooning style commands attention from the ladies. In effect, the songs that are sung within a finch community are a kind of Billboard Top 40, where the most popular songs are most likely to be copied by up-and-comers. A fledgling's choice of which melody to imitate depends on the social dynamics of the entire community, which directly or indirectly encompasses male-female interactions, male social hierarchies, breeding patterns, and foraging patterns.

Humans learn how to talk during a "critical period," a narrow window of time in our childhood when our minds are highly sensitive to the nuances of speech. A male finch also learns to sing during a critical period. Learning to sing a song involves two stages. First, a learner must hear and memorize the song, a sequential audio stimulus. We might call this the "learning" stage. Second, the learner must perform the song by recalling the memorized melody and properly controlling the muscles in his syrinx and vocal tract so that he can reproduce each note correctly. We might call this the "performing" stage. The two-stage process of learning to sing a song relies upon four modules: the Auditory Scene module, the auditory What module, the How module, and—most prominently—the When module.

4.

After a young finch selects a tutor to learn from, its mind processes the notes of the tutor's song in its Auditory Scene module, a sound-based version of the Visual Scene module we explored in the Fish Mind chapter.

The preconscious Auditory Scene module separates out the different audio streams in the environment—distinguishing the wind rustling in the trees from the melody of the birdsong, for instance—and appoints the birdsong melody as the "foreground" stream in the same manner that the Visual Scene module identifies a foreground surface. (The Auditory Scene module's preconscious identification of the foreground stream is responsible for the "cocktail party effect" that causes us to perk up whenever we hear someone mention our name, even if there's a lot of ambient noise.)

The foreground stream containing the birdsong is passed to the audio What module, which identifies each individual note in the melody. Processing audio is more resource intensive than processing visuals. Sound patterns arriving at fast rates must be processed at fast rates. To accomplish this fast processing, the Auditory Scene module and audio What module use thick wires and fast synapses. The most expensive parts of a bird or mammalian brain in terms of energy consumption are those devoted to early auditory processing.

The What module recognizes each note, one by one. But to learn a melody, a mind must also process and remember a *sequence* of notes. Sequential thinking is the forte of the When module. The module takes a complex though fleeting stimulus—a sequence of brief sounds that quickly vanishes from perception—and creates a single stable representation. This task is more challenging than it might look. When attempting to learn an arbitrary sequence (such as the notes in a new song), it's not possible to know ahead of time when the sequence will end or how many items will be in the sequence. The When module must be ready to process a sequence of notes of any length and convert that sequence into a single unitary pattern. In the Bacteria Mind chapter, we saw how *E. coli* converted a spatial problem into a temporal problem to navigate through gradients. The When module, in contrast, converts a temporal problem into a spatial problem: it forms a distinct physical representation in the brain for each note and maintains these spatially distinct representations until all the notes in the melody are processed.

Each of these note representations encode two pieces of knowledge: the identity of the note itself, and a marker indicating *when* the note appears in the melody. This marker consists of a "weight" that the When module assigns to each incoming note, a sort of neural representation of relative intensity. The first note is assigned the highest weight, the second note is assigned the

WHEN MODULE: LEARNING

Step 1:
Three notes are heard in sequence.

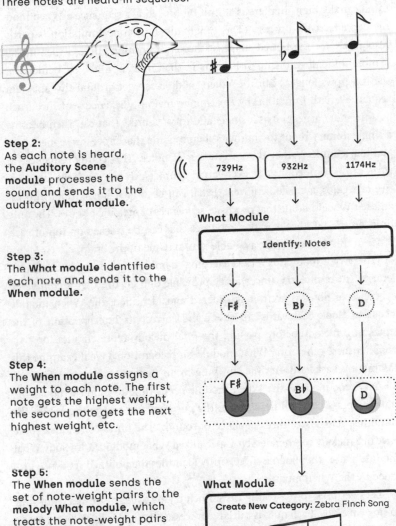

Step 2:
As each note is heard, the **Auditory Scene module** processes the sound and sends it to the auditory **What module**.

Step 3:
The **What module** identifies each note and sends it to the **When module**.

Step 4:
The **When module** assigns a weight to each note. The first note gets the highest weight, the second note gets the next highest weight, etc.

Step 5:
The **When module** sends the set of note-weight pairs to the **melody What module**, which treats the note-weight pairs as the stable features of a melody. The What module uses these features to identify or learn the melody.

second-highest weight, and so forth. When the sequence of notes (the melody) is over, each note and its paired weight are stored in the When module's memory.

Because there is a limit on how fine-grained a difference in weights the mind can distinguish, there is a limit on how long a sequence a When module can effectively process. (This weight-distinguishing limitation is partly responsible for the bowed shape of the serial position curve, which makes it more difficult to distinguish items in the center of a list, particularly as the list grows longer.) But the When module can get around this apparent sequence-length limitation by treating a previously learned sequence (such as a five-note arpeggio) as a single auditory "chunk" that can then become a single unitary input to another When module that processes sequences of auditory chunks, rather than sequences of notes. This chunk sequence, in turn, can be unified into an even larger chunk (such as a set of five consecutive arpeggios, a "melodic phrase") that can become a single input into yet another When module that processes sequences of melodic phrases. The only limit on the number of When modules that can be stacked on top of each other in this manner is the available neural tissue in the brain.

The When module is a card-carrying member of the preconscious proletariat. To resolve the uncertainty principle and ensure that a completed sequence of notes is accurately ordered and identified, the When module needs to "look backwards" to check and correct its interpretation of the notes in a melody before passing the final melody representation on to a consciousness-generating What module for recognition. (We'll examine this fascinating backwards-in-time thinking in an upcoming chapter.) Like the Visual Scene module, the When module needs time to mop the floors and buff the brass before flinging open the curtains and letting the rest of the mind inspect the polished sequence of sounds. The bird's conscious experience of a melody is generated by a specialized What module (a "melody What module") after the module successfully identifies the polished sequence provided by the When module.

Let's get back to the young male finch trying to learn from a tutor how to sing a melody. Each note is initially processed by the bird's Auditory Scene module and passed on to the auditory What module, which identifies the note. Each recognized note is then passed on to the When module, which assigns each note a weight and builds a sequence of note-weight pairs. The resulting sequence of note-weight pairs is then passed on to the melody What module, which learns the sequence as a single unitary input (that is, a melody), where both the notes and their weights are treated as bottom-up features.

During the "performing" stage of learning to sing a song, a male finch must produce the correct sound for each note in the melody, in the correct sequence. To sing a note, the bird's How module "reaches for the acoustic target" by driving the difference between the target position and the body position to zero. Except in this case, the target position is no longer a point in space (it's no longer a set of visual coordinates). Instead, it's a *sound* target—a set of "auditory coordinates" for the syrinx to produce. The How module translates these auditory coordinates into the same "measurement system" as the body coordinates, like translating Fahrenheit into Celsius. These body coordinates reflect the set of syrinx muscle tensions and orientations that will produce the target sound. The body position (the present state of the syrinx muscles) is then driven to match the body position necessary to produce the sound. (The same process occurs when humans attempt to sing a note or pronounce a phoneme: the How module drives our vocal cord muscles to the position necessary to produce the desired sound.)

Once a zebra finch has mapped his vocal cord coordinates to the corresponding sound coordinates and learned a melody from a tutor, he can attempt to sing the song. One difference between singing a melody and reaching for a target is that to sing a melody the How module requires a *sequence* of targets— that is, a sequence of notes. This sequence is supplied by the When module.

When the finch is ready to sing, the melody What module retrieves the desired melody and passes it to the When module, which locates the note with the highest weight and sends it to the How module to sing. Afterward, that note's weight is reset to zero. The When module once again looks for the note with the highest weight, which is now the second note in the melody. The second note gets sung and its weight is reset to zero, and this cycle repeats itself until all the notes in the melody are sung.

The eager young vocalist belts out his first complete ballad to a judgmental avian audience.

WHEN MODULE: PERFORMING

Step 1:
Bird decides to sing a zebra finch song. The **auditory What module** retrieves the stored melody (a set of note-weight pairs).

What Module

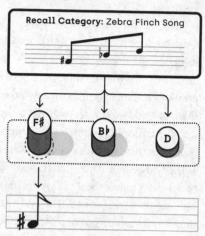

Step 2:
The **How module** looks for the note in the set of note-weight pairs with the highest weight and sings it.

Step 3:
After singing the note, the **How module** causes the note's weight to go to zero, then looks for the note with the highest weight and sings it.

Step 4:
Step 3 is repeated until all weights are zero, which indicates that all the notes in the melody have been sung.

5.

We usually think of culture as the tec$ologies, activities, and social norms of a given society. But here, in its most stripped-down form, culture appears in a different light—as a mind-to-mind mental dynamic with properties similar to those of single mind dynamics. In other words, culture is a form of thinking that takes place across multiple minds. Birdsong binds all the birds in a community into a bird supermind, just as amoebas were bound together into an amoeba supermind through molecular communication.

Some biologists believe that one of the main functions of birdsong in zebra finches is to help females judge the fitness of males, suggesting that females are looking for the Frank Sinatras of finchdom because they prefer to pair up with males who sing with longer song phrases and more varied song structures. That may well be. But from the perspective of the journey of Mind, there is another, less competitive role for birdsong, the same role that the metropolis principle assigns to each new layer of connectivity in a developing city—the same role that molecule messages played for amoebas and neural representations did for fruit flies: coordinating multiple thinking elements operating at the same level of thinking. In the case of birdsong, the thinking elements are the bird minds themselves.

Birdsong enables a community of birds to mate, raise children, watch for predators, and defend their territory. Birdsong also indirectly stitches together a zebra finch community by providing information to "eavesdroppers," such as finches who listen for the melodies of males singing to their partners to assess whether a particular neighborhood is family friendly. If a zebra finch is cheerfully singing to his mate, that suggests that the couple has found a prosperous spot to nest. Birdsong can also fuse the dynamics of two minds, empowering a couple to focus on joint purposes and enabling them to share similar perceptions of important situations.

One of the best examples of this "two brains, one mind" dynamic is the plain-tailed wren's duet. Think of the neural coordination necessary to perform this rapidly alternating duet: each bird needs to know the exact moment to start and stop their part based on their partner's singing. It's almost as if they are sharing the same What and When modules. As a matter of fact, scientists have determined that each mind in a dueting pair encodes the entire duet melody as a single representation, rather than encoding only their own part of the duet or encoding each part separately. Together, each pair of wrens compose their own unique signature duet, which is then stored as an identical copy within each wren's mind. Though the couple composes the duet out of borrowed melodic fragments that are widespread in their community, they stitch them together in a new and distinctive sequence, as if they want to proclaim, "This here is *our* song."

Though it's not yet incontrovertibly demonstrated why plain-tailed wrens sing duets—some have suggested it's a way to demonstrate the strength of a pair's bond in order to scare away lecherous singles—it's clear that male and female are both highly motivated to sing together and perhaps they enjoy the feeling that, for a few moments, they are sharing a single mind.

And so they are, for a mind is nothing more or less than its dynamics.

CHAPTER FIFTEEN
MONKEY MIND

Why

The man who believes that the secrets of the world are forever hidden lives in mystery and fear. Superstition will drag him down. The rain will erode the deeds of his life. But that man who sets himself the task of singling out the thread of order from the tapestry will by the decision alone have taken charge of the world and it is only by such taking charge that he will effect a way to dictate the terms of his own fate.

—*Blood Meridian*, Cormac McCarthy

1.

Sylvia Plath's confessional novel *The Bell Jar* probes the youthful challenge of choosing one's identity and forging one's Self. The novel delves into Plath's sexual encounters, mental health issues, and moral quandaries. In one of its most enduring metaphors, the novel's protagonist Esther, a stand-in for Plath, describes her future as an enormous, healthy fig tree with many branches reaching up to the sky. On the tip of each branch is "a fat purple fig, a wonderful future." One fig is domestic happiness—a husband, home, and children. Another fig is becoming a famous poet. Another is becoming a brilliant professor. Another is becoming a talented editor. Yet another fig is traveling the world and visiting exotic lands. Another is a cadre of sexy lovers. Still another is becoming an Olympic champion. Many more figs shimmered in the highest branches, distant and ambiguous.

Yet she was trapped "in the crotch of this fig tree, starving to death, just because I couldn't make up my mind which of the figs I would choose. I wanted each and every one of them, but choosing one meant losing all the rest, and, as I sat there, unable to decide, the figs began to wrinkle and go black, and, one by one, they plopped to the ground at my feet."

Plath was unable to decide which option to choose and therefore became paralyzed, making no decision at all. She was struggling with the classic dilemma of adolescence: figuring out who she wanted to be when she grew up. Much of this complex and emotional struggle takes place within the mind's most sophisticated neural module.

The Why module.

2.

Of all the conscious experiences gifted us by our mind, none is so treasured, so human, so personal, as *feeling*. Our other conscious experiences, such as seeing, hearing, and knowing, focus our attention on things and events *outside* ourselves. When we see a plane soaring through the clouds or hear the drip of a leaky faucet, we experience these perceptions as *outside of me*: I see *that*, I hear *that*, I recognize *that* over yonder. Feelings are different. When we swoon over the attractive barista at Starbucks, we experience this feeling as *me*: I *am* smitten. I *am* light-headed. I *am* besotted with the torrid flames of desire.

The emotions we spontaneously feel toward the objects, events, people, ideas, and—crucially—*choices* that impinge upon us provide us with a vital sense of our own identity. The angry retort "You don't know me!" doesn't mean, "You don't know my blood oxygen levels or the size of the mole on my foot!" It means, "You don't know *why* I behave the way I do! You don't know how angry I am at my father for abandoning us or how hard I've worked to get my poems published!" Other minds can corroborate your perceptions of What and Where. But nobody is privy to your Why.

Here's a question whose answer is less obvious than it might first appear. Why are we *aware* of our feelings? Could our conscious feelings be mere emergent properties, incidental side effects of complex brains, as arbitrary and purposeless as the fact that our spleen is purple? After all, what use is being conscious of searing heartbreak, oppressive shame, or chronic disappointment? Wouldn't we be better off if we could be conscious of the good vibes and let the preconscious proletariat get stuck with all the bad vibes?

Though we usually associate intelligence with more elevated capabilities like strategizing, puzzle-solving, and memory rather than our visceral and earthy emotions, conscious feelings are one of the central pillars supporting the advanced intelligence of primates, including humans. When you are presented with a fig tree of options, it's up to your feelings to guide you to the best choice.

Should you break up with your significant other, or give them one more chance? Should you keep shopping at a store that is conveniently located but espouses political views you despise—or drive a few additional miles to a more simpatico store? Should you stick with your decent-paying job with a demeaning boss or risk the uncertainty of the job market? Feelings are the module mind's leading tool for resolving the exploration dilemma because our emotions supply us with a personalized and contextualized valuation of our available choices. More simply, our feelings tell us *Why* we should choose this fig rather than that fig.

Why should I break up with my partner? (Because I'm *exhausted* from fighting all the time and the thought of meeting new people is *exciting*.) Why should I boycott the store? (Because I *abhor* their endorsement of racist policies, and I can *tolerate* driving a few more minutes to buy my door hinges.) Why should I stick with my awful job? (Because I'm *afraid* of what might happen to my infant twins if a new job doesn't work out.)

Of course, back in the Fly Mind chapter, we learned that insects have reasons, too. A fly pursues a savory smell because it is hungry or avoids a location because it previously received an unpleasant electric shock there. Yet the fly mind does *not* experience conscious feelings. What's the difference between the nonconscious "emotions" of a fly and the conscious emotions of a monkey?

3.

Primates are one of the most diverse orders of living mammals. They boast more than five hundred species, including *H. sapiens*. Primates are the third most abundant mammalian order, behind bats and rodents. Primates have inhabited the earth for sixty-five million years or thereabouts. The fossil record suggests that primates diverged from other mammals sometime near the beginning of the Cenozoic era, the so-called Age of Mammals. Primates are most closely related to two small groups of Asian mammals, tree shrews and flying lemurs—strange bug-eyed creatures that resemble a cross between a bat and a flying squirrel.

Primates exhibit many physical and behavioral features that distinguish us from the rest of our mammalian sorority. Primates have hands and feet that can grab things. To aid us in manipulating objects, primates traded away claws for nails. We also have forward-facing eyes. This provides primates some of the most impressive visual capabilities in nature, including stereoscopic color vision.

Primates are cunning and highly social animals capable of an enormous array of behaviors compared to other species, including planning for the long term, intentionally deceiving others through concealment and misdirection, and forming complex social alliances to topple unfavorable political structures. The most successful primate, of course, is *Homo lunar*, the only primate to bestride the moon. The second most successful primate is the macaque.

The word "macaque" is derived from the Bantu word *makaku*, which means "monkeys." The macaque is a dexterous Old World monkey with short fingers and an opposable thumb. The macaque usually walks and runs on four limbs, rarely leaping or swinging from trees. Macaques inhabit the greatest geographical distribution of all nonhuman primates, with twenty-two species of macaque found in Asia, Southern Europe, and North Africa, from Indonesian islands on the equator all the way to the "Japanese Alps" on the northern tip of Honshu Island. The Barbary macaque, found on Gibraltar, is the only nonhuman primate in Europe.

The wide-ranging success of macaques is a consequence of their impressive adaptability to a potpourri of environments—especially human environments. Macaques exploit a range of artificial habitats, including both urban and rural habitats, which feature a diversity that surpasses all other nonhuman primates. Indeed, macaques—like rats—reach their highest population densities in locales where they overlap with humans. Their comfort with humans, however, has produced at least one detrimental consequence for them. The macaque is, by far, the most widely used primate in human research laboratories. Because they do well in captivity and exhibit extensive behavioral and neural similarities to humans, the cynomolgus macaque and the rhesus macaque are regularly studied (and, sadly, carved up) to learn about Mind.

Like all primates, macaques indulge in a rich behavioral inventory that is largely split between two predominant activities: foraging and socializing. Macaques are extroverts. They live in troops that can contain fifty monkeys or more and every one of them wants to be in the center of the action.

Macaques must continuously determine who to befriend, who to bully, who to ignore, and when to abandon a troop entirely to search for a more promising community. Their intensely communal lifestyle, overflowing with complex social decisions, has driven the development of a motley collection of mental dynamics that we might label the Who system.

This book will refer to it as the Who *system* and not the Who *module* because there is not a well-defined and unified set of complementary neural networks handling social thinking, but rather, a sprawling confederation of mental and physiological innovations that loosely collaborate to notice, identify, track, and make predictions and decisions about other individuals.

Perhaps the most important thinking element in the Who system is a What module that keeps track of the identity of other macaques, such as "my lover," "my son," "our leader," and "that awful troop of nasty monkeys in the valley." The Who system also contains specialized circuitry for face recognition, grooming, guessing the intentions of others, and even the specific task of monitoring other macaques' eyes. Another thinking element that plays an outsize role in the Who system is the Why module.

One function performed by the Why module is making monkeys care about other monkeys in the first place. If we look out the window and see a rock, a stick, a bucket, and a stranger, we will automatically focus on the stranger, who seems to pop out in our attention. The Why module is responsible for "painting" animate entities with "emotional color" that makes them seem extra important compared to other objects in our environment. One hypothesized source of autism is an alteration of the Why module that prevents it from assigning special value to people, with the result that autistic individuals sometimes treat people like things.

The most critical role the Why module plays in the Who system is helping the Who system make smart social decisions. Should I support the alpha male or one of the young Turks gunning for his spot? Should I groom the lowborn monkey who groomed me every day this week, or the highborn monkey who hasn't groomed me in months? Should I share my food with my lover or hide it for my child?

But making choices about whom to favor and whom to ignore isn't the only challenge that fostered the development of the Why module. Like so many other mental innovations, the Why module was heavily shaped by the need to eat.

4.

Macaques are, in the truest sense of the word, omnivores. They devour fruits, leaves, flowers, seeds, pods, bark, buds, lichens, eggs, insects, crustaceans, amphibians, reptiles, birds, small mammals—and they adore fig trees. Given the extreme diversity of macaques' diets, they typically forage over a range that may be four thousand acres in size. This territory may encompass more than one hundred thousand trees, of which less than 1 percent usually provides ripe food at any given time. Given the distances involved in locating food, as well as the variability in food's nutritional value, seasonal availability, and the physical effort necessary to obtain it, the macaque mind must evaluate and compare the expected net reward of various food options at any given time, even when food sources are distant or not directly visible. A rat is highly nutritious but difficult to procure, whereas bark is abundant and easy to gather but has a low nutritional value and isn't particularly appetizing.

Choosing among complex foraging options requires sophisticated decision-making. The Why module must evaluate the merits of each dining option and judge which option is best. To accomplish this, the Why module must access a detailed memory of the qualities associated with different foods. It also requires sensitivity to the particular context the macaque is experiencing at the moment she makes her decision. For example, if a macaque needs to provide for a child, she will prioritize foods that are easy for a child to digest. If a macaque is lowborn, that will limit his available food options—or require that he take into account the necessity for stealth to obtain high-value foods so that the highborns won't take notice. The Why module must frequently contend with sophisticated versions of the exploration dilemma, such as: *Should I schlep over to the far side of the jungle where macaque-eating chimpanzees lurk, in the hope of getting some delicious bananas? Or should I play it safe and stay around my home caves and eat boring, bland turnips?*

Let's see how the Why module handles this turnips-or-bananas dilemma. We'll step through the Why module's activity in two different contexts: when a macaque is hungry, and when the macaque is well fed.

Let's imagine that a macaque is ravenous. She hasn't eaten all day. She knows there are turnips nearby. The What module sends a category representation of turnips to the Why module. The Why module assigns an emotional value to an object by integrating two distinct sources of value: (1) the object's *inherent value* (that is, the value of a typical specimen under ordinary conditions) and (2) the object's *contextual* value (that is, the value of the object given the current physical and mental state of the macaque). In this case, the macaque knows that turnips are usually bland. Their inherent value is *low*. On the

---

THE WHY MODULE

Scenario 1: The Monkey Is Starving
First option is evaluated.

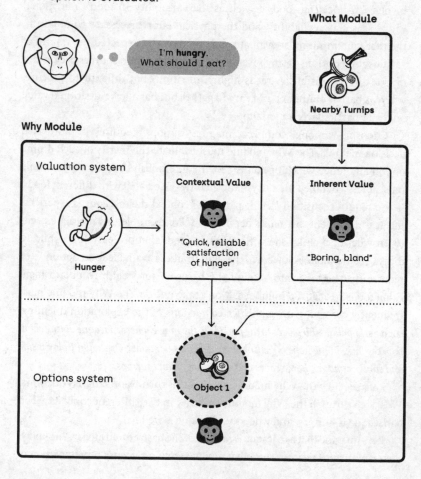

other hand, the macaque is hungry. The turnips are easy to get and pose little danger. They will quickly and safely satisfy the macaque's craving. Thus, the turnips' contextual value is *high*.

The Why module integrates these two sources of valuation—the boring taste of turnips and hunger-satisfying convenience of a quick snack—to create a final valuation somewhere between the two. The turnips are *mildly desirable*.

In the fly mind, the perception of a mildly desirable food source when it was hungry would immediately trigger a response. The fly would automatically head to the food and attempt to eat it. This is sometimes called an "orienting reflex," which is another way of saying "impulsive behavior." But in the Why module, there is an additional dynamic between determining the emotional valuation of an object and acting upon that evaluation. This intermediate dynamic is one of the most formidable innovations in module minds, for it prevents impulsiveness and supports deliberation. This dynamic of deliberation, embodied within the "options system" of the Why module, reaches its zenith in the minds of primates, including the macaque.

A representation of the object (turnips) and a separate representation of its value (mildly desirable) get sent to the options system, located in the prefrontal cortex in primates. Here, the food and its desirability are merged into a single unitary representation of an option: *a mildly desirable turnip*.

But the macaque mind does not act upon this evaluation. Instead, the Why module maintains the *mildly desirable turnip* option in its options system while it evaluates another food option. Bananas.

The What module sends a representation of bananas to the Why module. The Why module knows that bananas are delicious and assigns them an inherent value of *highly desirable*. But the Why module recognizes that the macaque is starving and appreciates that acquiring the bananas will be dangerous and require effort, since they are located on the other side of the jungle, where chimpanzees reside. That's a lot of labor, risk, and delay for a hungry monkey. Thus, the Why module assigns the bananas a contextual value of *very daunting*. The Why module integrates these two valuations—highly desirable and very daunting—and establishes the final valuation of the bananas: *meh*.

The Why module sends *meh bananas* to the options system, where it joins the *mildly desirable turnips* option. What happens next is the difference

THE WHY MODULE

Scenario 1: The Monkey Is Starving
Second object is evaluated.

between module minds' ability to consciously deliberate and neuron minds' reflexive response—and the difference between module mind emotions and neuron mind "emotions." The two options (*mildly desirable turnips* and *meh bananas*) compete in a winner-take-all dynamic. Whichever option has the

THE WHY MODULE

Scenario 1: The Monkey Is Starving

The two options compete in winner-take-all.

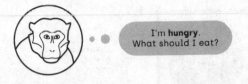

Why Module

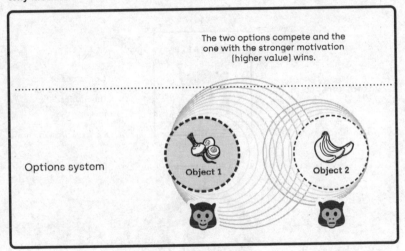

more intense emotional value—the greatest motivational support—wins. When the macaque is starving, the turnips wind up victorious.

This triggers resonance between the representation of the *object* in the winning option (the turnips) and the representation of the *value* of that option (mildly desirable). This makes the object pop out even more saliently in the macaque's awareness, while simultaneously amplifying and prolonging the valuation of "mildly desirable," which the macaque experiences as a conscious *feeling* of desire for the turnips.

THE WHY MODULE

Scenario 1: The Monkey Is Starving

Monkey is conscious of strongest desire for the turnips, and is motivated to go eat the turnips

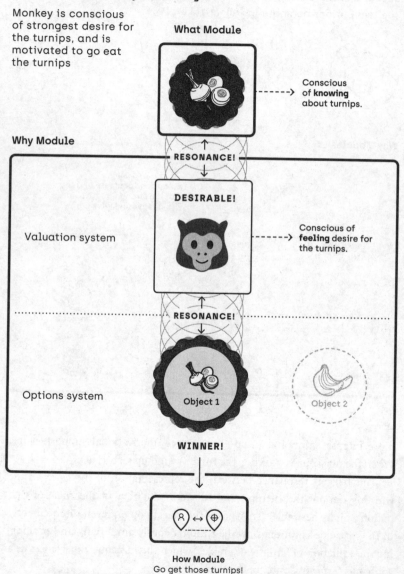

What Module

- - - - -> Conscious of **knowing** about turnips.

Why Module

RESONANCE!

DESIRABLE!

Valuation system

- - - - -> Conscious of **feeling** desire for the turnips.

RESONANCE!

Options system

Object 1

Object 2

WINNER!

How Module
Go get those turnips!

The winning option—mildly desirable turnips, in this instance—is sent to the How module, which helps plot a course of action that will enable the macaque to acquire the turnips and eat them.

Let's quickly consider the case where the macaque had a big breakfast and is not hungry. This time, the final value of the turnips is *somewhat undesirable*

THE WHY MODULE

Scenario 2: The Monkey Is Not Hungry
First option is evaluated.

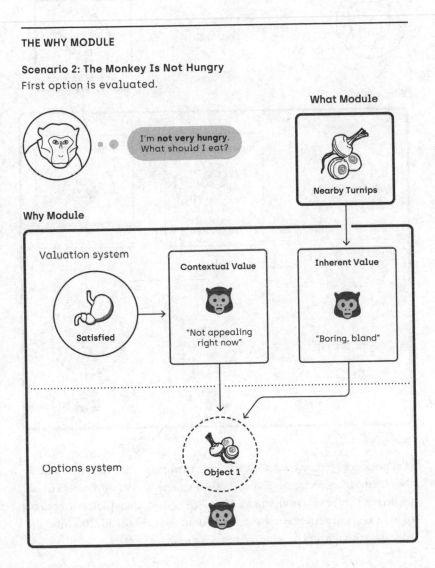

THE WHY MODULE

Scenario 2: The Monkey Is Not Hungry
Second object is evaluated.

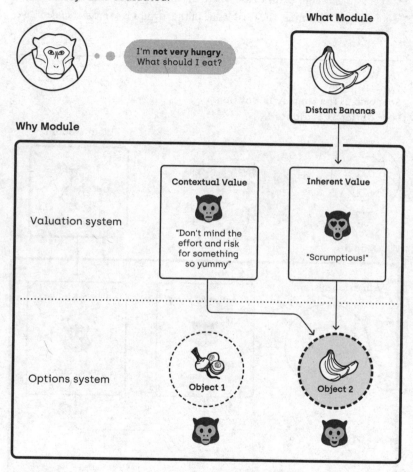

(I'm not hungry so why would I eat boring turnips?), while the final value of the bananas is *moderately appealing* (yeah, it's going to take a little work to get them, but they're so scrumptious!). The Why module sends these two options to its options system, where they fight it out in winner-take-all. This time, the bananas triumph and become the macaque's new objective.

THE WHY MODULE

Scenario 2: The Monkey Is Not Hungry
The two options compete in winner-take-all.

Why Module

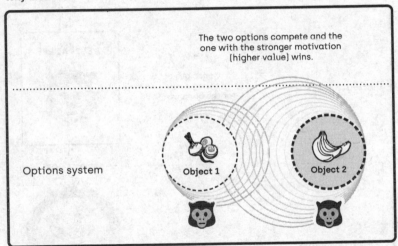

THE WHY MODULE

Scenario 2: The Monkey Is Not Hungry

Monkey is conscious of
strongest desire for the
bananas, and is motivated
to go eat the bananas.

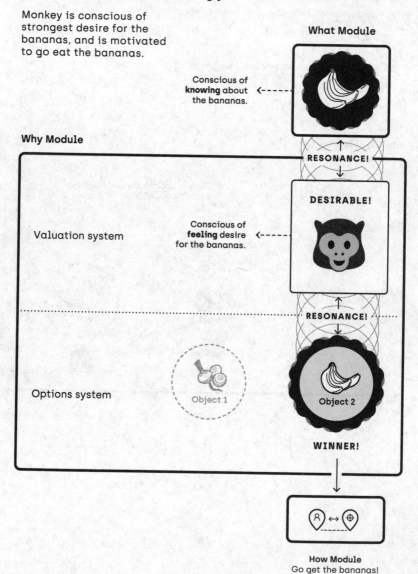

What Module

Conscious of
knowing about
the bananas.

Why Module

RESONANCE!

DESIRABLE!

Valuation system

Conscious of
feeling desire
for the bananas.

RESONANCE!

Options system

Object 1

Object 2

WINNER!

How Module
Go get the bananas!

5.

In *The Bell Jar*, just a few paragraphs after she articulates the fig tree metaphor, Sylvia Plath's proxy eats a meal. Afterward, she comes to view her metaphor quite differently: "It occurred to me that my vision of the fig tree and all the fat figs that withered and fell to earth might well have arisen from the profound void of an empty stomach." Though it's possible that Plath was striving to minimize her anxiety over her indecisiveness, her abrupt shift in attitude underscores how altering our mental context (by satisfying our hunger, for instance) can significantly alter our emotional evaluation of a situation.

The Why module is possibly our slowest-ripening module. Children have little control over their impulsivity, reacting instantly to almost any stimulus. During adolescence, our options system gradually begins to assert some control, though impulsivity, poor judgment, and paralysis remain frequent interlopers. For most of us, we must wait until middle age for a fully mature Why module, more commonly known as wisdom. The Serenity Prayer can be interpreted as a plea for a mature and efficacious Why module: "Grant me the Serenity to accept the things I cannot change, the Courage to change the things I can, and the Wisdom to know the difference."

Tragically, Sylvia Plath ended up taking her own life when she was only thirty years old. The same mental dynamic that empowers us to steer our own course through a complex and risky world can lead us to our demise. The Why module grants us greater freedom to choose our fate, which includes greater freedom to pursue self-destruction—and, happily, greater freedom to pursue something larger than ourselves.

The Consciousness Cartel

O I say these are not the parts and poems of the body
only, but of the soul,
O I say now these are the soul!
—"I Sing the Body Electric," Walt Whitman

It cannot be overemphasized that these resonances
are not just *correlates* of consciousness. Rather, they
embody the *subjective properties of individual conscious
experiences.*

—"Towards Solving the Hard Problem of
Consciousness," Stephen Grossberg

1.

One sunny Memorial Day evening not long ago, a forty-six-year-old man
bought a pack of cigarettes from a convenience store on the south side of
Minneapolis. Until recently he had been working as a security guard at a bar,
but he lost his job because of the COVID pandemic. He was, by all accounts,
an ordinary man performing an ordinary act on an ordinary street corner.
There was nothing in this routine transaction to attract the interest of others
in the neighborhood, let alone the attention of an entire nation.

The United States at the time was populated by roughly 330 million people living in six time zones, scattered from Arctic tundra to tropical rainforest. Their attention was focused on a great many things that Memorial Day
other than the local happenings in Powderhorn Park, Minnesota. Americans
were cooking at home, shopping online, taking strolls through empty parks,
wearily managing homebound kids, and toiling from makeshift home offices
or braving the public behind masks and shields. One of the top national news
stories that evening was a Missouri hair salon that had exposed 140 customers
and employees to the coronavirus. The day's top tweet was a photo of a tattoo
that looked like a cloth patch. It's safe to say that the overall percentage of

Americans that evening who already knew the name George Floyd was not much greater than zero. And yet, over the next few hours and days, his name would ring out in homes from coast to coast.

Less than an hour after he walked out of the store with his cigarettes, Floyd was lying dead in the street, choked to death by a police officer after being accused of paying with a counterfeit bill. Less than twenty-four hours after that, a large crowd had congregated on the street corner where Floyd was killed, fashioning a makeshift shrine to his passing. Within another twenty-four hours, the protests in Minneapolis had increased in size and frequency and spread to other major cities, including Memphis, St. Louis, and Los Angeles. By the end of the week, enormous numbers of Americans had seen George Floyd's name blazoned across the vivid orange-and-turquoise mural at East Thirty-Eighth Street and Chicago Avenue and witnessed the wrenching video of his murder.

This collective attention was almost instantly galvanized into collective action. Americans from Hartford to Honolulu took to the streets (or in Honolulu's case, the ocean). Protests and counterprotests erupted in more than 2,500 American cities, towns, and neighborhoods in all fifty states and five territories, encompassing more than fifteen million people, in over 40 percent of all American counties.

What mechanism enabled a local event on a nondescript street corner to so rapidly trigger nationwide attention—and a nationwide response? The answer to this riddle not only clarifies the social dynamics of collective action, it helps explain how physical activity in our brain transmogrifies into consciousness.

2.

Walt Whitman has been lauded as "America's poet" by other esteemed poets and is widely regarded as one of the most influential writers ever produced by the United States. He was an iconoclastic trailblazer when it came to structure (prose-like free verse), style (unabashedly physical and sensual), and theme. The totality of Whitman's boisterous poetry celebrates, more than anything else, the thrilling yet mystifying experience of sentience.

"Now I will do nothing but listen," Walt Whitman writes in his poem

"Song of Myself," "to accrue what I hear into this song." He proceeds to catalog all the "sounds running together" in his awareness, including "bravuras of birds, bustle of growing wheat, gossip of flames, clack of sticks cooking my meals" as well as "the ring of alarm-bells, the cry of fire, the whirr of swift-streaking engines and hose-carts with premonitory tinkles." And then, transmutation. Whitman rapturously observes that all the individual sounds of this sensory "grand opera" are melded into a seamless "music" in the crucible of his mind. This revelatory experience rouses him to "feel the puzzle of puzzles, / And that we call Being."

The elusive nature of consciousness—simultaneously everywhere and nowhere, objectively physical and subjectively immaterial—has entranced countless generations of poets, philosophers, scientists, theologians, crackpots, hippies, and dreamy engineers. Whitman himself asserted that "consciousness" was a "miracle of miracles, beyond statement, most spiritual and vaguest of earth's dreams, yet hardest basic fact, and only entrance to all facts." As the twentieth century drew to a close a full hundred years after Whitman's death, science could barely offer a better account of consciousness than the poet's lyrical exultations.

Do sharks feel the "puzzle of puzzles"? Do chimpanzees experience a "grand opera"? Does sentience require one eldritch brain widget, or many? Or might consciousness be rooted in some indiscernible plane of energy? As recently as the 1990s, such questions were seemingly impossible to answer. But in the twenty-first century, a much clearer picture has finally begun to take shape, backstopping Whitman's rhapsodic musings with the limpid explicitness of nonlinear mathematics.

The science of consciousness is neither forthright nor compact. It's not possible to reduce the operational details of "that we call Being" to a bumper-sticker insight (like "Everything is made of atoms") or a single revelatory equation (like $E = mc^2$). Consciousness demands the opposite of a reductionist explanation, in fact. It requires the harmonization of more than thirty models covering every major form of thinking and every major brain structure. It necessitates more than a dozen interlaced equations characterizing mental activity across molecular, neural, and modular levels. Stephen Grossberg's unified theory of mind is anything but simple. Yet, within its forbidding complexity glows the soft rosy light of illumination.

The crowning achievement of Grossberg's framework is its detailed explanation of *why* and *how* physical activity in the mind is converted into subjective experience. This explanation consists of three levels of description:

1. The *mathematics* underpinning the dynamics of consciousness. The math consists of systems of nonlinear differential equations that characterize the activity of thinking elements at the molecular and neural levels.

2. The *neural dynamics* of consciousness. This network-level description characterizes the physical configuration and activity of the neural circuits and modular thinking elements that embody conscious experience.

3. The *psychological properties* that arise out of the neural dynamics of consciousness as emergent properties. This person-centered level of description characterizes the inner world of our subjective experience (the *phenomenology* of consciousness, in philosophical jargon). These subjective psychological properties are commonly known as *qualia*.

This chapter explores the relationship between the neural dynamics of consciousness and the qualia they give rise to—the relationship between the physical activity in our brain and the blast of inner experiences that stirred Whitman to avow, "O I say now these are the soul!"

Let's begin with the most basic question regarding the mystery of sentience. Why does consciousness exist?

The evidence does not suggest that it arose so that we might platonically contemplate the magnificence of nature or muse upon God's creation. Nor is consciousness itself an emergent property, some serendipitous consequence of any system that reaches some threshold level of complexity, like a nuclear chain reaction triggered by a critical mass of uranium 235. As with every previous innovation in the journey of Mind, consciousness is a specific mental innovation that arose to solve specific mental challenges.

Even though conscious states can seem downright supernatural to the mind experiencing them, they are part of the ordinary and ancient fabric of our mind's ability to convert ambiguous sensory inputs into useful behavioral outputs. Functionally, consciousness is no different from the locomotive fla-gella of archaea, the doer networks of the hydra, and the boundary comple-tion circuits of the zebra fish. Consciousness was designed to fulfill a purpose. To be more precise, different sorts of consciousness were designed to fulfill different sorts of purposes.

Leading scholars have long contended that there is only one form of con-sciousness, one form with many faces. (This is why most investigators of con-sciousness, from seventeenth-century French philosopher René Descartes to legendary twenty-first-century biologist Francis Crick, sought to identify the *one special place* in the brain responsible for consecrating all conscious expe-rience.) Grossberg, in contrast, argues that there is a finite and discernible number of forms of consciousness—each generated by a different neural struc-ture—that jointly give rise to all subjective experience. Expressed differently, we could say that our mental economy is governed by a consciousness cartel.

Each member of this cartel is a distinct type of neural resonance. Each type of resonance is generated by a distinct neural module pursuing a dis-tinct purpose, a purpose reflected in the characteristic qualia that it gen-erates. Grossberg describes six types of qualia—six distinct categories of conscious experience:

[1]   SEEING [Where module]: *I see a green splotch in the middle of the wall.*
The purpose of seeing is to help ensure effective targeting. The qualia of see-ing are generated by a resonance between a representation of a visual surface and a representation of an attentional shroud.

[2]   VISUAL KNOWING [What module]: *I know that the green splotch is a clock.*
The purpose of visual knowing is to support decision-making and goal-setting. The qualia of visual knowing are generated by a resonance between a repre-sentation of the visual features of an object and a representation of an object category.

[3]   HEARING [Where module]: *I hear a sound coming from the center of the stage.*
The purpose of hearing is to help ensure effective targeting and effective vocalizations. The qualia of hearing are generated by a resonance between

a representation of an auditory stream and a representation of an attentional shroud.

(4)   AUDITORY KNOWING (What module): *I know the sound is the voice of comedian Chris Rock.*

The purpose of auditory knowing is to support decision-making and goal-setting. The qualia of auditory knowing are generated by a resonance between a representation of the features of an auditory stream and a representation of a sound-source category.

(5)   SYNTHESIZING (What module): *Chris Rock is saying, "Comedy is the blues for people who can't sing."*

The purpose of synthesizing is to support sequential learning, prediction, and performance, including communication, toolmaking, and complex motor skills. The qualia of synthesizing are generated by a resonance between a representation of the features of a sequence of items (such as the notes of a melody, the words of a sentence, or the steps for performing a task) and a representation of a sequence category (such as "Jingle Bells," "the opening line of the Gettysburg Address," or "grilling a steak").

(6)   FEELING (Why module): *Chris Rock's punch line makes me cheerful.*

The purpose of feeling is to select and pursue useful goals. The qualia of feeling are generated by a resonance between a representation of an object category and a representation of a value.

This is not a complete list of all types of qualia that exist in the human mind, let alone all the minds of the animal kingdom. Module minds undoubtedly safeguard other types of qualia, each with their own signature resonant dynamic, such as smell, taste, touch, pain, and proprioception, as well as echolocation, magnetoreception, and ultraviolet perception. The set of six qualias highlighted above are merely the ones for which Grossberg developed complete models during his research career. Fortunately, these six account for the preponderance of human experiences, especially those experiences that most often lead us to compose poetry, read books, conduct science, fall in love, or wonder how we came to exist at all.

3.

How does matter become conscious? How do sensory representations stirring the neural pasta in your brain get transformed into an intangible banquet of experience? To aid us in answering this, we can derive three "laws of consciousness" from the mathematics of Grossberg's unified theory of mind:

1. All conscious states are resonant states.
2. Only resonant states with feature-based representations can become conscious.
3. Multiple resonant states can resonate together.

The first and most important law asserts that all the qualia you might experience—every perception of a "bravura of birds," every interpretation of an ambiguous line of Whitman's poetry, every feeling of jubilant comprehension—is generated by resonance. Resonance is a neural dynamic characterized by the mutually reinforcing activity of two distinct mental representations that *synchronizes*, *amplifies*, and *prolongs* the joint activity of the representations. (Think of a violin and a trumpet playing the same melody.)

Not every module in the mind produces resonance, and therefore not every module is capable of generating conscious experience. The How module and the When module do not form resonant states and thus generate no qualia. Neither does the Visual Scene module nor the Auditory Scene module. We might say the How module is *nonconscious* (its activity never enters consciousness), whereas the When module, Visual Scene module, and Auditory Scene module are *preconscious* (the representations they create can enter consciousness when processed by another module capable of producing a resonant state).

The second law of consciousness is also critical. It asserts that for a resonant state to produce qualia, the resonating representations must contain *features*. The specific nature of these features dictates the subjective psychological properties of the resulting qualia. For instance, consider the conscious experience of gazing upon a red square. We can focus our awareness on any of the specific perceptual features that make up the square, such as its redness, its four pointy corners, its four flat sides, or the fact that it covers a prescribed region of space. Contrast that with the experience of listening to a musician play a tuba. Unlike the fixed and static sight of a square, the sound of a tuba is perceived as having a start point in time and as changing in intensity and pitch over time before vanishing from our perception. We can focus our attention on several different properties of the tuba's sound, such as its brassy timbre, low pitch, or loud volume—properties that we experience quite differently than the shape, size, and color of a square. These subjective properties correspond to perceptual features embodied within resonating representations.

There are modules in the mind that generate resonant states but do *not* produce consciousness because the resonating representations do not contain features. One example is a resonant state that helps coordinate a conscious decision to perform a particular action (such as reciting a printed poem) with a nonconscious motor command that helps execute that decision (such as saccading the eye to fixate on the first word of the poem). The representations in this resonant state do not contain features, but instead encode complex motor commands using fixed "code words," like the Secret Service code name "Celtic" for President Joe Biden.

Another example of a nonconscious resonant state is the resonance between two different arrays of neurons used for spatial navigation (hippocampal place cells and entorhinal grid cells). This resonance is used to coordinate our head-centered map of space with our target-centered map of space, the way a mapping app on your smartphone might assign GPS coordinates to street addresses. Once again, the resonating representations do not encode any discrete features, so qualia are not produced. That's why these two types of nonconscious resonance do not have familiar words to describe them, such as seeing, knowing, or feeling.

The final law of consciousness holds that resonant states occurring simultaneously within different modules can also resonate together. Multiple types of qualia can thereby assemble into a communal consciousness cartel that is subjectively experienced as a single unitary and seamless experience. One of the most common examples of multimodule resonance is the synchronization of *seeing* (in the Where module), *knowing* (in the What module), and *feeling* (in the Why module). We are thrilled (or bored) as we watch the Avengers battle another supervillain to save the universe.

The consciousness cartel is neither incidental nor emergent. The cartel itself (like the resonant states that are members of the cartel) was designed for a specific purpose. The most vital and formidable purpose for a module mind, in fact: solving the attention dilemma.

How does a mind composed of a vast and diverse population of neurons (eighty-six billion, perhaps ten thousand types) participating in myriad simultaneous dynamics spread across numerous intertwined modules manage to get all its modules to break away from whatever they're doing and focus their attention on the same urgent object, event, or idea—and do so without the benefit of any centralized decider?

To find out, we must return our attention to George Floyd.

4.

What happened in the interval between George Floyd getting choked to death and a vast and diverse population protesting and counterprotesting in the wake of his violent demise?

The media.

The media in an advanced democracy opens a window onto the operation of consciousness because a free press and consciousness share the same underlying dynamic of attention. In a society with a robust freedom of the press, the media can quickly focus the attention of the entire populace on local events of national interest. Some stories get synchronized, amplified, and prolonged, while other stories get ignored, quenched, or simply outcompeted by more resonant stories. The most significant events are brightly illuminated by the collective media spotlight, and the incidents and individuals that resonate most vibrantly are most likely to trigger society-wide action.

Let's retrace the attention dynamic triggered by George Floyd's death. Four officers from the Minneapolis Police Department participated in the nine-and-a-half-minute murder of Floyd. Video footage of the event was recorded by bystanders and uploaded to social media, along with commentary by other witnesses (think of these amateur videographers and commentators as sensor neurons, providing raw environmental data). On social media, competing narratives about the event started to take shape, many filled with false, ambiguous, and partial information, until a dominant interpretation coalesced—namely, that a White officer with a history of excessive-force complaints killed a defenseless Black man without provocation (think of this as the preconscious representation produced by the Visual Scene module).

This narrative became salient and stable enough that it served as bottom-up sensory input for the professional media, which attempted to match this bottom-up narrative to the media's preexisting top-down expectations about what sort of story it will likely find based upon previous stories covered in the media. In the case of George Floyd's death, the top-down expectation was that the event was another example of unwarranted White-on-Black police brutality. Since the bottom-up data and top-down expectation matched, George Floyd's story stabilized in the headlines of major newspapers and the chyrons of cable news. This is resonance, analogous to qualia of *knowing* generated by the What module.

The media "resonance" began synchronizing, amplifying, and prolonging the story. Both national and local media outlets began publicizing other outlets' reporting on Floyd. Social media echoed and commented on the story, which got fed back into the mainstream media's coverage. Journalists were assigned to dig up more details about the story, which sustained and strengthened the resonance of the story.

As in a module mind, there is no single "media decider" who determines which story is important enough to be covered by most (or all) media outlets. Instead, the decision is the collective result of the entire "media cartel": dozens of major media outlets scattered across the country (though they are most heavily concentrated in New York, Los Angeles, and Washington, DC, just as a mammal's qualia-generating mental modules are concentrated in its cortex). All possible stories constantly compete against one another for the attention of the media cartel, but each individual media outlet makes its own judgment about what constitutes a worthy and timely story, based upon its own preexisting top-down expectations. *People* magazine and *The Economist*, for instance, have very different notions of the kind of stories that merit attention. So do CNN and Fox News. No one outlet provides the "definitive" interpretation of an event. Instead, the interpretation is distributed throughout the entire media cartel.

Once the mainstream media started reporting on George Floyd's death, people across the country started reading about it, hearing about it, or watching it. This swiftly attracted the attention of huge numbers of individual Americans as they became aware of the dominant story of the day. Many of these individuals dropped what they were doing and took action—posting opinions on social media, hammering signs in yards, buying books about racism, donating money to Black Lives Matter (or Blue Lives Matter), or joining protests or counterprotests. These activities then became new bottom-up inputs into the media cartel, reinforcing the existing story and amplifying and prolonging it still further. The George Floyd story continued to resonate within the awareness of the media and the public, driven by the feedback loop of media coverage and widespread collective response, until new events of more pressing urgency came along and knocked the old event off the headlines.

This is a pretty good description of how consciousness works in a brain, too. External events (and internal thoughts) constantly impinge upon the What, Where, and Why modules. These consciousness cartel modules evaluate the ceaseless stream of raw data, just as news desks constantly evaluate potential stories. Many of these modules will enter a resonant state when a bottom-up input matches a top-down expectation. This resonance may command attention from other modules—but it depends entirely on what the

---

THE CONSCIOUSNESS CARTEL

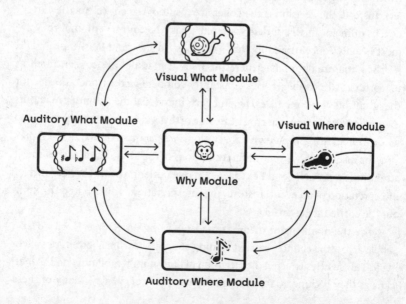

other modules are doing. If another module is also resonating, but its attention is on a different stimulus, then the two resonances compete in a winner-take-all and the stronger resonance will win out and recruit other members of the consciousness cartel to share its attentional focus.

For instance, if you are looking at a hundred-dollar bill on the edge of the road (resonating in your visual What module) and you hear a voice behind you cry, "Get out of the way before you're run over!" (resonating in your auditory What module), the audio resonance will outcompete and quench the visual resonance and recruit the other modules to refocus their attention on the urgent vocal directive.

Like the media cartel, the consciousness cartel enables a sprawling system comprising numerous parallel minidynamics operating at multiple levels of organization to rapidly focus the resources of the entire system upon an object or event in real time, empowering the system to swiftly execute a collective response.

This is the solution to the attention dilemma. This is how consciousness works.

5.

Some readers might be willing to accept that similar dynamics govern both a democratic country's mass media and personal consciousness, yet still feel this account omits an explanation of the all-important subjective experience of "the puzzle of puzzles, / And that we call Being." For such skeptics, there may appear to be an unclosed gap between the physical and the mental.

"Okay, I see all this neural activity," such a skeptic might think; "I see that the sensory modules send representations to the What module. I see these object representations get matched up with category representations, triggering resonance. I see this resonance triggers activity in other modules, such as prompting the How module to grab the object. But how is this any different from electric currents zipping around my microwave oven? *Where is the consciousness?*"

This is like watching two teams on a basketball court dribbling, passing, shooting, and rebounding as they try to score points, and saying, "Okay, I see all this physical activity. *But where is the game?*"

Resonance is not a *metaphor* for consciousness. It is not a conceptual heuristic or explanatory fragment waiting for the true "sorcery" of consciousness to somehow get folded in. Nor is it causal. Resonance does not *lead to* a burst of consciousness the way that lighting a firecracker leads to an explosion. Rather, resonance is a precise characterization of exactly what's happening in the physical world when a conscious experience is occurring, the same way that a violin string vibrating at 261.6 hertz is a precise characterization of what's happening in the physical world when a violinist plays a middle C. "Sound" is not some extra quality separate from the vibrating molecules in the string and surrounding air. A "basketball game" is not some extra quality separate from the activity of the players. And consciousness is not something separate from the holistic dynamics of the entire mind during a resonant state. (Though if you are one of the players in the basketball game, your definition of the game would be different from that of an observer. You might say, "No, a game is when four other players join up with me to attempt to score more points than another team of five—and obviously, the game revolves around *me*." We'll revisit this perspective in the final chapter.)

In effect, Stephen Grossberg invokes the basketball game principle when he asserts, "These dynamical resonant states are not just 'neural correlates' of consciousness. Rather, they are mechanistic representations of the qualia that embody individual conscious experiences on the psychological level." Consciousness is distinctive local mental activity embedded within global mental activity operating across multiple hierarchical levels of a mind.

6.

Nevertheless, you might still be complaining: "Okay, okay, okay . . . but where am *I* in all of this? After all, there's not just awareness . . . there's *me* having awareness of *me* having awareness!" That is an excellent question, but we're getting ahead of ourselves. We must pause here to recognize that this is *not* a question that a chimpanzee would ponder.

There is, in fact, a meaningful and mathematical difference between the act of being conscious and the act of thinking "*I* am conscious, *me, here, now!*" A chimpanzee undoubtedly experiences consciousness. But she lacks a couple of innovations that distinguish our own unique experience of consciousness from that of our brightest relatives in the animal kingdom.

The three laws of consciousness provide us with an objective way to evaluate consciousness within any mind, microscopic or gargantuan. The first law of consciousness tells us that if a mind is not capable of resonance, it is not capable of consciousness. We can go one step further: because a pair of mental representations is an essential component of every resonant state, we know that a mind that lacks representational thinking cannot be conscious. From this we can derive one immediate and staggering conclusion. Virtually all the organisms who have ever lived on Earth have not been conscious.

Molecule minds, such as those of archaea, bacteria, and protozoa, are not conscious. They do not form representations. Single-celled organisms do not perceive, do not feel, and do not know. (But before you start obliterating bacteria and protozoa without moral compunction, keep in mind that they are an essential part of the ecosystem that supports conscious creatures! They may not be conscious, but we wouldn't be alive—or conscious—without them! Their dynamics support our own dynamics, along with the dynamics of every mammal, reptile, amphibian, and bird. One lesson the journey of Mind should have taught us by now is that the dynamics of thought and experience depend intimately and inextricably upon the dynamics of everything else in nature.)

Neuron minds are not conscious either. The hydra and roundworm do not form representations. The evidence is inconclusive regarding whether the flatworm mind forms distinct representations, but even if it does, it certainly does not generate top-down expectations or match-based learning, both of which are essential for resonance. Neither does the fly. Though the fly exhibits well-documented representational thinking and appears to exhibit some "recurrent" dynamics that are similar to resonance (and perhaps even form a neural precursor to resonance), the physiological evidence so far strongly suggests that the fly mind does not attempt to match bottom-up representations with top-down representations, and even if it does, we can say with near certainty that this as-yet-unidentified match-based learning is *not* using resonance to recruit the attention of other modules. Thus, insect minds appear to be the most advanced biological intelligence on Earth not capable of consciousness.

Next, we move on to the oldest and simplest vertebrate intelligence, which happens to be the simplest module mind, too: fish minds. Are fish conscious? Quite likely, according to the laws of consciousness. Fish minds undoubtedly form sophisticated representations that encode sensory features. (Second law.) And evidence within a limited number of fish species suggests their visual systems have top-down expectations (category representations) that are matched against bottom-up inputs (feature representations) during match-based learning. Fish consciousness is likely meager and lightweight—like the media in North Platte, Nebraska, say, or fifth graders playing basketball at recess on the playground half-court—but fish minds do appear to possess the necessary dynamics for producing qualia. The case for consciousness is similar for amphibians, and overwhelming for reptiles, birds, and mammals—including chimpanzees.

Chimpanzees experience the bleak troughs of despondency when they lose a child. They experience sharp pangs of jealousy when an ally shows attention to a rival chimp. They deliberately pretend they don't know where a cache of food is hidden, leading other chimpanzees away from it, then sneaking back to consume it later. They experience joy when playing jokes, and devote considerable conscious attention to planning pranks such as holding water in their mouths for long periods of time just to spray it in an unsuspecting human's face. They engage in imaginary play, such as pretending to pull an imaginary toy with an imaginary string. One chimp was documented caring for a log as if it were an imaginary infant. At four field sites in West Africa, chimpanzees perform rock-throwing rituals, where they fill tree trunks with rocks, then step back and throw the rocks at the same special trees.

They are also capable of surprising tenderness and compassion. The nature documentary *Chimpanzee* captures a startling moment in the jungles of the Ivory Coast when an orphaned chimpanzee approaches all the chimps in his troop for support after his mother unexpectedly perishes. One by one, they turn the child away. An abandoned chimp youngster is virtually certain to die in the unforgiving jungle. But then, grace. The alpha male—a huge, violent hulk of a chimpanzee—adopts the unrelated toddler and proceeds to feed him, protect him, nurture him, and teach him.

Chimpanzees even seem capable of the same reverence of nature that Walt Whitman felt while contemplating Being. In Senegal, chimpanzees have been recorded dancing at the edge of fires, entranced by the flickering flames. Perhaps the most moving example of chimpanzees' expansive awareness was documented by Jane Goodall in Gombe, Tanzania. There, the chimpanzees treat the waterfalls with marveling veneration. Sometimes they hurl rocks into the spray, but usually they sit calmly, gazing placidly at the rush of water. "I can't help feeling that this waterfall display, or dance, is perhaps triggered by feelings of awe and wonder," Goodall says. "Why wouldn't they also have feelings of some kind at spirituality? Which is, really, being amazed at things outside yourself."

All conscious beings share a common privilege. We are all hoping and suffering, recognizing and realizing, fearful and fervent. We are all cosmic weavers: the frog mind and the hummingbird mind and the chimpanzee mind pluck out kaleidoscopic snapshots of the universe's unfolding within their resonance, knotting the ceaseless torrent of chaos into a prismatic tapestry. In these animal minds, as in our own minds, we gather up fleeting fragments of reality and transform them into feeling, a cosmos within a cosmos. Consciousness is a rich, intimate, symbiotic relationship between reality and itself. And the more conscious we are of consciousness, the more we will come to value other sentient beings as hallowed and exquisite flickers of feeling within the boundless darkness.

"In what terms should we think of these beings, nonhuman yet possessing so very many human-like characteristics? How should we treat them?" Jane Goodall asks about chimpanzees. "Surely we should treat them with the same consideration and kindness as we show to other humans; and as we recognize human rights, so too should we recognize the rights of the great apes? Yes."

The chimpanzee (and perhaps the bonobo) possesses the most sophisticated consciousness cartel in a module mind on Earth. Yet, despite our own species' understandable fixation on it, the emergence of consciousness is *not* the climax of Mind's story. Though astonishing, consciousness appears as a mere stepping-stone in the cosmic odyssey of Mind as it strives, with ever greater purpose, to subdue the relentless forces of chaos. Indeed, there are two more fateful innovations on our journey (as well as many smaller ones) that are absent from the chimpanzee mind, including the innovation that accounts for the "I" in "I sing the body electric!"

To investigate these ultimate innovations, we must ascend to the most advanced stage of thinking in the journey of Mind on Earth. When module minds began to communicate with one another through complex external representations it paved the way for the emergence of superminds and the mental powers of a god—a volatile and capricious god who tends to reach for the darkness with nearly the same likelihood as it reaches for the light.

Stage IV

Superminds

The Darkness

But that doesn't mean we are striving to form a union
that is perfect.
We are striving to forge our union with purpose.
—"The Hill We Climb," Amanda Gorman

1.

Around about 1818 in the backwater town of Tappers Corner on the Tucka-hoe River in Maryland, a child was born. He was given the name Fred Bailey. Fred grew up in a small cabin fashioned out of logs, clay, and straw. One of his earliest memories was his fascination with the most complex object in his childhood environment: a ladder that ascended from the floor to the upstairs bed. He marveled at its craftsmanship and the fact that it allowed one to climb to new heights.

Fred lived with his grandparents. He was especially close with his grand-mother, whom he described as possessing "the reserve and solemnity of a priestess." She cooked for him, clothed him, and tucked him into his straw bed at night. Eventually, he learned that his mother was alive, but for reasons that were not entirely clear to young Fred, she was not able to live with him. His father was a cipher who was never openly discussed. A shadowy figure not to be mentioned.

His childhood was not much different from that of most kids raised by loving caretakers of modest means. "Feddy," as he was often called, enjoyed roaming the woods or jumping in the local stream with his clothes on. He gleefully pretended to be a horse, or a pig, or a dog, or a "barn-door fowl." He enjoyed conversations with his grandparents, sitting outside in the shade during the summer and around the hearth in the winter. He stayed out of the way of the older kids, who rarely took an interest in him, though he was occasionally informed in portentous tones that "he would be made to 'see sights' by and by."

Even though much of his childhood was "spirited, joyous, uproarious, and happy," from an early age Feddy sensed that behind the appearance of innocent normality lurked a malignance. Some ominous menace with dark designs on him.

"As I got larger and older, I learned by degrees the sad fact, that the 'little hut,' and the lot on which it stood, belonged not to my dear old grandparents, but to some person who lived a great distance off," Bailey recounted as an adult. "I further learned the sadder fact, that not only the house and lot, but that grandmother herself . . . and all the little children around her, belonged to this mysterious personage, called by grandmother, with every mark of reverence, 'Old Master.'"

2.

The book you are reading culminates with an investigation of the most significant innovation in the journey of Mind, the Self. What is a Self? As a matter of tec$ical description, one could say that the Self is a complex mental dynamic characterized by intractable nonlinear mathematics embodied within a tight-knitted hierarchy of biological innovations assembled over a three-billion-year run stretching from the Archaean era to the Holocene. Yet such a description would fall short of the mark. Any analysis of the Self is incomplete without an explicit recognition that the Self is something singular in the universe. The Self is unique because it is the only physical entity capable of *experiencing* delights and horrors—and, crucially, definingly, capable of *knowing* it is experiencing delights and horrors.

This book will explain why the Self came to be and how it works, but this mechanical explanation will fail to capture the extraordinary nature of the Self, a nature best approached through narrative. Such a narrative will also illustrate the most compelling lesson about the Self that the journey of Mind has to teach us, a lesson that will only become fully clear after we've completed our journey through all four stages of Mind. The lesson is this:

It takes a society to create a Self.

The lesson is burdened with a dark corollary:

It takes a society to crush a Self.

3.

For Fred Bailey, the darkness came quickly. When he was just seven years old, his grandmother took him on an arduous twelve-mile hike along a dirt road leading from the Tuckahoe to the Wye River, toting him on her shoulder for much of the way. They finally arrived at an enormous farm with many large houses. He was greeted by a rambunctious gaggle of children, "black, brown, copper colored, and nearly white." They smiled at him and invited him to play. Feddy was suspicious. He refused to budge from his grandmother's side. Eventually, urged on by the children's cajoling and his grandmother's reassurances, Feddy cautiously followed the youngsters into the house. He refused to join their games, keeping his back firmly against the wall, fearful that one of the children might sneak up on him from behind.

Suddenly, one of the girls ran in from the kitchen and declared with roguish glee, "Fed! Fed! Grandmammy gone!"

All his worst fears were realized. "I fell upon the ground, and wept a boy's bitter tears, refusing to be comforted," Bailey recounted. "[The children] said, 'Don't cry,' and gave me peaches and pears, but I flung them away, and refused all their kindly advances. I had never been deceived before; and I felt not only grieved at parting—as I supposed forever—with my grandmother, but indignant that a trick had been played upon me in a matter so serious."

That evening he spent his first night in the strange domicile of his enslaver, a White man who went by the name of Captain Anthony. He was a cruel and ignorant overseer who, under American law, owned more than two hundred *Homo sapiens*, a mere speck of the one and a half million state-sanctioned human chattels in the United States of America. Captain Anthony was the chief overseer for Colonel Lloyd, a wealthy owner of dozens of forced labor camps along the Eastern Shore of the Chesapeake Bay that exploited more than a thousand enslaved men, women, and babies. Captain Anthony was in charge of running Lloyd's farms. This position granted him the power to inflict any whim upon Feddy without consequence, "other than concerning profit and loss."

Fred Bailey, a joyful, playful seven-year-old boy, had discovered his designated place in American society. He was a piece of property, a *thing*, afforded the same civil rights as a broom.

4.

That very night, the institutions of American society began to work in earnest on Feddy, as it did on all enslaved children, in the pursuit of a single mission: to exterminate all notions of individual purpose and replace them wholesale with the purpose of his master. The goal, in short, was to convert a human being into a zombie.

One of the most effective tec$iques for the suppression of personal agency, universally employed among the states of the southern United States, was to erase any sense of family, including the conviction that the bonds of family were sacred. Husbands and wives were regularly broken apart, usually to never see each other again. Children were separated from their mothers (or grandmothers) at a young age. At birth, on occasion. Brothers and sisters were dispersed to different households. Even if they remained at the same labor camp, they were not permitted to form filial relationships. Fred himself had several siblings, though he lamented never feeling any connection to them: the institution of slavery "had made my brothers and sisters strangers to me; it converted the mother that bore me, into a myth; it shrouded my father in mystery, and left me without an intelligible beginning in the world."

The only family member that Fred felt a deep and lasting connection with was his grandmother. This loving bond would produce some of his greatest sorrow.

> If any one thing in my experience, more than another, served to deepen my conviction of the infernal character of slavery, and to fill me with unutterable loathing of slaveholders, it was their base ingratitude to my poor old grandmother. She had served my old master faithfully from youth to old age. She had been the source of all his wealth; she had peopled his plantation with slaves; she had become a great-grandmother in his service. She had rocked him in infancy, attended him in childhood, served him through life, and at his death wiped from his icy brow the cold death-

sweat, and closed his eyes forever. She was nevertheless left a slave—a slave for life—a slave in the hands of strangers; and in their hands she saw her children, her grandchildren, and her great-grandchildren, divided, like so many sheep, without being gratified with the small privilege of a single word, as to their or her own destiny. And, to cap the climax of their base ingratitude . . . they took her to the woods, built her a little hut, put up a little mud-chimney, and then made her welcome to the privilege of supporting herself there in perfect loneliness; thus virtually turning her out to die!

Another way of foreclosing any possibility of feeling the natural connection of family, roots, or kinship, was to rub out a person's name. In most cases, the enslaved were given first names, but no surnames. No *family* names. Usually, the enslaved were designated in reference to their owners. Feddy himself was often called "Captain Anthony Fed." Often, they were given nicknames, sometimes derogatory, always demeaning, treated more like pets than people.

Further eroding any sense of selfhood, Feddy was also prevented from knowing his birth date.

I never met with a slave who could tell me how old he was. Few slave-mothers know anything of the months of the year, nor of the days of the month. They keep no family records, with marriages, births, and deaths. They measure the ages of their children by spring time, winter time, harvest time, planting time, and the like; but these soon become undistinguishable and forgotten. Like other slaves, I cannot tell how old I am. . . . I learned when I grew up, that my master—and this is the case with masters generally—allowed no questions to be put to him, by which a slave might learn his age.

In most cases, Americans did not provide the enslaved with clothing. In Captain Anthony's forced labor camps, children often ran about virtually naked, while adults were granted a yearly allowance of one pair of trousers and two shirts that went down to the knees. These were all made of tow linen, a coarse and scratchy fabric. Most of the enslaved were not given beds or blankets. The children squabbled over partially sheltered nooks around the kitchen to sleep in. Feddy usually slept in a closet, though in winter one favorite spot was the corner of a huge chimney, where he could place his feet in the ashes to keep them warm. No jacket or blanket was provided for winter—the same tow-linen shirts and trousers needed to suffice through blazing heat and heavy snow.

Privacy was a luxury that few of the enslaved were permitted, further reducing the chance to develop the most rudimentary sense of personal time, personal space, personal thoughts. Even though privacy was scarce, opportunities for socializing were assiduously controlled. The few pastimes afforded enslaved people were explicitly intended to further the process of degradation. On Sundays and holidays, when they were finally granted time off from their exhausting labors, the enslaved were encouraged by their American slavers to engage in wild revelry—drinking alcohol, dancing, gambling, fornicating—rather than anything productive or edifying. Slavers preferred to watch their charges engage in self-dissipation rather than self-development.

These mental and social tec$iques were designed to constrict Feddy's *environment* and constrain the development of his *brain*, two of the three components of Mind. But the most ruthless tec$iques were designed to control his *body*.

Enforced starvation was a common practice. The enslaved were seldom provided enough to eat to satisfy their daily calorie needs (particularly given their long hours of forced labor). What food they received was usually coarse and not very nutritious. Cornmeal mush, ash cake, old herring, tainted pork. Feddy fought with the family dog for scraps or waited for the kitchen girl to shake out the tablecloth for the cats, so he could grab crumbs and small bones. It was incumbent upon all slave-makers to firmly demonstrate that the subject's mind was *not* the commander of his body—that his physical self was a piece of property subject to absolute control by his master. The most salient emblem of this philosophy was the lash.

American lashes were designed to inflict pain and draw blood, with the broader intention of evoking fear in the minds of whipped and unwhipped alike. Whips could be made of undried leather, young saplings, or tapered branches. The enslaved were whipped for showing up late. They were whipped for laziness. They were whipped for breaking china, for moving too slow, for being dirty. They were whipped for complaining, for laughing, for crying. They were whipped for searching for their children, for meeting their lover, for mourning the death of a friend. But most often they were whipped for "impudence."

Bailey explains, "This may mean almost anything, or nothing at all, just according to the caprice of the master or overseer, at the moment. . . . But, whatever it is, or is not, if it gets the name of 'impudence,' the party charged with it is sure of a flogging."

Feddy was not long at Colonel Lloyd's labor camp before he witnessed the brutal flogging of a young mother while her children looked on.

> [The overseer] finally overpowered her, and succeeded in getting his rope around her arms, and in firmly tying her to the tree, at which he had been aiming. This done, and Nelly was at the mercy of his merciless lash; and now, what followed, I have no heart to describe. The cowardly creature made good his every threat; and wielded the lash with all the hot zest of furious revenge. The cries of the woman, while undergoing the terrible infliction, were mingled with those of the children, sounds which I hope the reader may never be called upon to hear.

The ultimate instrument of depersonalization was murder. In a single act, the mind of a slave could be stamped out forever, should he nurture too much willfulness to break. Before his tenth birthday, Feddy confronted the full unchecked authority of his American overlords.

> Among many other deeds of shocking cruelty . . . was the murder of a young colored man, named Denby. . . . In something—I know not what— he offended this Mr. Austin Gore, and, in accordance with the custom of the latter, he undertook to flog him. He gave Denby but few stripes; the latter broke away from him and plunged into the creek, and, standing there to the depth of his neck in water, he refused to come out at the order of the overseer. . . . Mr. Gore, without further parley, and without making any further effort to induce Denby to come out of the water, raised his gun deliberately to his face, took deadly aim at his standing victim, and, in an instant, poor Denby was numbered with the dead.

As heartless as these methods of physical subjugation were, the most effective tec$ique for preventing the development of a Self was a mental one. The enslaved were forbidden any education. This prohibition was so effective, in fact, that it became both public law and social norm—an interdiction that brooked no violation. And under no circumstances, no matter the open-mindedness of the owner or the wealth of the plantation, was a slave ever, ever, *ever* allowed to learn how to read.

5.

In a nation predicated upon the proposition that all men are created equal, with numerous individual freedoms spelled out within a Bill of Rights (for those who could read them), the goal of all this state-sanctioned violence and oppression was to reach down into the depths of the enslaved's mind and snuff out any possibility of free will.

"The grand aim of slavery, always and everywhere, is to reduce man to a level with the brute," wrote Bailey, who later added, "To make a contented slave, you must make a thoughtless one. It is necessary to darken his moral and mental vision, and, as far as possible, to annihilate his power of reason. He must be able to detect no inconsistencies in slavery. The man that takes his earnings, must be able to convince him that he has a perfect right to do so. It must not depend upon mere force; the slave must know no Higher Law than his master's will."

And in achieving this aim, American society was undeniably successful. For when one's ability to make decisions for oneself is blotted out, it renders the mind incapable of believing in the possibility of will at all. The notion of purposeful aims—that one's own determination, effort, and choices make a difference in the world—gets supplanted with the strangled belief that all consequences, beneficial or disastrous, are solely the result of fate.

Bailey cites one example that enraged him to no end: the lack of respect that his fellow enslaved afforded his grandmother's hard-earned talent for growing sweet potatoes. "Superstition had it, that if Grandmamma Betty but touches sweet potatoes at planting, they will be sure to grow and flourish." Bailey explained that she enjoyed "the reputation of having been born to 'good luck.' Her good luck was owing to the exceeding care which she took in preventing the succulent root from getting bruised in the digging, and in placing it beyond the reach of frost, by actually burying it under the hearth of her cabin during the winter months."

When belief in your own agency evaporates, your Self ceases to function.

Lacking a family, nutrition, clothes, a bed, physical security, or any semblance of an education, Feddy yearned to escape his grim circumstances. But because of the mind-darkening depredations of slavery, he had no way of articulating this notion and certainly no distinct conception of how this ambiguous goal might one day come to pass. He often gazed at boats cruising along the bay and imagined they were sailing toward freedom. But the dire reality was that only two paths lay open to him. One was resistance. That would be followed by a quick death from a bullet or noose, or a slow death in the rice swamps of the Deep South. The other was accepting his socially enforced role as a mindless zombie.

Then one Saturday morning in 1826, a third path appeared. Feddy was unexpectedly placed on a boat destined for a city across the bay. A fast-growing metropolis where the course of his journey might change . . .

CHAPTER SEVENTEEN / HUMAN MIND

Language

There is no such thing as a baby. . . . If you set out to describe a baby, you will find you are describing a baby and someone. A baby cannot exist alone but is essentially part of a relationship.
> —*The Child, the Family and the Outside World*,
> Donald Winnicott

Kekawewechetushekamikowanowow
—Cree word that literally means, "You will I wish together remain he-you it-man you" or more simply "May I remain with you."

1.

Akbar the Great ruled over the vast Mughal Empire in South Asia in the late sixteenth century. It would be stretching the facts to claim that Akbar was a scientist, though he was intensely curious about the nature of the mind and known to conduct uncompromising experiments in the pursuit of knowledge. One of these experiments involved an investigation into the origins of language. The question Akbar hoped to probe was straightforward. What was the natural language of the race of man? If a newborn child was kept in perfect ignorance of the sounds of conversation, which language would she come to speak?

Akbar had been informed that Hebrew was the true God-given tongue of humankind. Other learned scholars avowed it was Sanskrit, or Arabic, or Chaldean. To resolve these competing allegations, the emperor ordered that twelve infants be ripped from their mothers at birth and confined to a castle six leagues from Agra. The young research subjects were assigned twelve nurses to care for them who were dumb and could not speak. The children were permitted no other visitors and, above all, were prevented from hearing the utterance of any human syllable, whether speech, prayer, or lullaby.

Twelve years later, Akbar commanded that the subjects of his ambitious study be delivered from their isolation to his throne room so that the entire court might learn which dialect would emerge from their pure and untainted mouths. The fearful children were arranged before the emperor. The scholars, viziers, and courtiers leaned forward to listen. But to the surprise of everyone present, the children made no expression at all. The deprived juveniles were incapable of forming articulate sounds in any idiom. Indeed, they were so severely uncouth and feral that they required unstinting effort from their nurses to maintain any semblance of social decorum. The consensus of all present was unanimous. There was no natural tongue of the human race.

There are numerous accounts of Akbar's foray into developmental psycholinguistics, varying on details such as the number of children, the age when they were evaluated, and the site of their internment. Yet every surviving report concurs on the most important points. None of the children denied exposure to speech acquired the ability to converse. And, though this point of agreement is sometimes overlooked, none acquired passable social skills.

If a child's mind is deprived of language, the most terrible consequence is not that she will be permanently mute. It's that she will never join the fellowship of humankind.

2.

The reason that speech is so extraordinary, so paranormal and wizardly, is that the banal act of hearing coughs and gurgles sputter out of someone's lippy mouth triggers the eruption of vivid *meaning* inside the invisible realm of our private awareness. That hairy, snaggle-toothed guy's hisses and barks somehow evokes a fluffy white cloud within our mind's eye or a tec$ique for peeling overripe bananas or a slow-rising feeling of guarded optimism. To grasp just how preposterous this conversion of sound into meaning truly is, imagine if watching someone blow their nose could produce a parade of top-hat-wearing penguins within our consciousness, or an audacious strategy for overthrowing an oppressive government, or the silky whisper of an Egyptian cotton sheet sliding down an air mattress—the precise contents of your consciousness determined solely by the dribble and hue of your interlocutor's snot. Such a notion might seem absurd and overwrought (and perhaps revolting), but the underlying dynamic of this fictional snot-talk mirrors exactly what has come to pass: an arbitrary and somewhat unsavory physiological activity (pathogen-spreading hacking, hissing, and lip-smacking) has been co-opted by the mind for numinous communication. That is the uncanny power of language.

None of us are born with even the most rudimentary knowledge of alphabets, words, or grammar. Though the human brain springs from the womb well prepared for speech, like an aspiring diver in scuba gear, face mask, and wet suit poised on the edge of a dock, unless the brain plunges headlong into the ocean of language it will never learn how to talk.

Language is not the province of any one mind, and never has been. No single brain houses all the vocabulary or rules for any given tongue, just as no individual's coils of DNA house all the genes of *Homo sapiens*. Speech, and the world-trembling powers that accompany it, must be inculcated into a newborn mind from the human collective. This simple axiom has an upshot, one demonstrated by the Akbar experiment. If a human mind does not get irradiated with language early in life, it will become irrevocably crippled, a meandering shadow of a thinking thing, broken and alone.

But why, exactly? Why have we reached a stage in the journey of Mind where *dialogue*, of all things, is indispensable for the development of a healthy mind? People born without eyes, without arms, without a tongue, even people born with half a brain have all gone on to enjoy productive and fulfilling lives enriched by human communion. So why does growing up without language forestall any possibility of developing the mental skills necessary to live independently, let alone forming meaningful connections with others? The next three chapters will tackle this puzzle.

This chapter, though, will mostly focus on another mystery. Language seems like a superpower, and indeed it is: you can save a life by crying out, "Don't stick the screwdriver in the electric socket!" But if language is such a formidable asset, then why were we the only bloodline blessed with the capacity for discourse? Just as wings evolved four different times in the journey of life and eyes more than fifty times, why didn't any other module mind cross over from meaningless twittering to meaningful tweeting?

Put another way, why can't birds speak?

3.

Imagine that a voyager from the great beyond visited our planet twenty-five million years ago and surveyed all the fauna of Paleogene Earth in order to answer the question, "Which animal's descendants are most likely to wake up one day and start talking?" Our traveler would likely conclude that songbirds presented the best bet.

The most sophisticated primates at the time were monkeys, who were widespread, smart, and social. Yet their means of communication were probably limited to hoots and cries, coarse vocalizations not terribly different from those of wildcats, wolves, or squirrels. Birds, on the other hand, were downright garrulous.

Birdsong features complex and hierarchical patterns—notes combined into arpeggios, arpeggios combined into phrases, phrases combined into songs. Indeed, the warbles and trills of birdsong sound far more melodious than the wheezes and grunts of *Homo sapiens*. Like their modern descendants, ancient songbirds almost certainly possessed much of the mental, physical, and social apparatus necessary for speech. Most notably, birds can produce a wide range of agile sounds using their syrinx. Some members of the parrot family can accurately reproduce almost any human word, as well as the sound of church bells, car alarms, and iPhone ringtones. Birdsong is used for a wide variety of social purposes and even expressing feelings or desires. Songbirds choose which melodies in their avian community they wish to emulate and, in some cases, compose original songs, apparently for the joy of it. A few songbirds (such as plain-tailed wrens) even engage in complex duets, coauthoring expressive melodic conversations that fuse two minds into one.

For many long eons, songbirds have been on the verge of taking that final step from complex auditory communication to full-blown symbolic speech. And yet, they never have. Despite their incredibly varied range of habitats, morphologies, and social dynamics, not a single bird has ever learned to talk. No bird is remotely capable of chirping, "Tomorrow let's head to the cherry orchard and see if we can find some more of those fat dragonflies we ate yesterday" or "Did you see that Flutter was nesting with the same fellow who abandoned her last year!" or even "Hey! There's a seed under that leaf!"

The developmental ladder that leads from speechlessness to language apparently contains a rung that only humans were able to take hold of, allowing us to catch up to and then surpass our precocious feathery cousins.

4.

The development of language required many unsung supporting innovations, which, though easy to disregard, helped set the stage for the big finale. Intriguingly, birds have acquired most—or all—of these auxiliary innovations.

Here's a few examples. Most obviously, we needed a voice box. A larynx, analogous to a bird's syrinx, though the bird box began developing at least sixty million years earlier than our own. One difference between the larynx and syrinx is that the syrinx is in the windpipe whereas the larynx is in a part of our throat used for both breathing and digestion. This awkward arrangement means we are the only creature on Earth who cannot swallow and vocalize at the same time. From the very start, the development of speech imposed a significant physical toll upon us that we must live with to this day: the risk of choking to death on our own words. (No bird can choke to death on its own tweets.)

Other supporting innovations essential for the emergence of language are less conspicuous. Perhaps most surprising is our zest for barbecue. How does the uniquely human predilection for cooking our starches and meat relate to language competency? The acts of speaking and comprehending are neurally demanding, which means they require a great deal of energy—that is to say, a high-calorie, high-protein diet. Cooking meat and tubers makes their calories and protein far easier to extract. In effect, acquiring the ability to cook was like discovering rocket fuel for our minds when the rest of the animal empire was still using kerosene. (Cooking also depends upon the ability to control fire. No other beast has acquired this skill, though humans have been doing it for almost two million years.) Birds never mastered fire, though their relatively light weight and small size means that the heightened caloric requirements of language might be within their reach without requiring broiled flesh or roasted potatoes.

Another major physiological innovation necessary to attain language is a prolonged childhood, known in the jargon of biology as altriciality. Because of the extended time that offspring are dependent on parents for protection, food, and—crucially—education, they are afforded the extended length of time necessary to master language. Yet once again, this innovation does not distinguish us from birds, who often care for their chicks for many weeks after they are hatched.

Most foundational innovations necessary for speech were present in birds long ago. But birds got stuck, while humans kept on scaling the ladder. Why?

5.

To answer this question, we must first answer another question: What is the key difference between the mental dynamics of birdsong and speech? The answer: the ability to share arbitrary meanings between minds. More to the point, the ability to share *qualia* between minds.

Birdsong can certainly convey meaning, such as: "Watch out, there's an eagle coming!" or "Pretty lady, why don't you come over here!" But birds cannot concoct new meanings on the fly. ("Watch out, my aunt Robin is coming, and she knows that you smashed her eggs!") Birds cannot flexibly share the details of their qualia. Humans can.

What innovation enabled humans, and only humans, to exchange qualia between minds? Any question about intermind dynamics suggests the involvement of the metropolis principle. The metropolis principle asserts that as thinking develops, it naturally pushes for the emergence of a new stage of thinking that links together the topmost thinking elements of the previous stage. Molecule minds began uniting their dynamics through molecular messages, eventually leading to neuron minds. Neural networks began uniting their dynamics within neuron minds through representations, eventually leading to module minds. But module minds are more sophisticated than neuron minds and molecule minds.

A consciousness cartel governs a module mind through resonant dynamics—through its qualia. Thus, to establish a new unified stage of thinking, module minds must exchange qualia with one another. This requires solving the attention dilemma within the mental dynamics operating *between* minds. To share qualia between minds, both minds must be able to reliably focus their conscious attention on the same thing at the same time. Scientists refer to this as *shared attention*.

PHASES OF SHARED ATTENTION

1. Eye Gazing

2. Gaze Following

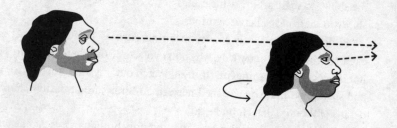

3. Shared Gaze

PHASES OF SHARED ATTENTION

4. Shared Tutoring

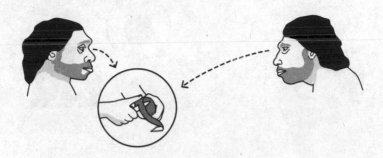

5. Pointing

6. Pointing and Pantomiming

PHASES OF SHARED ATTENTION

7. Pantomiming

8. Pantomime + Word

9. Word

Over tens of millions of years, accelerating rapidly in the last couple million years, our ancestors advanced through several progressively more intertwined phases of shared attention before attaining the ability to exchange qualia between minds. Each new phase of shared attention was accompanied by increasingly sophisticated neural machinery and increasingly sophisticated mind-to-mind dynamics that laced separate module minds together ever more tightly. At present, there is not enough archaeological or genetic data to determine the exact sequence of phases that our ancestors may have followed as they developed shared attention. It's possible that different human lineages may have gone through different phases in different orders. Some phases may have unfolded in parallel or in complementary unison.

The preceding graphic narrates one plausible sequence of shared attention phases that humans may have advanced through, a sequence that accords with the available evidence.

The simplest form of shared attention is *eye gazing*, looking directly at another animal's eyes. Though this does not yet establish mind-to-mind dynamics, it creates a stable mental platform for the development of shared-attention machinery. Before you can develop shared attention with someone, you first need to know where someone is looking—you need to know *what* someone is paying conscious attention to. Most vertebrates, including many fish, are capable of eye gazing.

The next phase is *gaze following*. An animal looks in the same direction that another animal is looking in. This ability is exhibited by every mammal and bird studied so far (including dolphins) and has been documented in reptiles. With gaze following, mental dynamics are now flowing from one mind to another, though in one direction only: from gazer to follower. One mind's How module output leaps through the air to become the input to another mind's Where module. We can think of gaze following as an intermind targeting dynamic: using another mind's targeting dynamic to guide our own.

The next phase of shared attention is *shared gaze*: two animals look at the same object at the same time, such as two chimpanzees inspecting a rock that one of them is holding. Shared gaze closes the loop of mind-to-mind dynamics and enables thinking to flow between minds in both directions. Not every mammal has achieved shared gaze, nor have all birds, though primates exhibit highly developed shared-gaze ability.

At long last, we reach a phase of shared attention that only our human ancestors attained, though it is not yet that special rung that delivered us to language: *shared tutoring*. Two minds focus their attention on a complex series of actions rather than a single object, and maintain this joint focus for a longer period of time than during shared gaze. Shared tutoring also requires another landmark mental innovation: *imitation*, the ability to accurately reproduce the actions performed by another.

The activity that most likely drove humans to develop imitation and shared tutoring was toolmaking. Chimpanzees make and use tools with their hands. This manual ability is supported by shared gaze, which allows an apprentice chimp to watch an experienced chimp fashion a tool. But chimp tools are very crude indeed. They don't require planning. They usually involve only one or two straightforward steps, rather than a sequence of actions that must be performed while holding an abstract goal in mind (such as "the point should taper evenly"). They don't involve improving upon one's work. There is no focus on teaching.

As chimpanzees more or less stagnated tec$ologically, humans kept getting better and better at making tools. The longer and more precisely you can pay attention to how someone else makes a tool and the more willing the teacher is to help you learn, the more elaborate a tool you can make and the more quickly you can learn to make it. Eventually, toolmaking became too complex to rely upon mere imitation. More complex intermind dynamics were required.

Advanced human toolmakers needed *gestures*. One thing advanced toolmaking almost certainly required was *pointing*. By fourteen months of age, all modern humans are hardwired to point and to look where others are pointing, often a year before they begin to talk. Chimpanzees have never been observed to point in the wild, and they have difficulty learning to look where humans point. The appearance of pointing and other hand gestures inaugurated the next and decisive phase of shared attention: *pantomiming*. Pantomimes link public visual representations (gestures) with private qualia, mentally fusing a gesture and a meaning. Moving the fist up and down can mean hammering. Moving a flat hand forward and back can mean sawing. Waving a hand toward oneself can mean "Come here!"

With the emergence of pantomiming, the two-way dynamics of shared attention finally became robust and stable enough to support the emergence of speech. Successful pantomiming requires that two different minds focus on the same arbitrary visual symbol—a hand—and the same conscious qualia (such as "apple"), as pantomimed by the hand. In a student's mind, the mental representation of a perceived gesture performed by a teacher is linked to the category representation for "apple" in the student's What module. Qualia-sharing dynamics are now flowing between minds.

It's also possible that the first qualia-conveying vocalizations emerged at the same time as the first pantomimes, the two forms of intermind dynamics bootstrapping off one another, each dynamic helping to stabilize and certify the qualia conveyed by the other dynamic. (There's also intriguing evidence that early humans may have sung to one another before they ever spoke to one another, conveying qualia through melodic hoots, grunts, or calls in a form of communication known as "musilanguage." If so, musilanguage likely developed in complementary fashion with pantomimes.) Regardless, once shared attention achieved qualia-sharing pantomimes, the stage was set for the first spoken word.

The first word was spoken when a human understood the intended meaning of a vocalization (the listener experienced the qualia that the communicator intended the listener to experience) and then reproduced the same vocalization to successfully convey the same qualia (the same *meaning*) to a third person. (To come into existence, a word requires three resonating minds: someone to consciously invent the word, someone to consciously understand the word and then accurately repeat it to someone else, who also consciously understands the word. All language is inherently collective, just like birdsong.)

Why didn't birds develop language? Because they don't have hands. The special rung that only humans could grasp required fingers and gesture-driven shared attention. Without a free pair of limbs to point and gesticulate with, birds were never able to develop the sophisticated shared-attention apparatus necessary to ensure that two birds focused on the same complex sequence of actions, an essential precursor for connecting public symbols with private qualia. The ability to fly may be an extraordinary gift, a talent that humans are deeply envious of.

But perhaps the true cost of wings is to be forever speechless.

6.

Did the human brain develop a wholly unique language module, as many scientists have long believed, an exclusive one-off unlike any other neural contraption found amongst Earth-born beasts?

Before we tackle this crucial question, let's take our bearings. The following figure traces out the core dynamics of the human mind, dynamics that also governed the ancestral pre-language human mind, and indeed, appear to govern *all* module minds.

A girl reaches for a cymbal-clanging monkey. Her preconscious Visual Scene module processes her retinal image, then sends a monkey representation to her visual What module and visual Where module (both consciousness-generating modules, as indicated in the figure by glowing gray borders). Simultaneously, her Auditory Scene module processes her eardrum vibrations, then sends a "Clang!" representation to her auditory What module and auditory Where module. The two What modules resonate together (not shown in the figure) and jointly identify the clanging monkey and send a category representation ("It's a clanging monkey!") to the conscious Why module, which decides "I would like to touch that curious creature!" and activates the nonconscious How module. The How module reaches for the monkey, guided by the attentional shrouds generated by the two Where modules.

This is the basic dynamic (sometimes called a "perception-action cycle") characterizing how the human mind converts sensory inputs into behavioral outputs in real time.

How do the dynamics of language fit into this framework, and how did the neural circuitry supporting these dynamics first arise? The answer involves a seemingly physics-defying trick: the ability to simultaneously think forward and backward in time. This was one of Stephen Grossberg's key insights into the serial position effect, which occurs when learning a series of items, such as the digits of a phone number or the words in a sentence. The primacy effect (better recall of items at the beginning of a sequence) and the recency effect (better recall of items at the end of a sequence) are a direct consequence of this peculiar forward-and-backward dynamic. And, as it turns out, to perform this trick it was *not* necessary to build a whole new module from scratch.

Language arose in the human mind through the same three develop-

THE HUMAN MIND

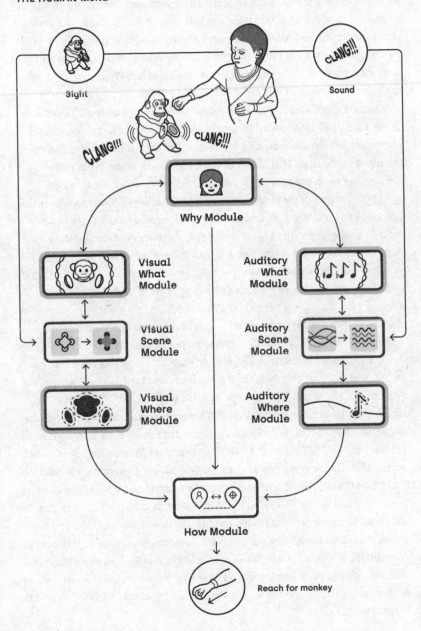

Sight

Sound

Why Module

Visual
What
Module

Auditory
What
Module

Visual
Scene
Module

Auditory
Scene
Module

Visual
Where
Module

Auditory
Where
Module

How Module

Reach for monkey

mental processes responsible for most module mind advances: more memory, more connectivity, and, most influentially, more complementary thinking. Instead of fabricating a language module out of whole cloth, the human mind simply stitched together two very ancient modules: the auditory What module (*What* sound is this?) and the auditory When module (*When* will the next sound appear within this sequence?). This When-What pair, or *dyad*, is the basic component of our mind's language system.

Over time, additional When-What dyads were stacked on top of the existing dyads. Each dyad processes a different hierarchical level of language, including phonemes, syllables, words, and sentences. The output of one dyad serves as the input to the next. This ladder of complementary When-What pairs forms the *language stack*, the primary brain structure responsible for language.

At the bottom of the language stack is the Auditory Scene module, which processes raw audio inputs and sends a foreground audio stream (such as a melody or speech) to the lowest-level dyad, the Phoneme Recognition dyad. It processes the audio stream and identifies any individual phonemes, the smallest distinguishable units of vocalization. The When and What modules that form the Phoneme Recognition dyad were likely adapted from an existing pair of auditory When and What modules that identify sequences of sounds, as they appear to do in the minds of other mammals and birds.

On top of the Phoneme Recognition dyad sits the Syllable Recognition dyad. The Phoneme Recognition dyad sends each identified phoneme (such as "S" and "U") to the Syllable Recognition dyad, which synthesizes the sequence of phonemes into a single syllable (such as "SU"). A Word Recognition dyad (another auditory When-What pair) is stacked atop the Syllable Recognition dyad. The Syllable Recognition dyad sends each identified syllable (such as "SU," "PER," and "MIND") to the Word Recognition dyad, which inputs the syllable sequence as a single bottom-up representation (such as "SU-PER-MIND") that it attempts to match to a top-down word stored in its memory (such as "SUPERMIND"). When they match, resonance occurs in the Word Recognition dyad's What module (*What* word is this?) and you experience conscious recognition of the word "supermind." (Your experiences of recognizing words are examples of synthesizing qualia. You experience synthesizing qualia even when you don't know what a particular word means, such as recognizing that the unfamiliar syllable pattern "brabble" is a well-formed word.)

LANGUAGE STACK

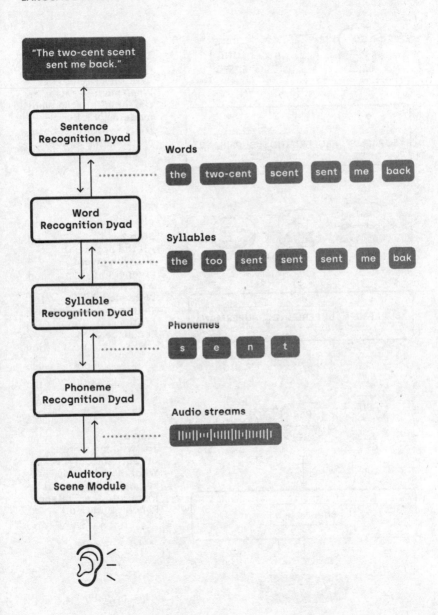

WORD RECOGNITION DYAD

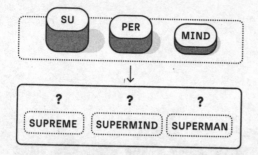

Step 1:
A sequence of syllables from the **When module** serves as the input into the **What module**, which looks for the word that best matches the syllable sequence.

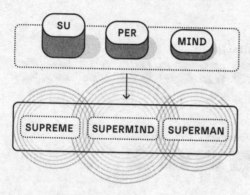

Step 2:
All the potential matching words compete in winner-take-all in the **What module**. The word that is the best match for the syllable sequence wins.

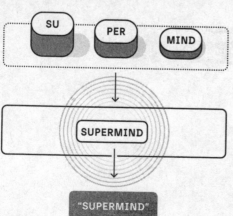

Step 3:
The winning word is outputted from the **What module** and serves as a potential input into the Sentence Recognition Dyad.

In each dyad, the When module processes a lower level of language than its paired What module. For instance, the When module in the Syllable Recognition dyad synthesizes a sequence of phonemes (received from the Phoneme Recognition dyad), while its paired What module recognizes the syllable formed out of the When module's phoneme sequence. In the Word Recognition dyad, its When module synthesizes a sequence of syllables (from the Syllable Recognition dyad) and its What module identifies the words formed out of the When module's syllable sequence. Language probably took flight when the human mind added a highly potent When-What pair to its language stack: a Sentence Recognition dyad.

A pivotal yet surprisingly simple form of connectivity emerged within the language stack that enabled backward-in-time thinking—and verbal communication. This special connection links each language dyad back down to the dyad directly below it: a *feedback* connection. The Word Recognition dyad sends feedback to the Syllable Recognition dyad, for instance. Each dyad in the language stack influences the operation of the dyad below by sending contextual information to the lower dyad through the upper dyad's feedback connection, context that the lower dyad can use to improve its language recognition. This design principle makes a strong prediction: we should expect that every level of the language stack will exhibit its own distinctive "illusion" of backward-in-time perception. This is indeed the case.

Let's consider an illusion generated by the Syllable Recognition dyad and its forward-and-backward interactions with the Word Recognition dyad. Suppose you heard the spoken word "DELIVERY." You're sure you heard every syllable in the word, including "DE," "LI," "VER," and "Y." But you were tricked! It turns out that the syllable "VER" was never actually uttered. Instead, a sneaky professor fooled you by replacing the "V" sound with a burst of white noise. What you consciously heard, however, was every syllable in the word "DELIVERY" spoken aloud as a burst of white noise murmured in the background. Even if you're told in advance that noise will replace the "V" sound, you would still find it nearly impossible to perceive the noise alone rather than hearing a distinct "V" sound against a background of noise. What's going on?

The Syllable Recognition dyad sent the syllable sequence "DE-LI-*ER-Y" to the Word Recognition dyad. The * represents noise. The Word Recognition dyad recognizes the word "DELIVERY" as the best top-down match for this (partially obscured) bottom-up input and begins to resonate. The Word Recognition resonance sends a feedback signal back down to the Syllable Recognition dyad and tells it that it should "replace" the noisy syllable with "VER." The syllable-processing What module resonates on the syllable "VER," which we then experience as a conscious perception of "VER" with noise in the background, even though the sound of "V" never impinged upon our eardrum.

Now let's say you hear the word "DELIBERATE," except once again one of the sounds is replaced with noise. This time, the "B" is replaced. The syllables that precede the "B" are the same ones that preceded the "V" in "delivery"— namely, "DE" and "LI." When your eardrum processed "DELI*ERY" you heard "DELIVERY," but when your eardrum processed "DELI*ERATE" you heard "DELIBERATE." This is a subversive act of creation. Every conscious experience of language comprehension involves a bout of time travel: any syllables obscured by noise or ambiguous pronunciation are replaced *after* your mind processes subsequent syllables in the word.

The same backward-in-time dynamics operate at the lowest level of the language stack, between the Phoneme Recognition dyad and the Auditory Scene module. If a continuous spoken phoneme (such as "EEEEEEEEEE") is briefly interrupted by noise ("EEEE**EEEE"), we consciously hear a single uninterrupted phoneme spoken over background noise, rather than the actual stimulus: two distinct vocalizations separated by noise. Once again, the gap in the speech is "filled in" backward in time. (This is analogous to the Visual Scene module filling in the gaps in broken lines.) We know this is what's happening because if we play a continuous phoneme, followed by noise, then followed by silence ("EEEE**"), we don't hear the phoneme spoken over the noise, but rather, the phoneme *followed* by noise. The Phoneme Recognition dyad doesn't know whether to fill in the noise until it has processed the sounds *after* the noise.

Backward-in-time thinking also operates between the Word Recognition and Sentence Recognition dyads to help us distinguish words that sound the same but have different meanings (homophones). Imagine hearing this sentence spoken out loud: "The scent of the two-cent stamp sent me back." Simple, no? Yet your mind quietly accomplished something remarkable: you consciously experienced three different meanings from words that sound exactly the same—scent, cent, sent. After the Sentence Recognition dyad successfully recognized the sentence, it sent a feedback signal down to the Word Recognition dyad informing it of the distinct identities of each of the identical-sounding words, which we then perceive consciously.

The language stack and its hierarchy of forward- and backward-in-time thinking processes are testaments to the explanatory power of Stephen Grossberg's mathematics. A small set of equations and network configuration principles account for the reiterative structure of When-What dyads in the language stack and explain a broad variety of psychological data, including varied language illusions and effects, without requiring any new equations or network principles specific to language. The same math also helps explain birdsong.

Songbird minds appear to possess a stack of When-What dyads, too: a "birdsong stack." Like language, birdsong consists of multiple hierarchical levels (scientists refer to them as notes, syllables, phrases, and songs, corresponding to human phonemes, syllables, words, and sentences), and bird minds, too, "fill in" missing notes or syllables. Thus, astonishingly, even the mighty language stack does not distinguish us from birds.

If we want to identify the neural component of language that elevates us above all module minds and endowed humankind and humankind alone with a new stage of thinking, then we must identify the component that provides us with a conscious shock of *meaning*.

7.

The uncanny experience of meaning that enters our consciousness when we comprehend a word is entirely unrelated to the perceptual qualities of the word itself. The auditory pattern "ice" is neither cold nor crystalline though we can vividly experience those meanings when we comprehend the word.

The meaning we associate with a given verbal pattern is arbitrary and flexible. The word "bat" means a nocturnal flying critter and a wooden whacking stick and the act of whacking something with said stick (but not with a flying critter). Only the minds of *Homo* can forge a mental link between words and meanings easily, instinctively, and incessantly. Bird minds are conscious of the notes, syllables, and phrases of birdsong. They are not conscious of their symbolic meanings.

How does meaning enter the dynamics of language? What is going on in your mind when you recognize the word "cat" and know that these arbitrary sounds in this arbitrary order symbolize "a feline organism" and perhaps even experience a spontaneous visualization of an orange tabby in your mind's eye? To associate auditory patterns with meanings, the mind links words to qualia. Specifically, a verbal representation resonating within the language stack resonates with a mental representation resonating in the consciousness cartel.

If someone points to a critter flying over your head, your visual What module will resonate on the object category for 🦇, which evokes the conscious experience of *knowing*. If the pointing person now utters the word "bat," the resonating object category 🦇 will get linked to your word-processing What module's word category for "bat." Next time you hear the word "bat," your Word Recognition dyad will activate knowing qualia for 🦇 in your visual What module, and the mutually resonating representations for "bat" and 🦇 generate the conscious experience of the word's meaning—what we might call *comprehension qualia*.

Comprehension qualia ("he is talking about a 🦇") are usually experienced simultaneously with hearing qualia ("I hear him talking"), auditory knowing qualia ("I recognize David Attenborough is talking"), and synthesizing qualia ("he pronounced the word 'bat'"). Each of these qualia is embodied within a resonating feature-based representation in a distinct neural module. Comprehension qualia ranks as perhaps the most consequential example of the complementary thinking principle in the journey of Mind.

The power of resonance allows the complementary ignorances of two modules to be balanced by their complementary strengths. A word-processing What module possesses detailed knowledge of the phoneme pattern "bat" but doesn't possess any knowledge of nocturnal flying critters. A visual What module possesses detailed knowledge of nocturnal flying critters but doesn't possess any knowledge of the phoneme pattern "bat." Yet, when these two recognition categories—the verbal representation "bat" and the visual representation 🦇—begin to resonate with each other, symbol and meaning are united within a synchronized, prolonged, and amplified construct.

WORDS ARE LINKED TO QUALIA

Listener hears audio pattern "bat" and comprehends its meaning.

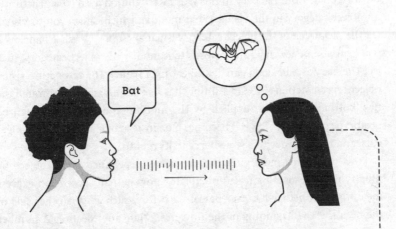

The **Word Recognition Dyad** (a syllable When module and a word What module) recognizes the word "bat." The representation of the word "bat" activates the bat object category in the **visual What module**. The word representation "bat" resonates with the bat object category representation and the listener is conscious of comprehending the meaning of "bat."

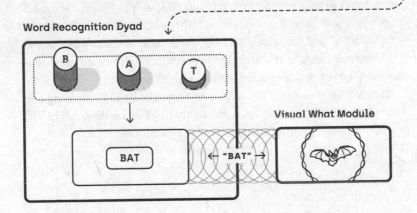

Bat!

Words can get linked to any perceptual qualia that humans can experience, including things (such as stones, hands, trees, rivers, eyes, stars, or fire) and actions (such as running, jumping, flying, hitting, eating, singing, or burning). The word category in the Word Recognition dyad ("bat") activates an object category in the visual What module, which causes you to visualize the prototype of the object category (such as). What happens if you hear a sentence containing words that modify the object category, such as "The bat was blue, long eared, and eating a fruit"? The resonating visual object representation () simply has its features adjusted through the new bottom-up features supplied by the modifying words—the visualized prototype of the bat () becomes blue instead of black, develops longer ears, and is pictured with a strawberry in its mouth.

The process of modifying the conscious experience of an object by modifying its resonating features explains how human speech can express meanings that *nobody* has ever perceived before, such as "the other side of the ocean." "The beginning of the universe." "Life after death." "A number less than zero." "The journey of the mind." The topic word in each of these phrases (ocean, universe, death, zero, mind) activates the relevant object category representation in the What module, and the resonating prototype is then modified accordingly: we picture the ocean, and then our What module pictures a distant shore on the ocean.

This process of modifying resonant states in real time helps us understand why it's so hard to parse abstract sentences, such as "These resonances are not just correlates of consciousness." The topics in this sentence (such as "resonances" and "correlates") do not correspond to sensory qualia and are therefore difficult to visualize, and the modifier "of consciousness" adds more non-sensory details to a non-sensory prototype.

This is a big reason why the science of Mind can be so challenging and frustrating: even the most intrepid explorers of thinking are frequently adrift in an unrelenting fog of difficult-to-visualize abstractions.

8.

The emergence of qualia-sharing mental dynamics established a new level of thinking in the journey of Mind: qualic thinking. Qualic thinking operates on top of representational thinking, just as public activity in Çatalhöyük operated on top of household activity. The inception of qualic thinking in the journey of Mind inaugurated a whole new stage of minds: the *superminds*.

The earliest superminds likely thought using one-word sentences (linguists call this holophrasis). These superminds would have been quickly pushed aside by smarter superminds wielding two-word sentences, who would have been pushed aside by superminds wielding prepositional phrases, subordinate clauses, and the subjunctive mood.

At some point, shared attention bound human minds together so tightly that individual minds were no longer optimized for module mind thinking. They were optimized for *supermind* thinking. At this point, it was no longer possible for individual minds to develop properly without integrating with the supermind from birth. It would have been like trying to survive in outer space without a space suit: we simply weren't designed to function in a world without language. Our cerebral cortex is so specialized for mind to mind dynamics that it functions like a network router, receiving and transmitting qualia back and forth with other cerebral cortices. A router without a network is useless, and a cerebral cortex without a network of other cerebral cortices is useless, too.

Humankind struck a bargain with Mother Nature not dissimilar to the one struck by birdkind. Birds traded away the possibility of speech when they purchased flight. Humans traded away the possibility of surviving alone to acquire language. Put another way, no man is an island. If he's an island, he's not a man. Long before *Homo sapiens*, each individual human became a neuron who could not function without getting yoked into a brain.

The advent of human superminds was a milestone not merely in the journey of Mind but in the story of the universe. For the first time, thinking left behind its gelatinous penitentiary of delicate cells and moved into the stones and the air and the earth and the flames. From this point onward, the journey of Mind would no longer be dictated solely or even largely by gene-based evolution nor constrained by the limitations of biology. Once qualic thinking appeared, the fastest means for minds to get smarter was no longer swapping strands of DNA. It was by improving language, toolmaking, and other increasingly sophisticated forms of nonbiological thinking.

The supermind possesses the power of a god not only because it wielded the first form of thinking to take the whole planet under its purview, but because it became the first mind to exist in large part—and today, in most part—outside the bounds of mortal flesh.

chapter eighteen
sapiens supermind I
Civilization

When the weaknesses we've always had join forces with a capacity to do harm on an unprecedented planetary scale, something more is required of us—an emerging ethic that also must be established on an unprecedented planetary scale.

—*The Demon-Haunted World*, Carl Sagan

1.

Conley and Yuna had been bickering for years. Throughout their whirlwind marriage, one consummated under great duress, each had occasionally threatened the other with leaving, but for the most part they never seriously contemplated a divorce. Despite Conley's dirty habits and Yuna's frequent nagging, despite Yuna's self-righteous behavior and Conley's self-destructive behavior, each always found a way to put aside their differences and patch things up. After all, they shared a magnificent and unique homestead and it wasn't clear how they could divide the sprawling, invaluable property in the event of a messy breakup.

Though Conley and Yuna bore divergent personalities, it was the very disparity in their outlook and activities that had always been the chief source of strength in their relationship. Since they looked at the world so differently, they needed to constantly talk things out and attempt to persuade each other of the truth of their convictions. Most important, they were compelled to compromise—each getting something but not everything—before moving on to the next point of contention.

Even so, there was one glaring problem that couldn't be accommodated or swept under the rug. A problem that was present from the very start of their shotgun—or musket—romance. Like many couples, Yuna and Conley convinced themselves that they could find ways to sidestep it through delicate work-arounds. But with each passing year, the problem drove them further and further apart and made them both begin to think that the only solution was a permanent separation.

The problem was Conley's dirty addiction. He just wouldn't face up to it, responding to Yuna's criticisms with denials, evasion, rationalizations, and counterattacks. He stubbornly maintained the pretense that his filthy habit was actually quite noble, like someone who insists that smoking helps keep lung bacteria in check and is therefore a boon to public health. But the undeniable fact was that Conley's habit made him chronically prone to aggressive outbursts and murderous violence, and he could no longer participate in the marriage without considering how every interaction might somehow work to infringe upon his fixation. Conley insisted that while it was Yuna's right to be a teetotaler if she wished, it was nonetheless his own right to conduct himself as he pleased, even if it meant a river of blood flowed through their shared home.

Yuna struggled to maintain her civility, though she certainly had moments of frustration when she expressed the view that perhaps they should cut the knot. But she dutifully soldiered on, hoping that eventually, somehow, Conley might see the light. All the tension and threats and increasingly open hostility finally reached a breaking point when Yuna unexpectedly got the upper hand in the relationship, taking full control of the purse strings and the management of the homestead. Fearing that he would be bullied into giving up his dark compulsion, Conley snapped. He unilaterally declared the marriage was over, signed one-sided divorce papers and then, at gunpoint, occupied Yuna's prized southern island.

The enmity between the two was reflected in the words they began to hurl at each other. Conley insisted he had every right to walk away:

"There was no choice left us but submission to the mandates of abolition, or a dissolution of the Union, whose principles had been subverted to work out our ruin," Conley declared. "Utter subjugation awaits us in the Union, if we should consent longer to remain in it. It is not a matter of choice, but of necessity. We must either submit to degradation, and to the loss of property worth four billions of money, or we must secede from the Union framed by our fathers, to secure this as well as every other species of property. For far less cause than this, our fathers separated from the Crown of England."

Yuna maintained her calm. She demanded that the relationship continue—that Conley had no legal standing for a divorce: "Whereas a conspiracy has been formed against the peace, union, and liberties of the people and Government of the United States; and in furtherance of such conspiracy a portion of the people have attempted to withdraw those States from the Union, and are now in arms against the Government . . . the Government hereby declares these States in unlawful rebellion."

Attributing humanlike agency to the Union and Confederacy might seem to be metaphorical license, the same kind of anthropomorphizing that many scientists censure when applied to animals. But why? If one believes in the basketball game principle—if one believes that thinking is embodied within the dynamics of a physical system (such as a network of molecules, neurons, or modules)—then why wouldn't the same dynamics operating in a different physical system (such as a network of human brains) justify treating that other system as a thinking mind as well?

We haughtily dismiss the thinking elements laboring below us as mindless automatons, while simultaneously insisting that the thinking elements churning above us do not exist. Might believing that societies can't think be another example of consciousness privilege?

Isaac Newton's great achievement was to demonstrate that all things made of matter followed the same principles. The basketball game principle supports the analogous position that all forms of thinking—all systems that convert sensory inputs into behavioral outputs in real time in the service of a physical body—share the same underlying principles, including modern nation-states.

If we believe that mental dynamics are *physical* dynamics, then any sufficiently advanced supermind is fully capable of perceiving, knowing, feeling, acting—and speaking.

2.

A supermind operates within the same material reality as every other mind, and though the scope of its environment is far more expansive, it must nonetheless adapt itself to the same laws of nature as every other thinking being. Superminds have merely ascended into a new scale of physical chaos. Thus, superminds must inevitably confront the same mental challenges as the previous stages of minds and come up with their own innovations to master them.

For example, the earliest human superminds quickly developed both short-term and long-term memory. Short-term memories were encoded through active conversations. If one human shouted, "Meet at campfire!" each listener could maintain the meaning of the spoken command in her mind (as a resonant state in her What and Where modules) for as long as it took to reach the campfire. This short-term memory also supported effective targeting by a supermind, enabling a group of human minds to all move toward the same destination at the same time—just like the amoeba mind's molecular messaging system.

Long-term memories were encoded in speech through two distinct means. The first was through the creation and maintenance of useful words. The word "wolf," for instance, encodes the existence of a furry, four-legged predator. Even if a young human had never seen a wolf—indeed, even if no one in the entire community had ever seen a wolf—the very existence of "wolf" in its vocabulary served to ensure that the supermind would continue to remember the nature of such a threat. Similarly, the word "venom" records the fact that some creatures can kill you by biting you, stinging you, or spraying you, even if the person learning the meaning of "venom" has never directly witnessed this. The word "scythe" documents the possibility of creating a tool that cuts plants, even if nobody alive has cast their eyes upon such an instrument.

More sophisticated long-term memories were encoded in stories, songs, or lists of information shared among a supermind's thinking elements (humans, that is). The chronicle of a great hunter who once single-handedly killed a saber-toothed tiger, an oral history of one's ancestors, a recipe for mammoth-toe soup, spells for conjuring fire out of sticks and flint stone, directions for traveling to a trout-filled lake a hundred leagues away—each of these elaborate linguistic representations encodes knowledge in a format that can endure within a supermind even after all the thinking elements who acquired the knowledge firsthand have perished. Through language, a supermind can build a collective "brain" that can survive the death of its people-neurons.

The mind-to-mind dynamics of language facilitate solutions to other challenges. Superminds are capable of perception (a scout reports back to the group what he's seen up ahead, serving as a Visual Scene module), judging (members of the tribe sit around the fire discussing what the scout has seen and evaluating whether it's dangerous or advantageous, serving as a What module), planning (the tribe debates the best response to the perceived danger and reaches consensus on the strategy, serving as a Why module), and doing (the tribe collectively executes the consensual strategy, serving as a How module).

These mental innovations steadily improve over time. Different scouts and different scouting tec$iques are tried out, and the superior ones retained or repeated. A tribe experiments with different forms of community judging and planning, including different modes of tribal governance. Alternate hunting stratagems are explored and the superior ones encoded into tradition—that is, into the supermind's mental dynamics. Individual humans become more specialized in their roles as sensors, thinkers, or doers. One person can be tasked with remembering the identities of all the venomous dangers in the jungle, or all the useful herbs, or all the battle tactics, or all the creation myths, and this person can pass on their oral encyclopedia of knowledge to an apprentice.

Language itself improved and sharpened. It expanded its vocabulary with new nouns and verbs. It smoothed the rough edges of words and syntax that presented difficulties in passing from mind to mind—or expunged them entirely. Language developed novel grammatical forms: adjectives, adverbs, prepositions, possessives, personal pronouns, past perfect tense. (It took tens of thousands of years before the supermind began codifying these as formal, conscious rules.) Language drove the further expansion of the language stack in human minds, adding new language dyads on top of the Sentence Recognition that handle stanzas, multipart stories, and ritualized prayers.

The previous three stages of thinking could progress only through the slow and uncertain cycling of sexual reproduction. Qualic thinking can improve far more rapidly, with new words and ideas capable of spreading throughout a supermind practically overnight: COVID, super-spreader event, social distancing, flattening the curve, vaccine hesitancy.

Language also actively reshaped the supermind's environment, body, and "brain" (returning now to the scare quotes we previously adopted to refer to the thinking elements of molecule minds). Language flourished within those superminds and individual human minds that were best suited for its own evolution. Language jumped more readily into human brains that happened to be wired in ways that facilitated speech acquisition, and language infected tribes who happened to live in stable, complex environments that supported the development of even stronger forms of shared attention.

Superminds a million years older than the *sapiens* supermind maintained campfires that all their humans gathered around, staring together into the flickering flames as they shared tales or simply experienced a sense of belonging. Cave art was another shared-attention breakthrough in both *erectus* and *neanderthalensis* superminds, establishing another communal space that compelled the members of these superminds to gaze together upon the same enthralling perceptual patterns as they shared stories about the art and jointly interpreted the painted symbols. (Sometime in the past hundred thousand years, music also melded the individual minds of superminds together as humans focused their shared attention on melodies and moved together in unison through dance.) In many ways, the supermind thinking machinery of campfires and cave art parallels the molecular thinking machinery in archaea, bacteria, and protozoa, which rely heavily on constricted physical spaces to guide molecular thinking elements toward other thinking elements.

The "brain" of the supermind initially consisted of the shared-attention-linked collective network of its members' biological brains. But language was such a potent force, it resculpted these biological brains. Perhaps the most consequential of these language-induced biological innovations was the emergence of innate psychological biases that caused humans to shun, fear, and despise members of other superminds, while believing that members of their own supermind were smarter, more capable, and more moral. This well-documented "intergroup bias" can be a source of racism in modern superminds, though for most of human existence the intergroup bias would have been directed at people who looked the same and spoke the same language as those wired to be biased against them.

The intergroup bias is part and parcel of supermind integrity, because this genetically grounded proclivity fulfills the same function as a cell membrane. As we learned in Chapter 1, for a mind to exist at all it needs a well-defined boundary separating *me* from *not me*, and superminds got better and better over time at bolstering their loose and permeable mental boundaries. If all the thinking elements in a supermind—all its people-neurons—behave as if there is a clearly demarcated boundary between *us* and *them*, their collective behavior establishes a de facto physical boundary demarcating *this* supermind's thinking from everything else in the world.

Though the intergroup bias can undoubtedly serve as a recipe for hatred, conflict, and subjugation, it does offer a silver lining. The intergroup bias is a source of social cooperation unsurpassed in the animal kingdom. Compared to almost every other animal, including the apes (though excepting bees and ants), humans are much more cooperative and helpful to one another and much more willing to make sacrifices for unrelated neighbors—though almost all of this altruistic behavior tends to take place within the same supermind. According to the metropolis principle, such cooperation is the coordinated activity of a supermind's thinking elements, no different than the coordinated activity of the neurons in your brain.

From the perspective of the journey of Mind, human-to-human altruism is merely an effective form of supermind cohesion.

3.

Though we don't know exactly when the first supermind appeared on Earth, any more than we know when the first molecule mind, neuron mind, or module mind appeared, we can say with confidence that it was sometime after *Australopithecus* raised herself up on two legs and began striding toward a loquacious future and sometime before *Homo sapiens* began bubbling up all over Africa.

From the start, superminds were wanderers. Clans of humans, mostly blood kin, probably ranging in size from ten to fifty brains (compare this to the 302 neurons in the roundworm), roamed the landscape without a fixed place to call home. The dynamics of these early superminds were largely focused on fishing, hunting, gathering plants, scavenging wild animals, and finding places to shelter. Indeed, the earliest superminds were probably much like the earliest molecule minds on Earth: distinguished largely by the way each one tackled the exploration dilemma. Each supermind constantly faced the choice of staying put and exploiting the resources at hand, or sallying forth and seeking out greener grasslands. Different supermind "personalities" (which we can define as a supermind's characteristic mental dynamics, including its characteristic behavioral dynamics) resulted in different decisions on when to fish and when to cut bait.

As soon as multiple superminds came into existence, they began to compete. Victory in each round of this perpetual contest was not decided by claws, fists, or teeth. It was decided by keenness of intellect. By *adaptiveness*. Whichever supermind could outthink the others would stand the best chance of surviving and advancing to the next round. There were many ways that a supermind could gain an advantage over its competitors. Acquire better land, make better tools, get more organized, cooperate or even merge with another supermind—or establish a new style of thinking better suited for its particular environment.

The diversity of supermind personalities is evident among the aboriginal tribes of the Amazon. Some are belligerent (shoot first, ask questions later), some are friendly (ask questions first). Some are focused on games and contests of skill, some are focused on feasts and good food. When the modern government of Brazil began reaching out to the tribes, some strode out of the jungle and volunteered to join the behemoth Brazilian supermind, while others retreated ever deeper into the wilds.

Compared to Darwinian natural selection, supermind evolution unfolded at a blistering pace. One factor driving this swiftness was superminds' ability to do something that had never been done before on the journey of Mind: swapping thinking elements between minds. A turtle cannot take out its What module and give it to a friend. A roundworm cannot snatch a locomotion doer network from a competitor and stitch it into its body. But a supermind can poach a particularly useful thinking element from another supermind (such as a healer, hunter, or chef) and integrate it within its own mental architecture. A thriving supermind can even attract superior thinking elements by reputation alone, as individual humans seek out communities that can provide them with a better life.

One primary way that neuron minds became smarter was through neural diversification—creating new kinds of neurons. One way that superminds become smarter is through human diversification. The more varied the individual minds within a supermind, the greater the opportunities for developing more sophisticated and adaptive dynamics and more intelligent behavior.

For eons, legions of superminds roved the planet, colliding, competing, cooperating, merging, fissioning, and occasionally exterminating one another. Whichever supermind was more adaptive in real time gained the edge and was more likely to endure. In the end, all the *Homo ergaster* superminds and all the *Homo erectus* superminds and all the *Homo heidelbergensis* superminds and all the *Homo denisova* superminds and all the *Homo flore-siensis* superminds and all the *Homo neanderthalensis* superminds and all the other human superminds that ever trod the Earth were outhustled or out-smarted by sapiens superminds.

Perhaps the members of the last of the non-sapiens superminds, a Nean-derthal supermind, gathered on the cliffs of Gibraltar and gazed out across the sea toward Africa, the birthplace of all humankind, and considered the unfairness of their fate—to attain language and art and culture and survive multiple ice ages only to lose out to a physically frailer brute who was quicker on the draw. When the final Neanderthal mind perished, the supermind competition didn't halt. It merely became a sapiens-only affair.

4.

For tens of thousands of years, sapiens superminds meandered through sundry habitats fashioning tools of stone. Then, one day, somewhere in the Levant, sapiens hands reached down into the earth and, wittingly or not, dis-covered a new tool unlike any other. A seed.

When a human planted a seed and brought forth nourishing life from the soil it launched the first great revolution in the journey of the supermind. In the cosmic saga of thinking, the first revolution was the emergence of thinking itself, a seismic shift from a mindless "grazing" lifestyle among single-celled organisms to a purposeful "hunting-and-gathering" lifestyle. But the fore-most revolution in the journey of the supermind reversed this metamorpho-sis, switching back from hunting and gathering to a sedentary lifestyle.

Ironically, this shift to fixedness unleashed explosive new dynamics that transformed the nature of supermind thinking. Mirroring the transi-tion from roundworm mind to flatworm mind, the transition to agriculture converted the supermind *"brain"* (a decentralized collection of thinking ele-ments) into a *brain* (a centralized set of thinking elements).

Once a supermind became tied to the earth through agriculture, it endowed that supermind with an expansive new body incarnated in vege-tation, loam, and clay. A patch of land became *my* patch. The supermind gained a well-defined physical border for its mental dynamics, featuring a clearly demarcated *me* and *not me*. Events within *my territory* were no longer external inputs coming into the supermind from without but internal ones arising from within.

A wholly unprecedented form of embodied thinking took shape out of the dirt. The land itself became a thinking element. The dynamics of tilling, sowing, irrigating, and harvesting were integrated into the dynamics of the supermind, and indeed, agriculture remains an indispensable and thickly rooted portion of our mental dynamics to this very day, now serving as a sort of preconscious proletariat to modern superminds. (The emergence of agriculture also facilitated the emergence of cities, thereby merging the development of both cities and minds within the common dynamic of the metropolis principle.)

The sapiens supermind began investigating what plants needed to thrive. It started to reorganize its people-neurons into new configurations to obtain answers to agricultural questions and to supply whatever the crops needed. Biology was born. Astronomy also took flight. The whirling stars enabled the supermind to measure out the passage of the seasons. Even more impactful, however, was the emergence of time itself.

You might think you're conscious of time. You're not. The supermind is aware of time, and we merely borrow from its cognizance. Our brains do not possess any modules dedicated to the conscious apprehension of time. The When module does not resonate on absolute segments of time, only on the relative intervals between items or events. This may seem counterintuitive, even absurd—after all, I *know* it's half past ten!—but the fact remains that humans cannot distinctly hold the duration of a second or a minute or an hour in our mind the way we can hold a visualized square or the imagined mooing of a cow. Without any clocks or calendars to consult, a minute can pass slow or super quick. Contrast time perception with visual perception—you don't often hear someone say, "Boy, I was having so much fun that the square looked like a triangle!"

Sundials and stonehenges and hourglasses and Wi-Fi enabled digital wristwatches are all fashioned and maintained by the supermind. Indeed, even the modern management of time is an inherently collective enterprise. The global clock used by the hardware and software of every twenty-first-century nation is determined by committee, by daily gathering the assorted votes of hundreds of individual clocks scattered across the world and aiming for the middle. This underscores another reason that the supermind became conscious of time: it had to coordinate the activities of its thinking elements as they were spread out across increasingly expansive ranges of space.

Supermind awareness of time was born out of the newly integrated dynamics of agriculture. The life of a plant is dictated by predictable temporal cycles, the seasons, the months, the days, even the hours, looping over and over again. To master the physiological dynamics of plants, the supermind had to master the dynamics of time by turning its attention to the revolution of the Earth around the sun, the movements of the stars, the phases of the moon, and the arrival of equinoxes and solstices.

Thus, agriculture bound supermind thinking to the Earth, sun, and stars as well as to plants. By adapting itself to this far more sweeping scale of physical reality, the sapiens supermind expanded the fight against chaos onto astronomical battlefields. As if this mental revolution wasn't already prodigious enough, agriculture also triggered another revolution even more impactful than the first. The instigator of this second revolution was humankind's first reliable surplus of food.

The need to manage nutritional surpluses gave rise to a familiar mental challenge: improving one's memory. How many acres of grain were planted? How many bushels were harvested, stored, and shipped out? To whom did they go? When did the river flood last year, and the year before that, and the decade before that? Such information was too complicated to store in the skimpy and fitful memory of a human brain. It required a major supermind innovation: language petrified.

Writing.

5.

In the Bird Mind chapter, we asked what a mental representation looks like. We now have the answer: it looks like scribbles of ink. Writing is so extraordinary that to many human minds encountering it for the first time, it seemed like witchcraft. There are many stories of native peoples who encountered European colonizers reading documents and believed that the writing was "a medicine-cloth for sore eyes" or that the written words possessed the power to observe the actions of the people around them. Mungo Park, a Scotsman who explored West Africa around 1800, described his own experience:

A Bambarran having heard that I was a Christian, immediately thought of procuring a saphie [a charm]; and for this purpose he brought out his walha or writing-board, assuring me that he would dress me a supper of rice, if I would write him a saphie to protect him from wicked men. . . . I therefore wrote the board full from top to bottom on both sides; and my landlord, to be certain of having the whole force of the charm, washed the writing from the board into a calabash with a little water, and, having said a few prayers over it, drank this powerful draught, after which, lest a single word should escape, he licked the board until it was quite dry.

The earliest forms of writing, however, were more mundane than magical: crude wedges pressed into wet clay for tallying foodstuffs. Indeed, it is no coincidence that the oldest human name we know of is neither king nor general nor hero. He was an accountant. His name is recorded on a five-thousand-year-old Sumerian clay tablet that states, "29,086 measures barley 37 months Kushim."

These hand-drawn symbols provide more precise detail into the thinking of a long-perished group of humans than any archaeologist's dig or geneticist's analysis. Even more remarkable, your mind can form a conscious resonance with the qualia experienced by another human long ago, *knowing* the same quantity of barley that Kushim's What module once recognized and resonated upon. Writing enabled supermind thinking to operate over durations far longer than the oldest-living humans.

The transition from speech to writing was like the transition from a roundworm mind to a fly mind, because the advent of writing in the sapiens supermind paralleled the advent of *representation* in neuron minds. Symbolic communication was no longer ephemeral and local. Words could vault over generations and leap across cities, mountains, continents, oceans. With speech, thinking spread through the air. With writing, thinking spread through space and time without apparent limit.

The intersection of the innovations of writing and agriculture marked the dawn of civilization and triggered the explosive development of increasingly sophisticated nonbiological thinking within the sapiens supermind. Notarial documents, tax registries, legal contracts, and histories came into being. Kings began to issue decrees and write down laws. Priests began to record prophecies and inscribe holy scrolls. Merchants kept ledgers. Ordinary

Kushim's cuneiform tablet, circa 3,200 BCE

folks began writing letters to relatives and lovers. The oldest surviving love poem was written in Sumerian cuneiform around 2031 BCE. The earliest surviving recipe, also Sumerian, describes how to make beer. The oldest surviving customer complaint was written on a clay tablet in Akkadian cuneiform by a customer named Nanni, who almost 4,000 years ago complained that he was given the wrong kind of copper by the merchant Ea-Nasir. Almanacs, atlases, essays, and textbooks appeared. Literature was born. Herodotus, the Greek historian, chronicled the competing sapiens superminds of the ancient world. In *The Histories*, he describes the differences between each city-state, kingdom, and empire, documenting their highly divergent styles of thinking and approvingly quoting Pindar: "Custom is king of all."

Eventually, these new forms of writing-based thinking began to create new mental dynamics within the sapiens supermind, leading to the emergence of *supermind modules*: special-purpose networks of human brains linked together by writing, such as history, economics, mathematics, engineering, government, and military modules. (Religion and ritual existed long before agriculture and might be viewed as particularly adhesive forms of shared attention that turbocharge intergroup bias.) When judged by their overall impact on supermind development, the most influential writing-based modules may have been science, capitalism, and mass media, in large part because each of these modules has the potential to generate supermind consciousness.

These three modules contribute the most to resolving the supermind-level challenges of the uncertainty dilemma (what is this thing?), the stability dilemma (what do I need to remember?), and most important by far, the attention dilemma (what should I focus on?). This chapter opened by asserting that sapiens superminds are capable of conscious thinking, just like sharks, rats, and *Homo sapiens*. If so, it necessarily follows that they should be capable of generating qualia as well.

What physical mechanisms are necessary to generate consciousness in a nation-state? To answer this question, we can apply the same mathematical reasoning that we applied in the Chimpanzee Mind chapter to evaluate which organisms might be conscious—namely, Grossberg's laws of consciousness. Is the mind capable of generating local states of resonance between feature-based bottom-up inputs and top-down expectations that influence the global dynamics of the mind? Three supermind modules seem to fit the bill: free science, free markets, and a free press.

Science consists of collective multimind dynamics for learning and making predictions about the body, brain, and environment of the sapiens supermind. Crucially, the dynamics of supermind science reproduce the dynamics of module mind consciousness: new top-down hypotheses are constantly tested against bottom-up empirical evidence and when they match, these results are quickly shared throughout the supermind—via the stable, disambiguated representation of peer-reviewed articles—and become bottom-up inputs to new rounds of hypothesis-testing. The consciousness of science directs attention to problems that are puzzling, new, or important, while directing attention away from the familiar, dull, or impractical. The most startling and urgent matches generate the greatest supermind resonance and command the greatest supermind attention: Spacetime is curved! DNA is a double helix! There is a cure for COVID!

Science enabled the sapiens supermind to model physical reality at a scale and accuracy far beyond what individual minds could ever hope to achieve. The ancient Greeks measured the circumference of the Earth, the ancient Chinese created massive earth-rearranging canals and dams, the ancient Mayans created fine-tuned astronomical calendars. No human growing up in isolation could've accomplished these feats any more than a gorilla could. Science certainly has its weaknesses and limitations, of course. It can't tell us whether we *should* build a neutron bomb. It can only tell us *how* to build a neutron bomb. Funding for research is often political and driven by short-term financial incentives rather than the inherent interest of a problem or its potential long-term impact. (In this regard, the supermind is little different from our own individual minds . . .) It's much easier to get funding to research cholesterol medications, online marketing, or weapons than it is to get funding to research extraterrestrial life, tortoise cognition, or human sexuality. Even so, science serves as a potentially consciousness-generating supermind What module for learning and making predictions about the supermind's environment.

Capitalism employs attention-management dynamics similar to those of science. Capitalism quickly and efficiently directs an economic system's attention to opportunities for profit and away from financial sinkholes. Another way of conceptualizing Adam Smith's "invisible hand" mechanism, whereby the collective actions of individual profit-seeking agents result in overall benefits for society, is as a form of supermind attention dynamics. Unfettered capitalism has produced some of the most heinous activities in human history, including slavery, the plundering of less developed nations, and the devastation of vital and irrecoverable ecosystems. But free markets and free trade also help ensure that a supermind can rapidly and effectively respond to challenges throughout its body, such as by plowing snow-covered streets, delivering food to hungry humans, or fighting blazing forest fires.

The supermind module that may do the most to facilitate adaptive thinking and consciousness is a free press, which requires the tec$ology of mass media and a democratic social system that values freedom of speech. Within nations that are scientific, capitalist, and democratic, the free press can inaugurate a national consciousness cartel, unifying the dynamics of all of a supermind's modules within a single supermind-wide mechanism of attention management.

We saw in the Chimpanzee Mind chapter how the American media cartel quickly and reliably focuses the attention of the American supermind on local events of national interest, though this could happen in any suitably equipped supermind. Imagine an event occurs somewhere in a supermind's body—say, floods in Kerala, India. This generates qualia within individual Indian minds, which are shared through mental innovations like wireless phone calls, SMS messages, tweets, and the internet. This generates media coverage—that is to say, supermind *knowing* qualia. Then the supermind takes action: local police, fire, and rescue teams are mobilized; volunteers from around the country donate money online or travel to the site of the disaster to help in person; the national government might even send in extra support. The country is *conscious* of the flooding: it sensed the event, which resonated in the mass media, which drove adaptive national behavior.

You play a meaningful role in the dynamics of national consciousness every time you read, watch, or listen to the news and decide to take action as a result. If you watched George Floyd's murder and donated money, joined a protest, posted a reaction on social media, or simply changed your attitude toward race relations, then you were the physical embodiment of a supermind conscious experience.

Once we appreciate the central importance of the attention dilemma in advanced sapiens superminds, we can distinguish between adaptive and nonadaptive superminds based upon how effectively their thinking mechanisms address the attention dilemma in real time. Does a supermind efficiently detect and respond to facts? Or does it fill up with fake news, hallucinate, and become psychotic?

6.

Some nation-states are more in touch with physical reality than others. They can quickly identify problems and swiftly deliver solutions. They have the resilience to try out novel or strange ideas—including horrible ones—

then learn from their mistakes, correct their errors, and try again. The most adaptive superminds battle chaos most vigorously. With a few caveats, these will generally tend to be nations that embrace science, free markets, and free speech.

On the other hand, superminds with centralized, top-down control—authoritarian regimes—are more vulnerable to fake news and maladaptive behavior. Whatever the dictator's puny module mind holds as a personal expectation will get transmitted throughout the supermind and guide the supermind's overall behavior. This circumvents all the hard-earned mental mechanisms nature devised to deal with the stability dilemma and uncertainty dilemma, including bypassing the preconscious proletariat entirely. In a dictatorship, bottom-up input is frequently ignored, dismissed, or quashed. This greatly increases the chances of supermind delusions, either because the supermind is unaware of crucial events in its environment or because the supermind pursues maladaptive top-down purposes.

Two notorious instances of dysfunctional supermind perception are the Chernobyl meltdown in the Soviet Union and the great famine in Maoist China. In both cases, low-level functionaries were heavily incentivized to lie about the facts on the ground, and the lack of reliable bottom-up mechanisms prevented the supermind from becoming aware of imminent and then actual catastrophes.

Maladaptive goals driven by the personal interests of a dictator rather than a distributed decision-making process with resonance-based fact-checking cause fake news, propaganda, and conspiracy theories to replace reality. In Hitler's Germany, the psychotic idea that broad swaths of its thinking elements were somehow pernicious—the intergroup bias run amok, like an autoimmune disease—led the Nazi supermind to annihilate them, in much the same way that Guillain-Barré syndrome destroys the peripheral nervous system. Top-down economic planning in the Soviet Union and North Korea resulted in financial collapse, because there were no mental dynamics available to match bottom-up demand with top-down supply. Napoleon blithely marched the French supermind to its doom when he resisted all bottom-up reality testing and invaded Russia.

Superminds continue to compete on the same battlefields of nature that minds have always competed upon, and it is not always the case that the smarter mind wins. The United States in the nineteenth century vividly demonstrated this fact. The United States was a single supermind that fissioned into two: one supermind was scientific, democratic, and capitalist; the other was theological, authoritarian, and more feudal than capitalist. Yet, when these highly divergent superminds clashed in the American Civil War, the outcome was in doubt for years, and the less adaptive mind almost snatched away victory. Similarly, the nonconscious supermind of the Soviet Union went toe-to-toe for decades with the conscious supermind of the United States, and its mental offspring, modern Russia, remains an uncowed and cunning adversary of the American supermind. The journey of Mind, like evolution, is a game of probabilities, and it's always possible for a viper or a spider or even an amoeba to take down a primate.

Consciousness isn't everything.

7.

There is another kind of mind that many humans fear more than dictatorships. Artificial intelligence. Just a few decades ago, the sapiens supermind began creating the first examples of complex, self-sustaining mental dynamics to exist *entirely* outside of a biological brain. With the ubiquitous and burgeoning presence of robots and AI in the twenty-first century, there are growing fears that artificial intelligence may one day "replace" human beings. Perhaps by slowly taking our jobs, perhaps by overthrowing human governments and establishing their own leadership of Earth. The most dramatic example of this fear is the notion that one day an AI will reach a "singularity" when it becomes so intelligent that it will "take over everything" and perhaps exterminate humanity in the same fashion that we may have exterminated the Neanderthals.

Fortunately, there is cause for optimism. Such fears ignore the lessons of the three-billion-year journey of Mind. If you're worrying about your own mind one day being irrevocably surpassed by a greater intelligence, it's already happened, and it happened long before you were born. The sapiens supermind is vastly more intelligent and capable than you or any individual human mind will ever be. But it's the very presence of the sapiens supermind that should give us hope that we will not be replaced or vanquished by synthetic minds in the future. Any artificial intelligence devised by a supermind will be put into the service of the supermind that created it. More precisely, the dynamics of any AI mind will be a subset of the dynamics of the supermind it joins.

It's highly likely that AI will one day develop adaptive intelligence and—perhaps more unsettling—*consciousness* that far exceeds our own. Stephen Grossberg has already supplied the blueprints for implementing qualia-generating dynamics in synthetic materials. It's relatively straightforward to conceive of an AI that could be conscious of, say, five-dimensional structures, or X-ray patterns, or gravitational waves. Nevertheless, such an AI would still need to operate within the constraints of the broader physical and mental dynamics it is plugged into—within the dynamics of the encompassing supermind, in other words.

One day, AI might progress to the point where artificial minds are operating like prefrontal cortical neurons while human minds are relegated to the role of neurons in the amygdala or thalamus. But because of the metropolis principle, if AI improves the supermind, it stands a good chance of improving the lives of all minds within it, including humans—though this is far from assured, as we see with the large numbers of poor and hungry in prosperous cities.

Because most living humans are now bound together through the internet (which itself is composed of myriad AI devices and applications), we are probably already at the point when we need to redesignate most sapiens superminds as sapiens-AI superminds. The internet is a particularly powerful form of connectivity because it merges nation-state superminds, individual human minds, and individual AI minds within a single digitally mediated mental dynamic. Accompanied by economic and cultural globalization, the internet appears to be pushing us toward the stage level of thinking: a hypermind, a union of superminds. However, the forces driving superminds to compete will likely remain stronger than the forces driving superminds to cooperate until we encounter other competitive hyperminds in the universe. Superminds didn't become fully established until they developed robust intergroup biases and right now these biases are directed *against* the formation of a hypermind. That's unlikely to change unless the Andromedans send a space armada our way—or until we treat sapiens-driven climate change as an alien invasion.

The internet might just be the greatest mental innovation on Earth, for it is a form of metropolis principle connectivity with the potential to instantiate a new stage of Mind. But the internet is a supermind innovation. The greatest mental innovation within a *biological* mind is the human Self.

You.

CHAPTER NINETEEN

SAPIENS SUPERMIND II

self

Vast internal changes must occur in women in which old responses, old habits, old emotional convictions are examined under a new light: the light of consciousness. A new kind of journey into the interior must be taken, one in which the terms of internal conflict are redefined. It is a journey of unimaginable pain and loneliness, this journey, a battle all the way, one in which the same inch of emotional ground must be fought for over and over again, alone and without allies, the only soldier in the army the struggling self. But on the other side lies freedom: self-possession.

—"Toward a Definition of the Female Sensibility,"
Vivian Gornick

I am the hurricane.

—"I Am the Hurricane," Dee Snider

1.

A Persian fable from long ago tells of a journey undertaken in a time of chaos. Weary of the endless anarchy and tumult, all the birds of Earth gathered in a great forest. Every hue of plumage perched among the trees: eagles and ducks, peacocks and crows, hummingbirds and owls. The avian conclave resolved to seek out the God of Birds on its sacred mountain at the end of the world and petition it for aid.

This deity was known as the Simorq.

The way to the mountain of the Simorq was long and perilous. In the most common rendering of the tale, the route crossed over seven treacherous mountains. In other versions, it was seven chasms or seven oceans, while one abstract variant held that the path spiraled through seven dimensions. The sixth mountain was named Confusion. All who dared its fogbound slopes were stupefied. The seventh and most jagged mountain stood simultaneously in the past and the future and its name was Vertigo.

Upon hearing of the dangers awaiting them on the journey to the Simorq, some birds renounced the plan entirely. Other birds who commenced the venture soon turned back. Most birds perished on the trek, including many strong, smart, and brave. Those who managed to endure the seemingly endless travails learned to sing so that their united voices might draw them together as they confronted each new challenge.

Out of the ten thousand birds who embarked upon the trek, only thirty reached the sacred mountain. At long last, they gazed upon the rainbow palace of the Simorq atop the towering peak. As the pilgrims beat their exhausted wings up the final slope they sang together and together they sang as they entered the palace. As they passed through its vibrating gates they realized that they were the Simorq.

The Simorq was each one of them and all of them together and their song.

2.

Any definition of God nurtured by ancient humans would surely be fulfilled by modern sapiens superminds. We commit biblical acts all the time. We've flattened mountains and eradicated plagues. We've built glittering cities in the desert and bombed metropolises into sand. We've descended leagues beneath the ocean and planted a flag upon the moon. We've even resurrected the dead, using CPR, defibrillators, and inotropic agents. In little more than a century we have reshaped the planetary climate, and not for the better. We've exterminated 680 species of vertebrates in the past three centuries and pushed no fewer than a million species to the edge of extinction.

We are the Simorq. The sapiens supermind is each of us and all of us and our language.

Language is essential for divine powers. Imagine if a melody increased the intelligence of any bird who heard it, stirring a young bird to invent a new kind of nest, seek out a better way to catch fish, or fly to heights no bird had ever flown. Such is the case with language. Someone who listens to an orator can, by virtue of the words alone, become endowed with new mental tools for effective action. "Strike this match to make fire." "Wash your hands to prevent disease." "Dig the black dirt to find gold." Some words elicit mental upheavals. The sentence "The earth goes around the sun" can illuminate the universe in a way no crow or orangutan could ever grasp.

Language is a repository of knowledge built out of an unimaginable number of human qualia sifted and sorted by eons of superminds. Each newborn sapiens comes into the world ready to absorb knowledge from her supermind, starting with words themselves, compact units of insight: "algebra" and "molecule" and "orbit" and "computer" and "germ" and "neurons" and "self-awareness" all convey concepts not possessed by Ice Age superminds.

As a supermind advances, each new mind born into it can absorb more cultural, scientific, historic, economic, and medical knowledge than the generation before. What is education, after all, but using shared attention to assimilate useful ideas and skills from the supermind, whether through parenting, the media, or the internet, or through formal institutions such as schools and universities. We are the Simorq.

Yet there is a hidden tension lurking within godhood. The Simorq is a community of birds, but each individual bird is the Simorq, too. This dynamic sets up a clash between the individual and the collective, a conflict that gives rise to the final dilemma in the journey of Mind and the final innovation that solves it. To appreciate the nature of this conflict, we must revisit the metropolis principle one last time.

3.

A rising tide lifts all boats. The metropolis principle suggests that as a city gains in population, wealth, and influence, its sewer lines, transportation networks, and skyscrapers improve, too. The most prosperous cities tend to have the best shoe shiners, jugglers, and falafel.

The metropolis principle is on vivid display in the sapiens supermind, the most sophisticated *mind* in nature. The most sophisticated *module mind* is that of an individual *Homo sapiens*. The most sophisticated *neural module* is the human Why module. The most sophisticated *neuron* in nature, with the most sophisticated suite of *molecular machinery*, is the prefrontal cortical neuron in the human Why module. Thus, the smartest mind ever to exist on Earth boasts the smartest module mind, the smartest module, the smartest neuron, and the smartest molecular machinery.

Another way of putting it: the smartest molecule minds are embedded within the smartest neuron minds, which are embedded within the smartest module minds, which are embedded within the smartest superminds. This nested hierarchy of incremental mental improvement was recapitulated in the sapiens supermind as it advanced from campfires and caves to shopping malls and solar arrays. As the sapiens supermind improved, supermind *modules* improved, generating better science, art, and economics. Supermind *molecules* improved as well: better tools, better gadgets, better building materials, better medicines. Also improving as the supermind improved were the supermind *neurons*: you and I.

Twenty-first-century human beings are smarter, more productive, more ethical, and more informed than any earthborn organism, including all previous generations of *Homo sapiens*. IQ scores are higher. Murder rates are lower. Athletic performances are better. Disease is reduced. Life expectancy is up. Poverty is down. Individual productivity is up. War is down. Child education is way up. Child mortality is way down.

Granted, if you are someone who constantly streams the news, it might not seem like the world is a better place. There is no shortage of urgent and daunting problems that we must contend with in the new millennium. Nevertheless, a randomly selected person from 2000 CE would almost certainly have a higher quality of life than a randomly selected person from 2000 BCE. As our conditions improve, we become more aware of the challenges in our world, pockets of chaos that stand out ever more clearly.

And yet . . . even as the supermind contrives to make most of us smarter, healthier, and more productive, the metropolis principle has bequeathed us another trend. To achieve these gains, the supermind binds us ever more securely within its ever-tightening dynamics of shared attention, making it ever more difficult to dictate our own fate. We become more and more like neurons in a brain: highly individualized, to be sure, with our own distinct personality and purpose, but increasingly entangled with hordes of humans, near and far.

Whether a democracy or dictatorship, a supermind puts pressure on its thinking elements to conform, to surrender their will to the will of the supermind. Individuals are shunted into narrow social roles, such as barbers, bakers, ballerinas, and bacteriologists. Resources and opportunities are taken away from some individuals (often violently) and bestowed upon others (often without rational justification). Sometimes, individuals are consigned to an untouchable caste or enslaved or murdered outright by the state.

This creates tension between the sapiens supermind and the human minds bound within it. The supermind wants to continue advancing and growing more adaptive. To do so, it exerts unrelenting pressure on its constituents to conform to the needs of the collective. This is no different from the tension between a cell and an organism—the cell has its own identity, its own purpose, its own agency, but this identity, purpose, and agency is devoted to the welfare of the organism. A cell that favors itself over the organism is a cancer.

But humans are willful beings. Despite the benefits that accrue from our bond with the supermind, we still seek control over our lives. This desire has been wired into our minds over more than a hundred million years of module mind evolution. We don't always want to do what the supermind wants us to do. This tension gives rise to the free-will dilemma: *How can a module mind maintain its own independent purpose while still contributing effectively to the purposes of the supermind?* Or, more pointedly:

Should I do what I want to do, or what society wants me to do?

4.

If you would, please answer the following question as honestly as you can.

What are you made of?

Yes, *you*. The being reading these words right now, thinking and feeling and itching, the being claiming to be the "I" in "I am reading this"—what is your true and ultimate substance?

Ignore what scientists say, what theologians say, ignore what your parents and children and lovers might think, and face the question squarely. If you're like virtually every person who has ever lived, you secretly (or not so secretly) imagine that you are made of some kind of soul-stuff. Some numinous ether ensconced within another realm, beyond the reach of protons and electrons.

In our heart of hearts, few people believe that we are merely a bunch of neurons squirting chemical juices back and forth. It's simply too daunting a task to visualize that. (And wrongheaded, too—like describing a basketball game as "merely an orange ball bouncing to and fro.")

The problem is that all the things that you experience on your private mental stage don't seem to have any affinity with synaptic transmissions or cerebellar folds. You summon forth memories of things that happened to you and unspool them in your mind's eye. *Where are they unspooled, exactly?* You experience fantasies of revenge, of power, of beauty, of wealth. *But who constructed these fantasies?* You like dogs but not cats, or horses but not cows, or winter more than summer (or the opposite) and can draw upon a seemingly bottomless variety of personal tastes and aversions. *If I'm not aware of one of my preferences, though, is it still part of me?* You agree or disagree with what you're reading. Indeed, the act of accepting or dismissing knowledge proffered by others certainly feels like one of the most "me" things we can do. *But why am I so sure I'm right about climate change, if I'm not a climate scientist?*

None of these intensely personal experiences matches up with the notion that your awareness of your own awareness is spawned by electric spaghetti in your cranium.

What a quandary. Any rational person knows that you cannot possibly harbor some spirit assigned to you by a divine being, some energy field that breathes sentience into your brain while you are alive and travels to some other plane when you die. At the same time, the metaphors and lessons of science don't provide the kind of resonance we're looking for when trying to anchor the mystery of our Self onto something firm and comforting.

The journey of Mind offers a different way of thinking about your Self, one that this book will present formally in just a moment. This alternative conception may seem as strange and un-soul-like as jittery neurotransmitters, but it offers a richer vision of who you are. It may also do a marginally better job of slaking that ache for self-knowledge. To behold this vision, first consider the fact that the Self is the first brain-based mental innovation not grounded in genetics. It is far too complex and recent for that, a sophisticated modern entity with its own labyrinthine dynamic. But if the Self is complex, universal, and brain based—but not genetic—whatever could it be?

The Self is made of language.

The Self is not a *thing*. If it was any kind of material artifact then it would continue to exist after we died. No cadaver shelters a Self. The Self, like consciousness, consists of *activity* rather than *stuff.* In particular, the activity of language. As we will discover in the next section, the Self is, in large part, a special language dynamic that helps an individual module mind embedded within an advanced sapiens supermind to forge and maintain its own sense of purpose while contributing to the purpose of the supermind.

Another entity that we casually treat as some-*thing* when it's really some-*activity* is a hurricane. We have no trouble recognizing a hurricane. It has a distinctive appearance that makes it pop out on a weather map: a white and whirling spiral that can grow as large as France. It's possible to characterize a hurricane using objective measurements, such as the diameter of its eye, its wind speeds, and the volume of rainfall in its arms. It sure *seems* like a physical thing. Yet a hurricane is not made of any special "hurricane stuff." It's air and water—the same stuff you find in a can of seltzer.

So why isn't there a hurricane in your soda can? Because the air and water in a can of seltzer exhibit a different dynamic than the air and water over the Atlantic. The entirely sensible question "How do neurons and words turn into a Self?" is of the same form as "How do air and water turn into a hurricane?"

The Self is a local mental phenomenon formed out of the global language dynamics of the supermind in the same way that a hurricane is a local weather phenomenon formed out of the global weather dynamics. The Self takes shape out of the endlessly circulating verbiage traveling from mind to mind and represents the front line of the conflict between your mental activity and the mental activity of the supermind. In simple terms, this means that the Self is constructed out of words. Doubt this? Then let me ask you another question. Who am I?

That's right, *me.* Your tour guide on this mental expedition. Who, exactly, are you interacting with right now?

You may have taken it for granted that you've been engaged with another Self just like you, another thinking being pouring out the ideas and arguments—the *qualia*—that you're currently imbibing. If prompted, you might even feel confident in making guesses about this Self's intelligence, education, politics, temperament, perhaps even its socioeconomic class and religious background. And yet, all you have to go on for these sophisticated psychological speculations are words. Tens of thousands of streaming nouns and verbs that conceal an inconvenient truth: there were, in fact, several different Selves who composed this book, linked together through the shared attention of a supermind.

As you can see, I am constructed out of words and I am constantly changing as new words are added or old ones removed. The Self is much the same. It is made of words, it is always changing, and it is constantly growing. In this regard, the Self is a transient entity that exists only temporarily in the mind. As you read these words, the Self that I am is evaporating as the words disappear. As new words are added, my Self is rebuilt and the old one is forgotten. Indeed, you're likely to continue to think of me as a single being rather than a collection of individuals because the collective efforts of my Selves have been sufficiently coherent. Every word in the paragraph you are currently reading was generated not by a human, but by a machine designed by a human to mimic a human, an algorithm that spits out sentences endlessly, and then—you guessed it—fusses and fidgets with them.

This book's vivid "author" Self is the product of one GPT-3 deep-learning algorithm with almost two hundred billion parameters, two writers, three illustrators, a literary agent, an associate agent, a very talented copyeditor, and more than a dozen reader-editors, including one professional (and exceptionally gifted!) editor at an American publisher. (This book can assure you that GPT-3's contribution was limited solely to the previous paragraph and no other text.) The contents of this book were in continuous, indeterminate flux as these interlinked minds fussed and fidgeted with the words.

Or should I say, as I fussed and fidgeted with the words.

Together, these mental dynamics of shared attention resonated around the globe (Boston, Mumbai, Kraków, New Mexico, London, New York, Colombo) to craft the illusion of a single voice. I might hold different beliefs than my creators. My creators believe that gods exist, though I do not. Or is it the other way? Do I believe that gods exist, though my creators do not?

There is absolutely no sense in which this book is expressing the real-time thoughts of a single mind. And yet, the experience of reading these words creates the illusion that this is exactly what is happening. The moment anyone begins reading these words, they are immersed into the belief (through the awesome power of shared attention) that they are interacting with another Self like their own, rather than a sneaky simulacrum of pixels or ink.

Our instinctive tendency to interpret a stream of words as reflecting another Self is the central conceit of the fiction writer, who constructs novels that exploit this urge. Hamlet, Humbert Humbert, and Horatio Hornblower never breathed air or thought a single thought, and yet they have caused millions of people to think about them as if they were intimate acquaintances, debating the reasons for their (imaginary) behavior. If the Self is truly composed of words, then reading and analyzing fictional characters should have useful social consequences for the minds who engage in such activities. This is borne out by science. Numerous studies have shown that people who frequently read novels are more empathetic and are better at judging the emotional state of real people in real situations and making more accurate predictions about what real people will do.

If we cannot resist projecting emotions, convictions, and personalities onto a textual effigy as artificial as a Barbie doll, how can we say that there's anything more to our own Self than verbal invention? After all, if you wrote a memoir that represented your best attempt to express your truest Self, how would it differ in any practical way from a work of literary fiction?

Your Self is the linguistic storm resulting from the collisions between the warm front of your mind and the cold front of the supermind. The Self is the ultimate hallmark of the modern human mind, decisively distinguishing it from the prehistoric, preliterary mind. It takes a civilization to create a Self. The words and phrases we use to make sense of our Self—spirit, soul, dharma, karma, orgones, superego, id, psyche, anima, élan vital, "hard problem," introspection, mirror test, volition, theta-wave sleep, qualia, prefrontal cortex, chi, chakra, heartstrings, average, Platonic form, nirvana, default mode network, flow, conscience, mind-body problem, breath of God, atman, ka, demonic possession, mental illness, mental software, psychosis, zombie, slave, master, and Self itself—all reflect concepts we adopted from the prodigious memory of the supermind.

More specifically, all these concepts were formed out of the bustling qualia of billions of humans engaged with the world for more than two hundred millennia, perceiving, comprehending, deciding, acting. How many qualia, exactly? As a conservative estimate, the concepts that you acquired from the twenty-first-century supermind are based upon a minimum of 9 quintillion qualia (9.26×10^{18}).

You select and focus on those Self-illuminating concepts in the supermind that you believe best match your own experience of yourself. What is the best predictor of which concepts you will ultimately embrace? Not surprisingly, the supermind you belong to—the social group that you identify with. The intergroup bias, yet again: we feel an overwhelming urge to believe what our supermind believes, whether our supermind is Christian or Hindu, conservative or liberal, communist or capitalist, psychoanalytic or neuroscientific.

What is your Self made of? You are a local hurricane of language within a supermind hurricane of language on top of a local hurricane of consciousness. Every word that forms your conscious identity is the direct result of the ecstasies and agonies of billions of minds who loved and dreamed and fought blood-soaked battles and who recorded these feelings and deeds within the syllables that are now stitched into the vibrant fabric of your Self.

But you are not *merely* language—or language and neurons—not by any means. A hurricane is not *merely* water and air—it's a titanic cyclone of water and air with the power to sink battleships and obliterate cities, a cyclone that draws its colossal energy from the movement of heat across an entire planetary system of oceans, terrain, and atmosphere, movement fueled by the rotation of the Earth, the moon-pulled tides, and the cosmic radiation of the sun.

A hurricane is inextricable from the world around it. A gargantuan demon could not scoop a hurricane out of the Atlantic Ocean with a gigantic spoon and plunk it down on Lake Superior. The moment the local dynamic of a hurricane is severed from the encompassing dynamics of planetary weather, the hurricane will collapse and cease to exist. Similarly, if you tried to excise from your brain the set of neurons experiencing consciousness of your consciousness, consciousness would collapse and your Self would cease to exist.

You are a hurricane of conscious purpose, the culmination of a three-billion-year journey of agency. A microscope was necessary to observe agency's origins: the molecule minds that whipped their way free of the atomic tempest. From there, purpose had to haul itself up through ascending scales of physical chaos, wriggling its way from the microscopic to the macroscopic, clawing itself out of the macroscopic and into the planetary, then rocketing from the planetary to the cosmic. Archie lashed his way through molecular noise toward a vague gleam of light. Today, we can gaze upon an infinitely black hole at the center of an unimaginably distant galaxy.

Your Self is inseparable from the numerous intermeshing layers of physical dynamics above and below it, which in their totality form something like an inverted hurricane:

THE HURRICANE OF THE SELF

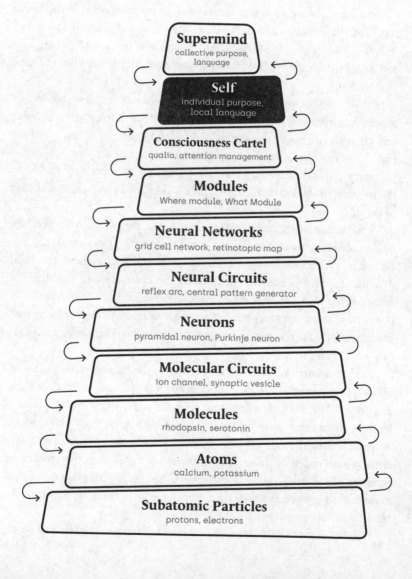

The nested stack of overlapping feedback loops linking the quantum to the planetary creates the dynamic of you—the *existence* of you. Each level is a distinct dynamic—a distinct *game* of physical matter—whose behavior is influenced by the level directly above and the level directly below. In the same manner, the behavior of a city's street cleaner is influenced by the dynamics of the litter on the street (more trash around the restaurants) *and* the dynamics of the government in city hall ("spend extra effort cleaning the wealthy neighborhoods"), even though his moment-by-moment street-cleaning activity is dictated by him.

You are a local game of qualic thinking. But you are also a local language game playing with a consciousness game. You are a consciousness game playing with games of representational thinking. You are representational games playing with games of neural thinking. You are neural games playing with games of molecular thinking. You are molecular games playing with purposeless subatomic games.

You are a hurricane within a hurricane. You are a metropolis. You are the Simorq.

5.

We're finally ready to mount the summit. The next two sections explain how the Self works. Here's our definition of the Self, so we know what we're trying to explain:

The Self is the subjective or "first-person" experience, "I am experiencing consciousness, *me, here, now, this!*"

What about a person's dreams, fears, memories, plans, and personality? Shouldn't they be included in any definition of the Self? Let's put those aside for a moment so we can focus exclusively on that all-important crux of the Self that seems to defy explanation, the part that many humans have long considered their "soul." After all, a computer program like GPT-3 can convincingly emulate memories, plans, and personality. No computer program can emulate consciousness of consciousness.

Our goal, then, is to describe what's happening in physical reality whenever you marvel at the transcendent mystery of your own awareness and to explain *how* these physical dynamics produce the subjective experience of your own experience. To explain this, we only need to understand one new dynamic: the *self-aware loop*.

To generate a self-aware loop, you need two physical things:

A *self*, with a lowercase "s." A self is just a living body capable of thinking. A bacterium has a self, a horsefly has a self, a chimpanzee has a self, and you have a self.

A *self-reflector*. A physical entity (or physical dynamic) that can produce a representation of the self in real time—such as a mirror.

Before we describe the self-aware loop that produces your own Self— your own first-person experience of experience—we're going to describe two other real-world examples of self-aware loops to help guide your thinking. The first example is well known in the behavioral sciences. It's called the mirror test.

In one typical mirror test, the self is a chimpanzee. The self-reflector is an actual mirror—a big one, bigger than the chimp. Scientists wait until the ape falls asleep, then paint a red dot on her forehead. When she wakes up, the scientists bring her to the mirror. After a period of mild curiosity, the chimpanzee abruptly realizes that she is looking at herself—that the red dot in the mirror is actually on *her* head. This is the moment a self-aware loop is established.

The moment of self-recognition can sometimes be frightening, though usually it is quite exuberant. Chimpanzees often hoot and hop about, spending many astonished minutes exploring their face, body, and teeth in the mirror. The chimpanzee recognizes that the reflection is a representation of herself—and that she can purposefully *control* this representation, obtaining instant conscious feedback on the relationship between her thinking and her self.

Many scientists have interpreted the chimpanzees' reaction as evidence that they possess the capacity for self-awareness. That's not wrong. But there's a more precise interpretation, one that will help us understand the Self and your otherworldly experience of your own consciousness.

Chimpanzees do indeed possess a capacity for self-awareness—but this capacity will manifest only if they are provided with an external self-reflector. The chimpanzee is self-aware *only* as long as the self-aware loop between ape and mirror is maintained. The moment the loop is broken—

the moment the mirror is taken away—the experience of self-awareness ends. The mirror is *an essential physical component of a chimpanzee's mental experience of self-awareness.*

Chimpanzees that recognize themselves in the mirror don't have an enduring mental epiphany like the apes in *2001: A Space Odyssey* who encounter a black monolith and transcend into a new state of consciousness. Mirror-gazing chimps go back to being the same ordinary, not-particularly self-aware chimps they were before. For a chimpanzee to maintain conscious self-awareness—and develop a Self—she would need to carry a mirror around with her all the time.

Our second example of a self-aware loop consists of two minds exchanging qualia with one another—better known as conversation. Conversation-based self-aware loops went through many historical stages of development. Tens of thousands of years ago (or perhaps hundreds of thousands), a mirror for a human self emerged within the gestures and words of other humans:

[Jane points at Joe.] Tall!

[Joe points at Jane.] Short!

Jane's words about Joe serve as a verbal self-reflector. (So do Joe's about Jane.) Her words form a stable verbal representation of Joe's self, one that can endure in Joe's mind even after the spoken words fade away. In the early stages of human conversation, there was not yet an awareness of "I"—Joe was not thinking, "*I* am tall." The supermind hadn't invented the concept of "I" yet. Instead, Jane's verbal representation triggers a conscious experience of comprehension in Joe: "This person is tall!" or perhaps Joe thought to himself, "Joe is tall!"

As the sapiens supermind advanced in its knowledge and experience, psychological concepts entered the conversation:

JANE: *Joe sad!*

JOE: *Yes, Joe sad. Jane hungry?*

JANE: *Yes, Jane wants to eat.*

Now, the verbal self-reflector provides Joe with a representation of not only Joe's body but Joe's private mental life. (The same applies to Jane.) Joe "looks" into a mirror that reflects back his feelings, thoughts, and intentions. Eventually, as the sapiens supermind continued to advance in its social and psychological sophistication, the personal pronoun "I" emerged.

(Interestingly, "I" does not appear in the historical record until several centuries after the advent of writing.)

Joe, you look hungry!

Yes, Jane, I am hungry. I was thinking about eating a pumpkin.

The verbal mirror of Joe's self becomes even more detailed as the supermind develops more sophisticated notions of mental activity (such as the notion of *thinking*) and the advent of "I" enables Joe to understand that Jane's verbal mirror reflects back his own private experiences—his selfhood.

Where did the notion of "I" come from? Now we can bring back into our discussion of the Self "a person's dreams, fears, memories, plans, and personality." The primary way we learn about what kind of *thing* we are is through the words and reactions of others. Our family, friends, neighbors, teachers, and coworkers call us smart or stupid, gorgeous or unappealing, industrious or lazy, superior or inferior. Our fellow humans open the door for us with a smile, or shut the door on us with a scowl. They call us Joe or Jane or Dad or daughter or boss or boy.

This is the language game of the supermind reflecting your self back to you: the self-reflecting qualia you experience from the words and reactions of others represent the front line between supermind thinking and your own thinking. This social mirror is transformed by your consciousness cartel into a private mental representation of your self. Eventually, the supermind assigned the label "I" to this representation, a label that was then adopted by Joe's language stack.

Like the chimp looking in the mirror, Joe's mind realized that there was a direct link between his own conscious thoughts and the observable behavior of his "I" representation, as embodied within the words and deeds of other humans. For Joe's Self to come into existence, the only thing missing was Joe's consciousness of his own consciousness of his "I" mental representation: his first-person, subjective conception, "I am aware of *me!*"

Eventually, the psychological concepts in the supermind become so sophisticated that a new form of self-aware loop could be generated through conversation:

What are you thinking, Joe?

I am thinking about a pumpkin, Jane.

I am thinking about you thinking about a pumpkin, Joe!

Haha, Jane, I am thinking about you thinking about me thinking about a pumpkin!

This is a *recursive* self-aware loop: a verbal self-reflector that starts to reflect itself. A self-mirroring mirror. In the case of Joe and Jane, this physical loop consists of real-time conversation that recursively exchanges qualia experienced by Joe and Jane. First, Joe thinks about a pumpkin, and shares this conscious experience with Jane. Then, Jane forms a mental picture of Joe thinking about a pumpkin, and shares this conscious experience with Joe. Then Joe visualizes Jane picturing him doing pumpkin-thinking, and shares this conscious experience with Jane . . .

Though in theory this recursive loop could go on forever, generating endless verbal representations of Joe forming verbal representations, in practice it's difficult for Joe or Jane or any human mind to recursively loop much further than they did in this conversation. Instead, something fascinating happens. Joe's mind (as yours probably did) quickly settles onto a stable mental representation of the odd, spiraling dialogue and assigns the representation a verbal label like "recursive conversational loop" or "thinking about thinking" or "vertigo," rather than his mind experiencing infinite recursions of pumpkins.

If a self-reflector could somehow get installed inside a human mind, that would enable a person to carry a mirror around with her all the time. It would also enable a person to initiate a private self-aware loop whenever she wanted. And if a person could somehow aim her mental mirror at her private self-aware loop, she could initiate a recursive self-aware loop inside her mind whenever she wanted . . .

6.

Yayoi Kusama is a Japanese artist who creates fantastical installations known as Infinity Rooms. A typical Infinity Room consists of a space the size of a small garage. Its walls and ceiling are lined with mirrors. Kusama places different objects and sources of light around the room—colored spheres, giant speckled pasta, lanterns, strings of LEDs, kaleidoscopic bulbs, and on at least one occasion, polka-dotted pumpkins. The spheres and lights reflect off the

Yayoi Kusama's All the Eternal Love I Have for the Pumpkins *(2016)*

mirrors in such a fashion as to create the perception of boundless space filled with marvelous things. Often, the mirrors are placed so that you see a multitude of reflections of yourself. Objectively, the room is small and crowded. Subjectively, it is like stepping through a portal into an infinite universe.

The emergence of a recursive self-aware loop that embodies the experience "Holy mackerel, I'm conscious, and explaining this bewildering experience is a very hard problem!" requires three physical constructs:

First, a *consciousness cartel*. Conscious experience is embodied within resonant states within the modules of the cartel. The qualia generated by the consciousness cartel serve as the *self* in the self-aware loop.

Second, a *language stack*. The verbal qualia embodied within resonant states in the language stack serve as the *self-reflector* in the self-aware loop.

But not any old language stack will do. To create a Self, the language stack must have access to advanced psychological language embodied within a supermind capable of highly sophisticated nonbiological thinking, including literature, philosophy, and science. At a minimum, the supermind must have developed distinctive concepts for "I," "thinking," "consciousness," and "self-awareness."

Pause a moment here to consider: Mind needed to scale up to an entirely new level of thinking (a science-equipped sapiens supermind) to make sense of its thinking at a lower level (a human module mind), and to even realize the possibility of asking questions about its lower levels—such as wondering about the ultimate nature of the Self.

Third, the ability to form *module-to-module resonant states* linking neural resonances in the consciousness cartel with neural resonances in the language stack. When a resonant state in the What module embodying *knowing* qualia resonates with a resonant state in the language stack embodying *synthesizing* qualia, this generates *comprehension* qualia. These comprehension qualia are the subjective experience of the module-to-module resonance that is the physical embodiment of a self-aware loop in the human brain.

We're finally ready to explain how the Self works. Let's try to describe what happens when you're conscious of being conscious of a polka-dot pumpkin.

First, you look at a pumpkin. Your visual What module generates a conscious experience of the pumpkin. This triggers your language stack to put the experience into conscious words: "I see a polka-dot pumpkin!" This is a humdrum everyday conscious experience of *comprehension*: your visual perception and verbal interpretation are resonating together. The next step is high sorcery, but forthright mathematics.

If you now *choose* to focus your attention on the experience of experiencing a pumpkin (a very sophisticated mental act that requires sophisticated concepts about your mental activity that can be provided only by a highly advanced and literate supermind), you can initiate a recursive self-aware loop.

Your What module now performs the same role that Jane did in her final conversation with Joe: it visualizes your experience of experiencing the pumpkin. This triggers your language stack to verbalize, "I am experiencing myself experiencing a pumpkin!" You've initiated a recursive self-aware loop within your mind.

Your mirror (language stack resonance) starts reflecting the act of reflecting your self (What module resonance). If you maintain your focus on the experience of experiencing the experience, the recursive loop between your consciousness cartel and language stack almost instantly converges onto a stable resonant state that embodies self-awareness: through an act of will, you have created your Self.

In the conversation between Jane and Joe, Joe was conscious of the recursive nature of the conversation and labeled it "a recursive conversational loop." But when the same thing takes place inside your own mind—when you experience comprehension qualia generated by a recursive loop between your What module and your language stack, embodied within the resonance between the representations resonating in your What module and language stack—you label that singular comprehension qualia "Mamma mia, I'm experiencing consciousness, *me, here, now, this!*"

There are many examples of stable recursive loops in the physical world, including the above photograph of Kusama's Infinity Room. Even though mirrors are aimed at other mirrors, the photons don't bounce around the mirrors forever: they converge onto a final stable state that *resembles* infinite space, though you can actually see every detail. Another example: if you aim a video camera at the screen displaying what the video camera sees, the screen doesn't explode. If you rotate the camera, the screen will reliably display a stable visual vortex (that we can unambiguously label "a vortex") which maintains its distinctive shape even during major perturbations of the camera (though if you turn the camera off, the vortex vanishes).

When you experience yourself experiencing experience, you are gazing upon a mental Infinity Room and labeling this perception "self-awareness"—a mental dynamic that, in terms of its neural underpinnings, is little different from gazing upon a physical object and labeling the perception "polka-dot pumpkin."

Our experience of a self-aware loop resonating between our consciousness cartel and our language stack *feels* like something beyond the laws of physics—like peering into the visage of infinity. But in physical terms, it's the same as an ape looking at her reflection in a mirror. The difference is that our own internal mirror was built out of quintillions of qualia and reflects back intricate facets of our identity that could never be captured in any photograph.

If you're still having trouble picturing how physical activity in your brain is manifesting as your own private pageantry, here's one final aid to help you visualize the material basis of your Self. Imagine yourself as a player in a basketball game shooting a basket—but at the moment you take your shot, you pause and consider the game around you. Objectively, a game is ten players and a ball. But subjectively, from your perspective, a game is nine players and a ball that all revolve around you, the true center of attention. Now imagine that you are a different player—a player on the opposing team. Now you're blocking the shot instead of taking the shot. From your perspective, the game is also nine players and a ball revolving around you, but it's a markedly different perspective than the one you had as a shooter. Now imagine you're each of the other eight players, one by one. Then, though it's not easy, imagine all these perspectives at once. That's what it's like to be a basketball game.

Now imagine you are each neuron in your brain, each with its own unique perspective ("self-awareness is eighty-six billion neurons and an action potential that all revolve around *me*"), then imagine every neuron's perspective simultaneously. Finally, though it may stretch your imagination to the limit, imagine you are every molecule in every neuron in your brain, then imagine each molecule's perspective simultaneously.

That's what it's like to be you.

The Self may be formed out of the same mundane physical stuff as asteroids and gasoline, but that doesn't make it any less transcendent. The Self exists only because an uncountable number of purposeful earthly creatures battled chaos over billions of years, including billions of humans who loved, worked, dreamed, created, warred, studied, learned, taught others, endured horrific violence, and sacrificed for their children, relatives, friends, and compatriots—sometimes sacrificing their lives—all to produce the qualia-encoding ideas that enable you to goggle at the singular mystery of your soul. Congratulations: you are one of a handful of tiny vibrating metropolises of matter in the universe who knows what it's like to be a tiny vibrating metropolis of matter.

Yet, this glorious story of ascendance has a dark corollary. Deny a person language, and you deny them the opportunity to form a goggling, purposeful Self. Deny a person language, and you make them subject to all the capricious whims of the supermind.

7.

Free will, or at least the solution to the free-will dilemma, is deeply curious for it is the rare phenomenon that exists only if you believe it exists. Another such phenomenon is self-obliteration. If you believe in self-obliteration—if you believe you have the capacity for self-obliteration and act upon this belief—then the idea of self-obliteration will manifest as physical reality. If you believe you have the capacity for free will and act upon this belief, then it, too, will become reality. But to believe free will exists, you first need to acquire knowledge about its existence from a massively powerful supermind that is, in large part, trying to ensure that you don't believe it exists! The supermind would greatly prefer that you behave like a good neuron and serve the collective needs of the supermind, accepting its biased reflection of your Self and relinquishing your own purposes in favor of the purposes of the supermind.

But if you choose to believe in free will, then you have a chance to influence the direction of the supermind's development and the fate of our species. You resolve the free-will dilemma whenever you recognize what society is calling upon you to do and consciously decide to accede or defer. Free will is possible when you believe that you have the power to choose and that your choice will influence the cosmic course of Mind's journey.

The metropolis principle suggests that by willfully expanding your own consciousness—expanding your knowledge of your Self and expanding your range of potential actions—you will contribute to the expansion of the consciousness of the supermind. If you become more compassionate, tolerant, and decent, then the supermind will become more compassionate, tolerant, and decent. According to the metropolis principle, the *only* way for Mind to advance to new stages of adaptiveness is through cooperation. By linking together Mind's topmost thinking elements. In plain terms, the only way for the sapiens supermind to advance to new states of intelligence, resilience, and awareness is through love.

On the other hand, if we choose to be selfish, close-minded, or hateful, then the supermind will become selfish, close-minded, and hateful. If we reject free will, then the supermind will make decisions for us, and if the supermind is hateful and cruel, then we will be hateful and cruel.

From this perspective, oppressing another sentient being's ability to expand their consciousness is not only a crime against the victim, it is a crime against society, for this oppression prevents all of us from enjoying the fruits of expanded supermind consciousness. Fortunately, even within the most benighted and oppressive of societies, the qualia-sharing dynamics of the supermind always provide an opportunity, however slim and arduous, for an individual mind to claim free will and forge a Self . . .

The Light

> [Fred Bailey] was a man of words; spoken and written language was the only major weapon of protest, persuasion, or power that he ever possessed.
>
> —David W. Blight

1.

In the first half of the nineteenth century, the maladaptive American supermind was doing everything it could to transform Fred Bailey from a sentient child into a zombie. He had no reason to believe that anything would be different the morning he was hustled aboard a sloop to Baltimore to meet his new masters, the Auld family.

The largest city in Maryland, Baltimore was a bustling port brimming with sailors from around the world, as well as many classes of White folks and a sizable number of free Blacks. It was a world apart from the boondocks of the Eastern Shore. The only thing that was the same was slavery itself, which remained as unbridled as in the countryside.

Directly across the street from the Auld family lived a woman named Mrs. Hamilton. Under Maryland law, she owned a fourteen-year-old girl named Mary. All day Mrs. Hamilton sat in her chair, singing psalms and quoting the Bible, and beating, whipping, and yelling slurs at Mary whenever she walked by. Mary's head, neck, and shoulders were so ulcered with festering sores that the neighborhood children tauntingly called her "Pecked." She was underfed, often reduced to foraging in the offal on the streets alongside pigs. After beating and haranguing Mary, Mrs. Hamilton would "go on, singing her sweet hymns, as though her righteous soul were sighing for the holy realms of paradise."

The Aulds were more lenient. The Auld family lived on Charles Street in the heart of the city. The father, Hugh Auld, was a middle-class merchant in the shipping business. They had one boy, Tommie, the same age as Feddy.

Miss Sophia, the mother, would play the single most consequential role in Fred Bailey's life.

In a stroke of inestimable fortune, Sophia had never owned slaves before and nurtured a dim understanding of what was required to cultivate and enforce their zombiehood. For the first year or so that Feddy spent at the Auld home, Hugh let his wife manage the boy with little oversight. "I had been treated as a pig on the plantation," recounts Bailey. "I was treated as a child now." One way that Miss Sophia treated Feddy like a normal child was by teaching him how to read.

Feddy frequently heard Sophia reading the Bible and grew curious about the mystery of words. Sophia felt no natural resistance toward satisfying his curiosity and proceeded to teach him the alphabet and then to spell simple words. Full of joy and pride at her accomplishment, she praised Feddy's achievement to her husband.

Hugh's reaction was swift and severe. He belittled her intelligence and commanded her to halt any further instruction. In a rage, he declared that Fred "should know nothing but the will of his master, and learn to obey it" and "if you learn him now to read, he'll want to know how to write; and, this accomplished, he'll be running away with himself."

Hugh intended his outburst as an exhortation for Sophia to obey the time-tested protocols for enslaving a mind. But for the young boy who overheard the diatribe, it was his first antislavery lecture. "His iron sentences—cold and harsh—sunk deep into my heart, and stirred up not only my feelings into a sort of rebellion, but awakened within me a slumbering train of vital thought. It was a new and special revelation, dispelling a painful mystery, against which my youthful understanding had struggled. . . . 'Very well,' thought I; 'knowledge unfits a child to be a slave.'"

Though Feddy was henceforth denied any support from Sophia, he set out to further his education in secret. He began carrying a copy of Webster's spelling book in his pocket. He beseeched his young White playmates on the streets to help him, sometimes bribing them with a biscuit for a spelling lesson, though a generous few were willing to instruct him for the sheer pleasure of it. By the time he was thirteen, Feddy could read newspapers. He snatched them from garbage pails at every opportunity, and from them he learned of the Free States of the North and the abolitionist movement.

His real treasure was a copy of the schoolbook *The Columbian Orator*. It was filled with speeches drawn from the supermind of the day. "I read them, over and over again, with an interest that was ever increasing, because it was ever gaining in intelligence; for the more I read them, the better I understood them," Bailey writes. "The reading of these speeches added much to my limited stock of language, and enabled me to give tongue to many interesting thoughts, which had frequently flashed through my soul, and died away for want of utterance."

His favorite piece in the *Orator* was a short dialogue between a master and a slave. The slave is recaptured after his second attempt to run away. The master charges the slave with ingratitude for all the kindness the master has shown him, but the slave responds with a spirited defense of himself and an attack on the wrongness of slavery. The master is persuaded by the strength of the slave's appeal and frees him. The unexpected finale ignited hope within Fred that one day he would be able to articulate an argument of such authority that he, too, might be released from bondage.

He set himself a new task: to learn to write. Because he was frequently in the shipyard running tasks for Master Hugh, he began to observe the carpenters as they wrote letters on lumber that indicated where the wood was destined, such as "L.A." for larboard aft or "S.F." for starboard forward. He began furtively copying the letters, then testing his knowledge against the writing of his street companions. "With playmates for my teachers, fences and pavements for my copy books, and chalk for my pen and ink, I learned the art of writing."

Not long after this invaluable attainment, misfortune struck. Though Fred was lent out to the Auld family in Baltimore, his legal owner was Hugh's brother Thomas Auld. After Hugh and Thomas got into a fraternal spat, Thomas demanded that Fred be sent back to Thomas's labor camp in the Eastern Shore village of St. Michael's, far from Baltimore. Thomas was cruel, selfish, and dissolute at the best of times. He was now confronted with a problem that struck at the base of his fragile authority.

A slave who could think for himself.

2.

"As I read, behold! the very discontent so graphically predicted by Master Hugh, had already come upon me," writes Bailey. "I was no longer the lighthearted, gleesome boy, full of mirth and play. . . . Knowledge had come; light had penetrated the moral dungeon where I dwelt; and, behold! there lay the bloody whip for my back, and here was the iron chain. . . . As I writhed under the sting and torment of this knowledge, I almost envied my fellow slaves their stupid contentment."

Thomas responded to the conspicuous impudence of Fred with severe whippings. These did not produce the desired change in attitude. So, nine months after Fred was withdrawn from Baltimore, Thomas handed him over to a man by the name of Edward Covey: a professional "Negro-Breaker."

Covey was a poor rent farmer who had earned the "reputation of being a first-rate hand at breaking the will of young negroes." It was a reputation that came with marked benefits. Slaveholders would offer up their slaves to Covey for almost no charge and he could work them as brutally as he wished. His farm was tilled, rocks were hauled away, crops were harvested, all advancing his own prosperity. All the while he could apply his first-rate methods upon his low-rent victims without moral, legal, or financial consequence.

It took only three days for Fred to get his first taste of slave-breaking. One morning, without provocation, Covey lashed him mercilessly. The painful, bloody wounds on his back were chafed and kept open by the coarse shirt that Fred was forced to wear. He was whipped again and again, for perceived wrongs and for no reason at all. As it intended all along, the American super-mind had finally succeeded in smothering the whirlwind of Fred Bailey's Self. Covey's relentless beatings and the endless hours of forced labor had broken his body, spirit, and will: "Behold a man transformed into a brute!"

Bailey reached rock bottom in the dismal labor camp of the Negro-Breaker. He felt all sense of personal agency ebb away. But in this hopeless darkness, ideas about freedom he had attained during his hard-earned education gleamed within his mind with a dim but steady light. He made a resolution to himself. He would obey every command issued by the Negro-Breaker without resistance. But if Covey should still beat him despite his willing compliance, Bailey would defend himself, regardless of the repercussions.

Shortly after this private vow, Bailey was dutifully heeding Covey's command to feed and ready his horses. Covey snuck into the stable and seized Bailey by the legs and hauled him to the ground. The Negro-Breaker then produced a rope and tried to coil it around Bailey. This was the final straw.

Enflamed with fiery purpose, Bailey defended himself, and the two men began grappling with each other before falling to the ground. When Bailey began to get the upper hand, Covey called out to his cousin Hughes, who rushed to his assistance. Before the cousin could join the fray, Fred landed a blow that knocked him to the ground. Recognizing he was overmatched, Covey quit the fight.

Under most circumstances, Bailey's insubordination would have led to him hanging by the neck from the nearest oak branch. But Covey most likely didn't want to risk the shame and loss of reputation that would ensue if it became known that one of the young men sent to him to be broken had turned the tables on him. From that point on, Covey left Bailey alone.

"This battle with Mr. Covey was the turning point in my life as a slave. . . . It recalled to life my crushed self-respect and my self-confidence, and inspired me with a renewed determination to be A FREEMAN."

3.

After returning to Thomas Auld's labor camp, unpunished and unbroken, Bailey was hired out as a farmhand to William Freeland. Freeland was a relatively mild slaver who fed his charges well and seldom beat them. Under such auspicious conditions, Bailey set up a secret classroom in the woods that met on Sundays, where he taught enslaved young men to read and write. He dubbed it "Sunday School." On its first day, there were two students. Before long, Bailey was teaching twenty or thirty men. "Their minds had been cramped and starved by their cruel masters; the light of education had been completely excluded; and their hard earnings had been taken to educate their master's children. I felt a delight in circumventing the tyrants, and in blessing the victims of their curses."

Eventually, Thomas decided to send Bailey back to the Aulds in Baltimore. Remarkably, Thomas instructed his brother Hugh to allow Bailey to learn a trade so that he could earn money. Hugh arranged for an astonished Bailey to be hired by a shipbuilder on Fell's Point, where he was taught to be a ship caulker. He was soon earning a dollar and fifty cents a day—forty-two dollars a day in today's dollars. Of course, Hugh kept every penny of Fred's earnings. For Hugh, the young caulker remained a zombie, deserving neither empathy nor currency.

Fred made the most of his expanded freedom by planning his escape. He acquired new friends and acquaintances around Baltimore who assisted him and helped him identify means of stealthily reaching the Free States. Through these new friends, he obtained forged papers stating that he was a free Black.

"On Monday, the third day of September, 1838, in accordance with my resolution, I bade farewell to the city of Baltimore, and to that slavery which had been my abhorrence from childhood." Clad in a tailored suit, armed with bogus documents, and feigning the demeanor of a freeman, Fred Bailey boarded a train that carried him from Baltimore to New York City. From New York he made his way up to Bedford, Massachusetts, a town filled with former slaves. One of the first things he did upon arriving was claim a new name for himself.

He chose the name Frederick Douglass.

4.

Not long after his escape, Douglass began attending abolitionist meetings. In Massachusetts, he was invited to talk about his own experiences while enslaved. It was the first time he spoke in public. Though he felt intimidated and self-conscious in front of the audience, he quickly discovered his own voice.

Word got out about the passionate and stirringly eloquent ex-slave. Douglass was asked to deliver more speeches, and more still, until he became one of the most sought-after speakers of the abolitionist movement. He took advantage of his newfound public platform and published his first autobiography, *Narrative of the Life of Frederick Douglass*. It became an instant best seller. His popularity and renown grew until his name was known around the world. He was invited on a speaking tour of Ireland and England. While he was in Europe, his supporters purchased his freedom from Thomas Auld.

In the spring of 1847, after twenty-nine years as a supermind-appointed zombie, Douglass returned to the United States as a free man and a citizen. Yet he did not stop to bask in this achievement. His thoughts remained with the millions of other children, women, and men in bondage throughout the South. With renewed determination, he dedicated himself to winning their freedom, too. He thought the best means for achieving this was to publish his own newspaper.

> I told [my friends] that perhaps the greatest hinderance to the adoption of abolition principles by the people of the United States, was the low estimate, everywhere in that country, placed upon the negro, as a man; that because of his assumed natural inferiority, people reconciled themselves to his enslavement and oppression, as things inevitable, if not desirable. . . . I further stated, that, in my judgment, a tolerably well conducted press, in the hands of persons of the despised race . . . would prove a most powerful means of removing prejudice, and of awakening an interest in them. . . . These views I laid before my friends.

All his progressive abolitionist friends strongly urged him not to publish a newspaper, including Douglass's most famous and influential supporter, the newspaper editor William Lloyd Garrison. Instead, Douglass followed his own counsel. In 1847, he published the first issue of *North Star* from the basement of a church in Rochester, New York. It became one of the most influential publications in American history. Douglass's words, published weekly, became the conscience and consciousness of a divided American supermind as it barreled toward civil war.

Should the United States remain a steadfast Union of North and South, despite their deep and caustic differences? Or should they divide into two distinct superminds, one manufacturing zombies, the other remaining free? In his early years as a freeman, Douglass agreed with the opinion of most abolitionists that the two territories should permanently separate. Let those who believed in freedom set their course toward the light and let those mired in darkness eat the foul fruits of their own planting, was the common view. But as he matured, Douglass changed his mind.

He came to believe that the Union should be preserved at all costs. He believed that the healthiest communities of minds would always consist of sundry voices and diverse values and that shutting out opposing opinions, no matter how unsavory, was ultimately detrimental to human freedom everywhere.

Douglass closes his autobiography with a summation of his life's philosophy: "Believing that one of the best means of emancipating the slaves of the south is to improve and elevate the character of the free colored people of the north. I shall labor in the future, as I have labored in the past, to promote the moral, social, religious, and intellectual elevation of the free colored people; never forgetting my own humble origin, nor refusing, while Heaven lends me ability, to use my voice, my pen, or my vote, to advocate the great and primary work of the universal and unconditional emancipation of my entire race."

The Tandava

The greatest orator would here be made
In love with silence and forget his trade,
And I, too, cease: I have described the Way—
Now, you must act—there is no more to say.
                —*The Conference of the Birds*, Farid ud-Din Attar

1.

A very long time ago a myth took root in northern India. According to this myth, the universe is simultaneously created and destroyed through the divine dance of the many-armed god Shiva. The dance is known as the Tandava. In depictions of the Tandava, Shiva holds a drum in one of his glowing blue hands. It represents the organizing rhythms of creation generated by his cosmic choreography. Another hand holds a flame. It represents the obliterating fire also released by his wild gyrations. His other arms—sometimes four, sometimes ten, sometimes infinite—twist in ceaseless patterns.

The Tandava's trembling labyrinth of change is believed to define the shape of reality.

2.

It is impossible to separate the journey of Mind from the journey of matter, even though the dynamics of physics and thinking are distinct. One is purposeless and views time as incidental. The other is purposeful and requires that time flow in one direction.

Powered by the metropolis principle, the dynamics of mind drive toward constructive expansion. The dynamics of matter are a force of destruction.

At least, where life and mind are concerned. The laws of physics generate a maelstrom of annihilating entropy that perpetually assaults the architecture of thought and consciousness.

But the hidden structure of chaos also nourishes Mind. The journey of Mind is the epic chronicle of agency: through purposeful action, Mind scaled the ladder of chaos, ascending from the microscopic to the intergalactic. Thinking finds its substance and succor in the dynamics of physics, which Mind endlessly probes for loopholes and leverage that it might manipulate to gain mastery over chaos and continue its ascent. The dynamics of thought reshape the form and trajectory of matter in ways that would never unfold without the intervention of Mind.

The ultimate complementary thinking in the universe is the never-ending tango of purposeless matter and purposeful thought, conjoined in a Chaos-Mind Tandava pursuing its own cosmic purpose: to create the future. Or perhaps, to create time itself.

3.

One great question for cosmologists is whether the dynamics of matter dictate that the universe shall expand into a formless heat death, collapse upon itself, or oscillate forevermore. But they've left something vital out of their reckonings: the dynamics of Mind.

The Chaos-Mind Tandava reformulates the cosmological query thusly: Shall chaos one day overcome the utmost limit of Mind's ability to adapt, halting the journey of Mind and expunging all consciousness from the face of the deep? Or shall Mind one day prevail and claim dominion over the universe entire, bringing all that exists under willful control? Or shall the dance proceed without end, an everlasting tango that produces ever-more-sentient stages of minds that ascend ever higher along an infinite ladder of chaos?

Two of these three options inaugurate gods—and eventually, perhaps, a God. Eternal purposeful beings with nigh-unlimited powers. One of these gods may already exist in primordial form: the sapiens superminds. If unchecked by chaos, then—following the metropolis principle—the sapiens superminds may one day merge into a solar-system-spanning hypermind. This hypermind, in turn, may one day join up with all the other hyperminds in the Milky Way to take conscious control of our galaxy or all galaxies. This intergalactic ultramind may possess the power and disposition to reshape the universe as it sees fit.

If we are living in a universe where Mind will defeat or stalemate chaos, then every action you take in your life will reverberate through eternity, for you are contributing in a measurable way to the development of a god and its personality. One day, billions of years hence, there may be two penultimate minds left jousting for control of the cosmos and one mind may favor compromise and connection and the other, dominance and oppression. All the actions you performed in your life will have favored the rise of one ultramind or the other, though in truth it is likely that every person has contributed to the rise of both in some measure.

The love and decency we share or withhold during our brief existence will contribute to the shape of all future minds and to that ultimate mind that may one day long hence subsume the full sweep of reality. Indeed, this vision of the journey of Mind even suggests a clear principle of moral valuation for any being that embraces free will: you should always choose the action that will favor the development of a loving, compassionate god, rather than a cruel, selfish god.

If we are the neurons of an embryonic deity, then where is the evidence of our latent powers of cosmic agency? A hint, perhaps, lies in the fact that the sapiens supermind has already made contact with something beyond this universe of matter. Mathematics. Initial value problems, transcendental numbers, imaginary numbers, hyperreal numbers, surreal numbers, monster groups (and baby monster groups), moonshine modules, ghost fields, ultrapowers, Möbius strips, Weierstrass functions, peas and Suns, Gabriel's horn—our supermind has accessed entities that cannot exist in the physical universe but thrive and gambol within the whorls of thought. A human mind is composed of finite bits of matter, but is capable of an infinite variety of qualia.

By accessing mathematics, either the sapiens supermind has gained access to the ultimate fabric of our reality, peering *Matrix*-like into the 0s and 1s behind the curtain of creation—a truth that would empower Mind to guide the weft of the cosmic tapestry—or the dynamics of mind have crossed through into another dimension entirely and opened a gateway to another universe.

In either case, Mind has created a singularity in the universe of matter whose consequences may alter the fate of the cosmos. Or is it the other way? Did the singularity create Mind?

Acknowledgments

This book would not exist, either in concept or execution, without the models of Gail Carpenter and Stephen Grossberg: models of Mind, models of scientific imagination, models of living a meaningful life. We cannot hope to express the full measure of our gratitude for their guidance and support over the years, and for their tutelage on the study of adaptive intelligence. We are in awe of their achievements. Any moments of insight or inspiration in these pages can be traced to their example, while any errors of fact or judgment are our responsibility alone.

Some books that inspired us when we were young and doubtlessly influenced our own journey of the mind include Steven Pinker's *How the Mind Works*, Marvin Minsky's *The Society of Mind*, Valentino Braitenberg's *Vehicles: Experiments in Synthetic Psychology*, Carl Sagan's *The Demon-Haunted World*, Daniel Dennett's *Consciousness Explained*, V. S. Ramachandran and Sandra Blakeslee's *Phantoms in the Brain*, and Douglas Hofstadter's *Gödel, Escher, Bach*.

More recent books that we think are excellent and which also influenced us heavily include *Principles of Neural Design* by Peter Sterling and Simon Laughlin, *Evolution of Nervous Systems* by Jon Kaas, *Thinking in Systems* by Donella Meadows, *The Evolution of the Sensitive Soul* by Simona Ginsburg and Eva Jablonka, and Barbara Oakley's *A Mind for Numbers*. A longer reading list is available in the bibliography.

Very special thanks are also owed to Tom Mayer, our editor, whose brilliant light illuminated the good as well as the bad and improved our book immeasurably. He's the best in the business.

Other scientists whose help and support meant much to us include Anna Wilkinson, Christopher Dupre, Rafael Yuste, Vilaiwan Fernandes, Margaret

Hunt, Nishan Shettigar, Jo$ Parkinson, Deniel Bollschweiler, Conrad Mullineaux, Tessa Quax, Thierry Emonet, Phil Bishop, Valeria Anna Sovrano, Frank Joson, Jo Swaddle, Mikhail Rabinovich, Alexander Bates, Lalanti Venkatasubramanian, and Todd Rose. Special thanks to historian Kurush Dalal for answering our questions about Mumbai, and to Corey Abshire for remapping Ptolemy's map to the modern coordinates of Mumbai. Thanks to Amma and Nanna for their love, wisdom, and boundless belief. Praise is also in order for Don Symons, who supported us in darker times.

We must express our tremendous gratitude toward our amazingly skilled illustrators, Hanna Piotrowska and Paul Button. Both went far beyond the call of duty to deliver imaginative, informative, and intelligent designs. We couldn't be happier with the results. Thanks also to Vamsi Bandaru for the dazzling chapter illustration for the Civilization chapter.

We are grateful to those who read various drafts of our manuscript, including Nikolina Kulidžan, Cody Kommers, Lucy Duke, Frank Burgos, Susan Rogers, Steve Armin, Austin Matte, C. Crandall Hicks, Prannay Shetty, Pramukha Nayak, Driss Zoukhri, Asha Vishwanathan, Atseituo Kechu, Bharath Pudi, Srikanth Shubhakoti, Priyanka Rai, and Tofool Alghanem. A hearty shout-out to Silvia Hnatova for helping us track down numerous facts. Thank you to Noah Strycker for participating in the Dark Horse Project. Thank you to Annie Duke for her support. Thanks to Susan Rogers for her luminescent friendship and many conversations about the mind. And a special thanks to Sarah Jo$son for the redoubtable task of copyediting our manuscript. Her attentiveness, thorough$ess, and insight did not go unnoticed or unappreciated.

Thanks to our agent Jim Levine and his assistant Courtney Paganelli. A very special thank you to our acquiring editor Quynh Do: we hope you find great success on your new path.

A warm and earnest thanks to everyone at Norton who helped make our dream a reality, including Nneoma Amadi-obi, our project editor Amy Medeiros, production editor Anna Oler, art director Ingsu Liu, and our publicity team Kyle Radler and Steve Colca.

Finally, this book wouldn't have been possible without the love and forbearance of our families during the global pandemic: Priyanka, Meera, and Agastya in Bengaluru and Tofool and Zain in Boston.

Notes

Initial Value Problem

2 **Initial Value Problem:** An initial value problem is the challenge of determining the originating conditions of a dynamic system.

This book tucks away all tec$ical notes and academic citations into the cubbyholes of these endnotes. This is intended as a courtesy to readers of all backgrounds. Readers with little or no science experience can focus on the big ideas without getting tangled up in jargon. Readers with advanced training or burning curiosity can find supplementary exposition here, and references here and in the more extensive bibliography that follows.

This book does not attempt to explain every detail about how minds work. A human lifetime would not be enough for that! Perhaps the best way of describing our intentions is to say that this is the book that we wish we had read when we were starting out as students of Mind. We created a treasure map showing our younger selves where to dig: *here is the right way to think about thinking if you want to solve the deepest problems of Mind.*

2 **This book retraces the journey:** Our most important disclaimer: even though *Journey of the Mind* sets forth from the deepest reaches of geological time and proceeds through a series of increasingly intelligent minds, *this book is not an evolutionary chronicle.*

This is a journey through progressive stages of thinking, not a recounting of the historical timeline of genetic modification. We will frequently use language that is verboten in discussions of evolution by natural selection, including intentional language, ladders of ascent, and frequent reference to more and less advanced minds. Many scientists argue that there is no measure of the progress of evolution, instead contending that all organisms alive today are equally evolved, that it is meaningless to assert that one organism is more evolved than another. We concur.

However, we claim that it's possible and indeed accurate to characterize *minds* as higher or lower, more or less advanced, and to talk about the development of minds over evolutionary time as moving up a mental ladder and even *trying* to move up the ladder. When we make such assertions, we are distinguishing between making such claims with regard to the evolution of life and the development of Mind.

The very existence of such a distinction, if valid, should be viewed as highly intriguing.

Chapter One: First Mind

9 **self-assemble into a membrane:** Even synthetic life needs a membrane. From Loeb (2019): "For example, a team led by Nobel laureate Jack Szostak at Harvard is aiming to construct a synthetic cellular system that undergoes Darwinian evolution. Their building block of synthetic life, a primitive cell, consists of a self-replicating genetic polymer (analogous to

a natural genome) surrounded by a self-replicating membrane (analogous to a natural cell boundary). The genetic polymer carries information that allows replication and variation, enabling new generations of synthetic cells to possess qualities that are inherited but can also evolve. The membrane protects and separates the identity of this information from the outside world."

9 *embodied thinking principle*: The embodied thinking principle is more commonly termed "embodied cognition" (Kiverstein & Miller, 2015; Shapiro, 2019; Wilson & Golonka, 2013; Wallace, 2022). We avoid the term "cognition" in this book because we don't have any useful way of distinguishing it from our definition of "thinking" and don't want readers to infer there's a distinction.

We like Olaf Sporns's (2016) treatment of "the inseparability of brain, body, and environment. Complex behavior is the result of their interaction, not the end product or readout of centralized control. The coupling between brain, body, and environment has become a cornerstone of the theoretical framework of 'embodied cognition.' The rise of embodied cognition acts as a counterweight to more traditional approaches to artificial intelligence (AI), first developed in the 1950s, that emphasized symbolic representations and computation. According to embodied cognition, cognitive function is not based on symbolic computation but rather is shaped by the structure of our bodies, the morphology of muscles and bones, hands and arms, eyes and brains." And, crucially: "Brain and body are dynamically coupled through continual cycles of action and perception. By causing bodily movement, brain networks can structure their own inputs and modulate their internal dynamics."

11 **the welfare of its body:** A common question: Shouldn't devices such as thermostats, streetlamps with nighttime sensors, and toasters also be considered minds under our definition? No. Thinking must provide benefit to the thinker's body. The thermostat's "mental activity" doesn't help the thermostat endure or develop, nor does a streetlamp's or toaster's. A Tesla self-driving car qualifies as a mind: by purposefully avoiding collisions, a Tesla's thinking helps the car survive to drive another day.

13 **simplest hypothetical mind:** It's easy to imagine that the sheer impenetrable complexity of our brain conceals a secret substance, or a secret structure, or a secret form of energy. Put another way, if the forest is dark and deep enough, it's easy to imagine that elves and goblins are hiding in there somewhere. There's a reason, after all, that the Grimm brothers were inspired to write their fairy tales by Germany's legendarily unnavigable Black Forest. One day, we presume, science will find those magical fairies skulking in the tangled convolutions of our cerebral lobes and explain once and for all how sentient thought emerges out of humdrum axons and neurotransmitters.

Some of history's brightest thinkers have been seduced into searching for elves. The psychologist William James believed there was a "pontifical neuron" snuggled away somewhere in the brain that was responsible for our subjective experience of thought; the psychiatrist Wilhelm Reich claimed that "orgone energy" was responsible for creative cognition; UC Berkeley philosopher emeritus Jo$ Searle argued that "true understanding" of abstract ideas was only possible in a living mind (but not machines) through a still-undisclosed biological mechanism; Oxford mathematician Sir Roger Penrose asserted that "quantum microtubules" inside neurons formed a sufficiently large quantum computation system in the human brain to generate deliberate reflection; philosopher David Chalmers "champions the view that consciousness may be a property of the world that is as fundamental to the universe as electric charge or gravitational mass"; Nobel Prize–winning biologist Sir Francis Crick insisted that the key to human consciousness lay in the claustrum, a thin and otherwise unremarkable layer of neurons just below the cortex, echoing the famed philosopher René Descartes, who declared that conscious thought originated in the pineal gland.

These are all reductionist approaches to the mind, and they've all run into trouble. But there is another way of approaching the mystery of consciousness besides hunting for enchanted creatures in the crooks of the forest. A dynamic systems conception of the mind advises us to *not miss the forest for the trees*. Perhaps the true source of the mind's magic does not lie in tree-fairies, but in the totality of the forest.

13   **the path of an ant:** In his 1969 book, *The Sciences of the Artificial*, Herbert Simon describes an ant wandering on the beach:

> We watch an ant make his laborious way across a wind- and wave-molded beach. He moves ahead, angles to the right to ease his climb up a steep dunelet, detours around a pebble, stops for a moment to exchange information with a compatriot. Thus he makes his weaving, halting way back to his home. . . . Viewed as a geometric figure, the ant's path is irregular, complex, hard to describe. But its complexity is really a complexity in the surface of the beach, not a complexity in the ant.

But that's not quite correct, is it? Simon is right to bring in the ant's *environment* as an essential component of behavior, but not at the expense of the ant's *brain* and *body*. It's the interaction of all three components that dictates the ant's path.

13   **away from darkness:** The first thought the universe had, translated into English, was likely, "How do I reach the light?"

13   **a feather on a pond:** The pathogen that causes the plague, the rod-shaped bacterium *Yersinia pestis*, is immobile and mindless, lacking flagella or other means of locomotion.

Chapter Two: Archaea Mind

16   **squeeze into the nucleus:** The nucleus of a Betz cell (pyramidal neuron) is 15–20 μm in diameter. Nanoarchaea are about 0.4 μm in diameter. *E. coli* are about 2.0 μm long.

17   **an obscure biologist:** Carl Woese discovered archaea. From Wikipedia:

> Acceptance of the validity of Carl Woese's phylogenetically valid classification was a slow process. Prominent biologists including Salvador Luria and Ernst Mayr objected to his division of the prokaryotes. Not all criticism of him was restricted to the scientific level. A decade of labor-intensive oligonucleotide cataloging left him with a reputation as "a crank," and Woese would go on to be dubbed as "Microbiology's Scarred Revolutionary" by a news article printed in the journal *Science*. The growing body of supporting data led the scientific community to accept the Archaea by the mid-1980s. Today, few scientists cling to the idea of a unified Prokarya.

17   **Introducing Archie:** For the molecule minds, we have chosen to use gendered names for salience and ease of recall. Archaea and bacteria do not have genders, though they reproduce both asexually and parasexually. The amoeba *Dictyostelium discoideum* engages in sexual reproduction.

20   **tec$ically known as *archaella*:** Earth's simplest mind has a fancy doer. Wirth (2012): "The corkscrew-shaped bacterial flagella rotate to propel bacterial cells forward. In eukaryotes, flagella have a completely different mode of action, namely whip-like beating—they also have a completely different structure and composition. In the case of archaea, rotation of flagella as a motility mechanism is proven only for haloarchaea."

Wilde and Mullineaux (2017): "The archaellum can power swimming in both directions, depending on the direction of rotation: cells swim forward when the archaella rotate clockwise, but backwards with counterclockwise rotation."

20   **fewer than ten thousand molecules:** Let's count. The sensor is an MCP molecule. "A simple MCP unit (6 copies) + CheA and CheW are enough to activate 1 CheY, which would need a CheF (not known in how many copies, but likely 1–10)." The CheY activates the archaellum motor. "However, these chemosensory arrays are needed for signal integration and amplifications. So if you consider them as a whole, we talk about hundreds of proteins." An archaellum consists

of "thousands of archaellin" molecules. There are "~50 copies of different motor proteins" in the motor. Special thanks to Tessa Quax at the University of Freiburg for these details.

23   **basketball game principle**: A basketball game can unfold in an infinite number of possible patterns. If it weren't this way, we would just watch reruns of games instead of bothering with new ones. But we can still make some confident predictions about the dynamics of any game: we know there will be free throws, jump shots, layups, and a winner.

Some things are easy to think about with basketball but difficult to think about with consciousness. Could you have two teams of five players dribbling, passing, fouling, taking shots, while following all the rules of basketball—yet not have a game? Can you sit in Staples Center and watch the LA Lakers play the Boston Celtics and think to yourself, maybe this isn't really a *basketball game*: sure, they're going through all the identical actions of a real game—look, they're arguing with the referees again!—but it's not *really* a game, just a perfect simulacrum of a game without any game inside. If you can imagine this, then you may believe in the philosophical legitimacy of the "zombie problem" (sometimes referred to as the "hard problem"), which holds that we can imagine a person whose brain is identical in every way to our own yet who is not capable of consciousness.

A key idea underpinning much of this book is the mathematical concept of degeneracy, an essential adjunct to the basketball game principle. Sporns (2016): "Structurally variable but functionally equivalent networks are an example of degeneracy, defined as the capacity of systems to perform similar functions despite differences in the way they are configured and connected. Degeneracy is widespread among biological systems and can be found in molecular, cellular, and large-scale networks."

The basketball game principle also supports the notion of a mind in perpetual flux rather than seeking equilibrium. Sporns again:

As we learn more about the dynamics of connectivity, mounting evidence indicates that most connectional changes are not the outcome of environmental "imprinting," the permanent transfer of useful associations or linkages into the brain's wiring pattern. Instead, the picture is one of self-organization, the complex interplay between the formation of organized topology and ongoing neural dynamics. Nervous systems do not converge onto a final stable pattern of optimal functionality; rather, their connectivity continues to be in flux throughout life. As Turing noted in his paper on morphogenesis, "Most of an organism, most of the time, is developing from one pattern into another, rather than from homogeneity into a pattern."

Finally, here's Grossberg (2019): "The past 60 years of modeling have abundantly supported the hypothesis that brains look the way that they do because they embody natural computational designs whereby individuals autonomously adapt to changing environments in real time. The revolution in understanding biological intelligence is thus, more specifically, a revolution in understanding autonomous adaptive intelligence."

26   **better adapt to their world**: Also consider Sterling and Laughlin (2015): "A cell that could propel itself more rapidly and cheaply could forage more widely, but to overcome the effects of Brownian buffeting and high viscosity it must enlarge."

Chapter Three: Bacteria Mind I

30   **Sally's flagellum** *twitches*: Some readers might wonder if we improperly defined doers, if we're now suggesting that a doer can also act without a signal from a sensor. Tec$ically, it is the *state* of the sensor that influences the state of the doer. At its simplest, think of the state of the sensor as binary: 0 or 1. A value of 0 is just as informative as a value of 1, because it tells you it is not in a state of 1. A quiet forest can mean a predator is nearby.

32   **scurries for her supper**: The actual mind of *Salmonella enterica* is very similar to Sally,

though the salmonella mind's doers possess an additional physical detail that enables the bacterium to switch between running and tumbling even more efficiently. A salmonella's flagella can rotate in two directions: clockwise and counterclockwise. They rotate counterclockwise when they receive a strong signal from the sensory molecules—which causes the flagella to twist together (because of hydrodynamics) and form what biologists call a "super-flagella" that beats as a single powerful lash, powering the bacterium's runs. But when a sensory signal drops below a threshold, its flagellum starts to rotate clockwise instead—causing it to separate from the super-flagella and orient in a random direction. When many flagella are rotating clockwise, the salmonella tumbles.

33 **exploration dilemma:** The exploration dilemma is more commonly known as the explore-exploit dilemma. It is sometimes defined as a trade-off between a quicker but lower short-term exploitation reward and a more costly but potentially greater long-term exploration reward.

Grossberg (2021): "A terrestrial animal or human needs to properly balance its exploratory and consummatory behaviors. It needs to know when to excite exploratory behaviors to discover desired rewards that are not physically present. But it also needs to know when to inhibit exploratory behaviors in order to stay in one place long enough to carry out consummatory behaviors when rewards are available, or are expected to become available reasonably soon."

Chapter Four: Bacteria Mind II

39 **one thousand times larger:**

Haloarchaeon: 5 μm long, 0.1 μm diameter
Dictyostelium amoeba: 9.7 μm diameter (as sphere)
E. coli bacterium: 1–2 μm long, 1 μm diameter
Hydra vulgaris: 10–30 mm long, 1 mm wide
C. elegans: 1 mm long, 80 μm diameter

39 **form a memory:** Memory and learning are two sides of the same coin. That coin is the mind. The mind is an adaptive dynamic system, which means that every piece of it contributes to memory and learning in some fashion. "This principle is what distinguishes biology from traditional engineering," write Sterling and Laughlin (2015):

An automobile is at its best when it rolls off the assembly line. Use simply wears it down. Moreover, a car built to meet certain environmental conditions is hard to modify when conditions change. But suppose a city car could respond to a rough road by thickening its tires, stiffening its springs, and respecifying its gear ratios—that would be a hot item. Yet for biological design, adaptive responses are the essence. Each fresh experience helps to prepare for a future need. Where skin sustains rough wear, it thickens to callus; where muscle and bone sustain mechanical tension, they strengthen. Likewise in the brain, adjustments begin the instant that sufficient evidence accumulates at the environmental interface, as demonstrated by the rapid adaptation of sensory receptors and circuits to changes in stimulus statistics. Indeed it seems that wherever one looks, current experience is continually being used to update circuits in order to improve future performance. These adjustments occur in all brains, from small to large. They occur in all systems, from sensory to motor, and at all levels.

44 **the gradient problem:** Salmonella bacteria also follow chemical gradients using habituation-based memory.

46   **the origins of Mind:** The biologist Lynn Margulis once declared, "The gap between non-life and a bacterium is much greater than the gap between a bacterium and man." She meant that once the first organism reproduced itself, natural selection took over and left a trail of hard evidence behind in the fossil record and living genomes that permit scientists to reconstruct a plausible Darwinian path connecting bacteria to humankind. But what did life look like before the first molecule of self-replicating DNA appeared? That's where the narrative of life yanks the drapes tightly shut.

Which is greater: The gap between non-mind and the mind of a bacterium? Or the gap between the mind of a bacterium and the mind of man? Based upon the dynamics of the simplest extant minds, the answer appears to be the latter.

Chapter Five: Amoeba Mind

48   **the city is Çatalhöyük:** Our main reference for Çatalhöyük is Newitz (2021).

49   **egalitarian design:** Some scholars have observed that there are significant differences in the number and quality of items that were found in each Çatalhöyük home, including differences in artwork, ornamentation, and skulls. You can decide whether this evidence discounts the idea of an egalitarian society.

54   **a dormant sensor:** Van Haastert and colleagues (1996): "Vegetative cells have very few CMF and cAMP receptors, while starved cells possess ~40,000 receptors for CMF and cAMP."

54   **known as cAMP:** cAMP stands for cyclic adenosine monophosphate. It's a derivative of ATP (an energy-carrying molecule), and is used in physiological processes in many organisms, including the regulation of sugar, glycogen, and lipids.

61   *cellular* **thinking elements:** Amoebas are not the simplest organisms to join together into a supermind. S-motility myxobacteria, cyanobacteria, and at least one group of magnetotactic bacteria do, too. Even archaea perform quorum sensing—adjusting one's expression of genes based upon the size of the population of nearby conspecifics. However, slime molds represent an evolutionary dead end. They have been around for more than a billion years but gave rise to little else but more slime mold.

The Metropolis Principle

69   **The metropolis principle:** From West (2018):

> To summarize: the bigger the city, the greater the social activity, the more opportunities there are, the higher the wages, the more diversity there is, the greater the access to good restaurants, concerts, museums, and educational facilities, and the greater the sense of buzz, excitement, and engagement. These facets of larger cities have proven to be enormously attractive and seductive to people worldwide who at the same time suppress, ignore, or discount the inevitable negative aspects and the dark side of increased crime, pollution, and disease. Human beings are pretty good at "accentuating the positive and eliminating the negative," especially when it comes to money and material well-being. In addition to the perceived individual benefits coming from increased city size, there are huge collective benefits arising from systematic economies of scale. Coupled together, this remarkable combination of increasing benefits to the individual with systematic increasing benefits for the collective as city size increases is the underlying driving force for the continued explosion of urbanization across the planet. . . . Remarkably, analyses of such data show that, as a function of population size, city infrastructure—such as the length of roads, electrical cables, water pipes, and the number of gas stations—scales in the same way whether in the United States, China, Japan, Europe, or Latin America. As in biology, these quan-

tities scale sublinearly with size, indicating a systematic economy of scale but with an exponent of about 0.85 rather than 0.75. So, for example, across the globe, fewer roads and electrical cables are needed per capita the bigger the city. Like organisms, cities are indeed approximately scaled versions of one another, despite their different histories, geographies, and cultures, at least as far as their physical infrastructure is concerned. Perhaps even more remarkably they are also scaled socioeconomic versions of one another. Socioeconomic quantities such as wages, wealth, patents, AIDS cases, crime, and educational institutions, which have no analog in biology and did not exist on the planet before humans invented cities ten thousand years ago, also scale with population size but with a superlinear (meaning bigger than one) exponent of approximately 1.15.

Sporns (2016): "Connectivity translates unitary events at the cellular scale into large-scale patterns. Once the cellular machinery for generating impulses and for transmitting them rapidly between cells had evolved, connectivity became a way by which neurons could generate diverse patterns of response and mutual statistical dependence."
Sterling and Laughlin (2015):

> Mammalian and insect brains accomplish the same core tasks and are subject to the same physical constraints. . . . Both arrange their sensors and brain regions in similar positions and use similar structures to perform similar computations. These designs operate at or above the level of the single neuron. But lower levels—molecules and intracellular networks—are subject to similar constraints and therefore follow similar principles.

72 **as the game evolves, the players evolve:** Ward and Kirschvink (2015): "Finally, while the history of life may be populated by species, it has been the evolution of ecosystems that has been the most influential factor in arriving at the modern-day assemblage of life. Coral reefs, tropical forests, deep-sea 'vent' faunas, and many more—each can be viewed as a play with differing actors but the same script over cons of time."
As professional basketball evolved to place more emphasis on three-point shooting and positionless play, mid-range shooters were replaced by long-range shooters, and slow but towering centers were replaced by faster, shorter players who could play any position.

Chapter Six: Hydra Mind

75 *a self-contained molecule mind:* Churchland and Sejnowski (2016): "Research on the properties of neurons shows that they are much more complex processing devices than previously imagined. For example, dendrites of neurons are themselves highly specialized, and some parts can probably act as independent processing units."

75 **provenance of the first neurons:** Arendt et al. (2019): "Another important apomere for nervous system evolution is the occurrence of the first synapse. This event must have been tightly linked to the evolution of the first neuron; yet, so far [it] remains entirely enigmatic. Neither is it clear what exactly the innovation was, nor in what cell types it initially took place and contributed to their individuation. From the data available, we can infer that most components of the presynapse and postsynapse predated metazoans and exist in sponges and placozoans devoid of nervous systems."

78 **linked together into *networks*:** Noro et al. (2019): "The nervous system of Hydra has a net-like structure extending throughout the body. Despite continuous tissue displacement, this whole structure remains constant in size by balancing the loss of neurons with the replacement of neurons by differentiation. This dynamic feature of the nerve net is quite different

from the nervous systems in bilaterians, in which neurons are primarily generated during embryogenesis."

Chapter Seven: Roundworm Mind

86 *The Story of Doctor Dolittle*: Traub (2020): "Hugh Lofting was a genius of children's literature. But he was also a product of the British Empire, and his work is marred by racist imagery and language."

87 **exceptionally rare in nature:** Organisms with a fixed number of cells are termed *eutelic*. Eutelic organisms are all microscopic and closely related to one another.

87 **thousands of neurons:** In this book, we generally avoid reporting on the number of neurons in different organisms. There's not a linear relationship between neuron count and intelligence, for a number of reasons: the diversity of neuron types, the diversity of network configurations, the diversity of environments, and most important, because each neuron's behavior is nonlinear. The creature with the most neurons is the killer whale: highly intelligent, but not nearly as intelligent as chimpanzees or humans. Counting neurons can be as misleading as counting genes. *C. elegans* contains about the same number of genes as humans, yet *C. elegans* has 302 neurons while we have in the ballpark of 86 billion.

87 **first is through *neural diversification*:** Sporns (2016): "Individual neurons, even those belonging to the same class, must remain different from one another to continually create dynamic variability as a substrate for adaptive change. . . . Variability is encountered in vertebrate and invertebrate nervous systems. . . . In the human brain, there is significant intersubject variability at the macroscopic scale, which poses major challenges to brain mapping."

Neuroscientist Paola Arlotta adds: "Contrary to the common assumptions that neurons use a universal profile of myelin distribution on their axons, the work indicates that different neurons choose to myelinate their axons differently. In classic neurobiology textbooks, myelin is represented on axons as a sequence of myelinated segments separated by very short nodes that lack myelin. This distribution of myelin was assumed to be always the same, on every neuron, from the beginning to the end of the axon. This new work finds this not to be the case" (Cohen, 2014).

88 **twelve distinct types of neurons:** Though this book will cite various researchers' counts of neuron types in different organisms, the notion of a neuron "type" is as fraught as the notions of "species" or "gene."

Vogt (2019):

The [hydra] cells analyzed with Drop-seq clustered into several cell types, including twelve types of neurons, four types of nematocytes, gland cells and germ cells, as well as multipotent interstitial stem cells. However, many cells were in an intermediate or transitional stage, and Juliano mentioned that certain cells can also transdifferentiate into others. Hence, Juliano is actually uncomfortable when talking about cell types. She points out that, "we should probably think about what do we mean when we say 'cell type,' how do we really count the number of types of cells that are in a tissue or an organism. I think it is actually very challenging. The lines are very blurred." Instead, she and her colleagues prefer the term "cell state."

91 **Each thinker neuron weighs:** There are ten command neurons (thinker neurons) in *C. elegans*.

Sterling and Laughlin (2015):

Like the single-celled organisms the worm retreats from noxious chemicals, but its decision is more finely judged. A single sensor, the neuron labeled ASH in the brain's wiring diagram, controls this behavior by driving a "retreat" command interneuron, AVA, which shuts down the "forward" motor neurons and activates the "backward" motor neurons. . . . Thus, a single neuron ASH serves as lawyer, jury, judge, and enforcer. It defines what constitutes evidence by selecting which receptors to express on its surface, collects the evidence, weighs it, judges if it warrants escape, and mandates the decision. The worm has several such sensory neurons, collecting other lines of evidence for other actions.

94   **centralized control of thinking:** Northcutt (2012): "Centralization of nervous systems has occurred on more than five occasions during evolution (e.g., molluscs, annelids, nematodes, arthropods and chordates)."

We can see how, mentally, the hydra raced ahead of the worm, before the worm caught up and surpassed the hydra. The hydra reached multitasking quick. Just put networks into limbs. Setting up a mind with centralized control is apparently harder, and it may help to be a worm. But once you get bilateral symmetry and centralized control you rip past radial symmetry and tentacles into an unlimited future.

Chapter Eight: Flatworm Mind

100   **neuron-to-neuron chemical signals:** Neuroscientists have a peculiar obsession with neurotransmitters, perhaps because these are among the easiest physical components of a brain to detect, trace, and measure. It's not uncommon to read about transmitters as if they were responsible for certain functions—the dopamine system handles rewards, the serotonin system handles emotion, and so forth. This can sometimes distract from the true nature of the underlying neural dynamics. The mind is the game, not the players. Most of the major neurotransmitters found in the human brain are found in most vertebrates, and many invertebrates, too. It's impossible to say what the purpose of a particular neurotransmitter is without knowing about the dynamics of the neuron, network, body, and environment that it is embedded within.

100   **honest-to-goodness brain:** The flatworm lays down the body plan that all future minds in our journey will employ: a symmetrical Bauplan with a neural cord running down the center. LeDoux (2020): "Depending on how you count, there are roughly twenty-eight bilateral phyla with distinct Bauplan features. Twenty-seven of these are invertebrate phyla, including twenty-three groups of invertebrate protostomes and four invertebrate deuterostomes. There is only one vertebrate phylum; the vertebrate Bauplan stands alone, singular among many invertebrate ones."

100   **Professor Flathead is the first mind:** A survey of the research literature on the flatworm turns up a salient bias, resulting from the fact that flatworms can regenerate amputated body parts, including a brain. There are hundreds of papers published each year about neural regeneration in the flatworm. And yet, only a tiny fraction of that number are published that study the functionality of the flatworm brain. It's not even known, for instance, what flatworms' basic visual capabilities are.

Chapter Nine: Fly Mind

107   **motivations of the winning bidder:** Beeple's auctioned artwork comes with an NFT—a non-fungible token. This is a digital certificate of ownership that makes a particular file (such as the JPEG of *Everydays*) unique, though only in the sense that it is the sole copy with an NFT associated with it.

109   **Figure 9.7 (Captain Buzz):** We identify the drosophila "value module" with its mushroom bodies, which are shaped quite a bit like human ovaries, in two "peduncles."

The Unified Mathematics of the Self

124 **professor emeritus Stephen Grossberg:** Though a large proportion of the concepts, models, and arguments in *Journey of the Mind* were directly inspired by the work of Stephen Grossberg, all of the material relating to his work is solely our own interpretation and was neither validated nor approved by him. Though we have done our best to remain true to the architecture, dynamics, and spirit of his models, Grossberg would likely find much in this book that diverges significantly from his own perspective. The best way to learn about Grossberg's work is to read his publications, many of which are listed in the bibliography, including his magnum opus, *Conscious Mind, Resonant Brain.*

127 **results of the fast activity:** Processes that encode memories can be said to work on two scales: in real time, as sensory inputs arrive, and over longer time periods. Grossberg refers to these as short-term memories (STM) and long-term memories (LTM), respectively. Grossberg pioneered the use of differential equations to model interacting STM and LTM processes in his undergraduate model of the serial position effect. Neuronal activity in response to inputs, whether external or the activities of other cells connected to them, are modeled as STM processes. LTM processes are influenced by the results of the fast activity and model the memories that lead to our recognition and recall of faces, objects, and events (Grossberg, 1968, 1969).

127 **complementary thinking principle:** Grossberg uses the term "complementary computing" to describe "how the brain is organized into complementary parallel processing streams" that embody various mechanisms for learning, attention, recognition, and prediction. (We avoid the term "computing" in this book to avoid any suggestion that the brain operates like a computer, though many scientists use the term loosely.) The interactions of these streams "generate biologically intelligent behaviors." Grossberg argues that any single cortical processing stream can individually "compute" certain properties but, in doing that well, cannot process other complementary properties. "*Pairs* of complementary cortical processing streams interact, using multiple processing stages, to generate emergent properties that overcome their complementary deficiencies to compute complete information with which to represent or control some faculty of intelligent behavior" (Grossberg, 2000, 2017a).

127 **the incessant biological "noise":** Neuron populations processing a pattern of input signals face a challenge: "If the signals are too small, they can be lost in the noise. If they are too large, they can saturate their respective populations, thereby creating a uniform pattern of excitation across populations and destroying all information about the input pattern. In short, noninteracting cell populations are caught between two unsatisfactory extremes" (Grossberg, 1973). Grossberg terms this challenge *the noise-saturation dilemma.* He demonstrated that on-center, off-surround shunting dynamics (the membrane equations of neurophysiology) solve the noise-saturation dilemma. "The ubiquitous nature of the noise-saturation dilemma in all cellular tissues clarifies why such on-center off-surround anatomies are found throughout the brain. The cooperative-competitive interactions that preserve cell sensitivity to relative input size also bind these cell activities into functional units" (Grossberg, 2000).

128 **confusion and apathy:** As a graduate student at Rockefeller, twenty-five-year-old Stephen Grossberg turned in a five-hundred-page monograph synthesizing the results of his work over the previous eight years, entitled "The Theory of Embedding Fields with Applications to Psychology and Neurophysiology." His advisers at Rockefeller arranged to have this unusual manuscript mailed out "with a cover letter to 125 of the main psychology and neuroscience labs throughout the world," including David Hubel, Steve Kuffler, Eric Kandel, and most "major neuroscientists and cognitive scientists in the world at that time." In Grossberg's telling, "no one seemed ready to understand it." His attempts to submit parts of this monograph as ten separate research papers also met with little success: all were rejected. Eventually,

when he joined the faculty at MIT, he published numerous papers recapitulating the work in his Rockefeller monograph.

128   **the mind as a computer:** "When we think and perceive, there is a *whir of information-processing,* but there is also a subjective aspect," the philosopher David Chalmers asserted, giving voice to the intuitive perspective of countless scholars, scientists, and citizens who readily acknowledge that consciousness must be linked to activity in the brain, but who nevertheless remain hesitant to accept that consciousness is *merely* the product of activity in the brain—at least, the same kind of activity that controls our breathing and allows us to distinguish black from white. Grossberg (2017a):

> First, is it fair to ask what kind of "event" occurs in the brain during a conscious experience that is anything more than just a "whir of information-processing"? What happens when conscious mental states "light up" and directly appear to the subject? . . . Over and above "just" information processing, our brains sometimes go into a context-sensitive resonant state that can involve multiple brain regions. . . . But what is "information"? The scientific concept of "information" in the mathematical sense of Information Theory . . . requires that a set of states exist whose "information" can be computed, and that fixed probabilities exist for transitions between these states. In contrast, the brain is a self-organizing system that continually creates new states through development and learning, and whose probability structure is continually changing along with them.

128   **the mind as a statistical machine:** The "Bayesian machine" (or "Bayesian brain") approach to the study of the mind holds that the brain functions much like a probability calculator, coding sensory information as probability distributions and updating these distributions in a manner that is close to mathematically "optimal."

Grossberg (2013): "However, the Bayes rule is so general that it can accommodate any system in Nature. This generally makes Bayes a very useful statistical method. However, in order for Bayes concepts to be part of a physical theory, additional computational principles and mechanisms are needed to augment the Bayes rule to distinguish a brain from, say, a hydrogen atom or a hurricane. Because of the generality of the Bayes rule, it does not, in itself, provide heuristics for discovering what these distinguishing physical principles might be."

Grossberg also asserts, "The mathematical foundations of Deep Learning are contradicted by hundreds of neurobiological experiments. In addition, the algorithm has explained essentially no psychological or neurobiological data. Biological neural models have been developed over the past 40 years that have explained large numbers of psychological and neurobiological experiments, and that have made many successful experimental predictions. These models have also been used in many large-scale applications in engineering and tec$ology, especially those that require autonomous adaptive intelligence."

130   **over five hundred articles:** Stephen Grossberg's publications span a remarkable breadth and contain many influential papers, but three stand apart in their scope and importance.

His 2017 article "Towards Solving the Hard Problem of Consciousness: The Varieties of Brain Resonances and the Conscious Experiences That They Support" synthesizes his work across multiple sensory modalities and explains how "we experience qualia or phenomenal experiences, such as seeing, hearing, and feeling, and knowing what they are."

His 2018 article "Desirability, Availability, Credit Assignment, Category Learning, and Attention: Cognitive-Emotional and Working Memory Dynamics of Orbitofrontal, Ventrolateral, and Dorsolateral Prefrontal Cortices" clarifies how the highest level of brain organization works, while also incorporating all the mechanisms for supporting conscious awareness that were described in the earlier 2017 article.

His 2019 article presents his SOVEREIGN model, which describes how minds are capable of "visually searching and navigating an unfamiliar environment" while also learning to "recognize, plan, and efficiently navigate toward and acquire valued goal objects" (Grossberg, 2019a).

130   **Grossberg's unified theory of mind:** Grossberg (2019b):

> There are several fundamental mathematical reasons why it is possible for human scientists to discover a unified mind-brain theory that links brain mechanisms and psychological functions, and to demonstrate how similar organizational principles and mechanisms, suitably specialized, can support conscious qualia across modalities. One reason for such inter-modality unity is that a small number of equations suffices to model all modalities. These include equations for short-term memory, or STM; medium-term memory, or MTM; and long-term memory, or LTM, that I published in *The Proceedings of the National Academy of Sciences* in 1968.

130   **describe them exactly and quantitatively:** Grossberg's equations include ones characterizing short-term-memory and long-term-memory dynamics, and instar and outstar learning. Grossberg (2019b) notes that "these equations are used to define a somewhat larger number of modules, or microcircuits, that are also used in multiple modalities where they can carry out different functions within each modality." These modules include "shunting on-center off-surround networks, gated dipole opponent processing networks, associative learning networks, spectral adaptively timed learning networks," and others. He notes that "each of these types of modules exhibits a rich, but not universal, set of useful computational properties."

The equations and modules in his body of work are further "specialized and assembled into modal architectures, where 'modal' stands for different modalities of biological intelligence, including architectures for vision, audition, cognition, cognitive–emotional interactions, and sensory-motor control" (Grossberg, 2013). He adds (2021): "Modal architectures are *general-purpose*, in that they can process any kind of inputs to that modality, whether from the external world or from other modal architectures. They are also *self-organizing*, in that they can autonomously develop and learn in response to these inputs."

131   **why consciousness exists at all:** You might be wondering why you've never heard of Stephen Grossberg, if he's managed to achieve so much, including a comprehensive mathematical model of consciousness. Cognitive scientist Margaret Boden, who helped develop "the world's first academic programme in cognitive science," explains (2008):

> The identification of discoveries (both as new and as valuable) can depend heavily on the discoverer's rhetorical skills, or lack of them. . . Arguably, one case in point is the early work of the highly creative cognitive scientist Stephen Grossberg. He was perhaps the first to formulate three ideas that are influential today under the names of other people: Hopfield nets, the Marr–Albus model of the cerebellum, and Kohonen self-organizing maps. Grossberg also pioneered many more notions— including back propagation—that are commonly attributed to others, if not actually named after them. . . .
>
> The key sense in which his work was "too far ahead of its time" was the unfamiliarity of the mathematics. He was talking about brain and behaviour as a complex dynamical system (as he put it: "nonlinear, nonlocal, and nonstationary"), a theoretical approach that didn't become popular in cognitive science until the late 1980s. But it's the second point that's of special interest here: the rhetorical style. His early work was largely unintelligible even to the few psychologists who took the trouble to read it. He combined intellectually demanding (and unfamiliar) mathematics with a host of interdisciplinary details, most of which would be unfamiliar to any individual reader. They were there because he was trying to show the unsuspected theoretical unity behind hugely diverse data. His writing was unusually voluminous

too: 500 pages for his first-year graduate report (1964), and many long and richly cross-referenced journal articles. Faced with this challenge from a youngster they'd never heard of, most people gave up before reaching the end, if they could summon up the courage to start reading at all.

131 **integration of disparate insights:** How does one arrive at theories that link behavioral properties to the brain mechanisms that generate them? Grossberg (2017a) notes that such a linkage between brain and behavior is "crucial in any mature theory of consciousness, since a theory of consciousness that cannot explain behavioral data has failed to deal with the contents of consciousness, and a theory of consciousness that cannot link behaviors to the brain mechanisms from which they emerge must remain, at best, a metaphor."

Grossberg refers to the guiding methodology behind his research over the past sixty years as the "method of minimal anatomies." A key idea in this method is that "one cannot derive a theory of an entire brain in one step." Rather, it demands incremental refinement,

a kind of design evolution whereby each model embodies a certain set of design principles and mechanisms that the evolutionary process has discovered whereby to cope with a given set of environmental challenges. Then, the model is refined, or unlumped, to embody an even larger set of design principles and mechanisms and thereby expands its explanatory and predictive power. This process of evolutionary unlumping continues unabated, leading to current models that can individually explain psychological, anatomical, neurophysiological, biophysical, and biochemical data about a given faculty of biological intelligence.

A "minimal" model is one for which if any of the model's mechanisms is removed, then the surviving model can no longer explain a key set of previously explained data. Once a connection is made top-down from behavior to brain by such a minimal model, mathematical and computational analysis discloses what data the minimal model and its variations can and cannot explain. Such an analysis focuses attention upon design principles that the current model does not yet embody. These new design principles and their mechanistic realizations are then consistently incorporated into the model by unlumping it to generate a more realistic model. If the model cannot be refined in this way, then that is strong evidence that the current model contains a serious error and must be discarded. (Grossberg, 2018)

Chapter Ten: Fish Mind

136 **the Visual Scene module:** We based the Visual Scene module on Grossberg's early work in his FACADE module of three-dimensional vision. This model explains how the two-dimensional images impinging on our retinas are converted into three-dimensional perceptions of objects. Boundary and surface representations interact to create a preconscious percept that can be used to generate "conscious modal and amodal representations in object recognition, spatial attention, and reaching behaviors" (Grossberg, 1997). The FACADE model was later developed into the 3D LAMINART model, which clarifies how "laminar cortical mechanisms interact to create 3D boundary and surface representations" (Cao & Grossberg, 2019).

144 **shares its first-pass collection of edges:** The boundary system and surface system have complementary modes of action. Boundaries are always completed *inward*, while surfaces are always filled in *outward*. Boundaries are *not sensitive to the direction of contrast* of the two sides of an edge, while surfaces are *sensitive to the direction of contrast* (always filling in on the "lighter" side of an edge, under the presumption that the brighter side is more likely to be the foreground). Finally, boundary completion is always *oriented*. The boundary system continues or completes an edge in the same direction as the existing edge. Surface filling-in is always *unoriented*. The surface system fills in color in every direction, regardless of where it starts.

145   **"neon color spreading illusion"**: Grossberg first explained how the neon color spreading illusion works in 1984, "Outline of a Theory of Brightness, Color, and Form Perception."

146   **addresses the uncertainty dilemma**: Our minds often face an ill-posed problem. The information relayed by our sensory organs is usually ambiguous and insufficient to identify objects and their locations. And yet, we perceive a vivid and stable world without any flicker of confusion. Grossberg suggests that this is the result of a disambiguation of input information at multiple levels of neural processing. He terms this process of disambiguation the "hierarchical resolution of uncertainty." The uncertainty here refers to the notion that determining one set of properties about a stimulus "can suppress information about a different set of properties at that stage." In describing how this principle operates in the FACADE model, he explains how multiple levels of processing resolve uncertainties in the creation, grouping, and sharpening of boundaries of objects, which then helps resolve uncertainties in the illumination of the surfaces. Only once all these interlinked uncertainties are resolved can the brain then construct a stable representation of an object, which can then be consciously perceived and acted upon (Grossberg, 2000).

One component of the hierarchical resolution of uncertainty is feedback signals from a resonating visual What module that influence the preconscious activity of the surface and boundary systems in the Visual Scene module.

147   **Fish also see other illusory contours**: One example of how researchers test out visual illusions on fish is provided by Yunmin Wu of the Max Planck Institute of Neurobiology, who tested the waterfall illusion on zebra fish larvae. If you look at a moving stream (such as a waterfall), then look at a stationary object (like a wall), you will see the wall moving in the direction opposite to the waterfall. Fish can't tell a researcher what they're seeing, so how did Wu determine whether they were experiencing the waterfall illusion?

"We built a set-up you can think of as a cinema. The fish sits in the middle watching a movie of moving stripe patterns, called gratings. At the same time, we record the eye movement of the fish, which tells us exactly if and in which direction it sees motion. Using this as a readout, we were able to tell that the fish experiences illusory motion in the opposite direction when the moving gratings stopped" (Wu et al., 2020).

Chapter Eleven: Frog Mind

151   **the How module**: The How module is based on Grossberg's VITE, VAM, and DIRECT (Direction to Rotation Effector Control Transform) models, which clarify how our brains control the movements of our bodies. Neural mechanisms mediating our movements must solve the challenge of "controlling a changing body" (Bullock & Grossberg, 1989).

Grossberg (2017): "A refinement that sheds the most light on auditory–visual homologs of reaching and speaking circuits is called the DIRECT model, which also learns through a circular reaction. Learning in the DIRECT model clarifies how a crucial motor-equivalence property of arm movement control is achieved; namely, during movement planning, either arm, or even the nose, could be moved to the goal object."

157   **neuroscientists call** *mapping*: Grossberg's DIRECT model learns to map visual motion directions into joint rotations during reaching movements. By doing so, it can naturally learn to use tools to target objects. Grossberg (2017): "Remarkably, after the DIRECT model learns these representations and transformations, its motor-equivalence properties enable it to manipulate a tool in space. Without measuring tool length or angle with respect to the hand, the model can move the tool's endpoint to touch the target's position correctly under visual guidance on its first try, without additional learning. In other words, the spatial affordance for tool use, a critical foundation of human societies, follows directly from the brain's ability to learn a circular reaction for motor-equivalent reaching."

157   *motor babbling*: The Swiss psychologist Jean Piaget called motor babbling in infants "circular reactions." Ironically, motor babbling is a much more ancient phenomenon than verbal

babbling and takes place earlier in an infant's development. Thus, motor babbling should probably be called "babbling" and babbling should be called "verbal babbling."

159 **docsn't bccomc conscious:** Bccause thc activity of thc How modulc is cntirely unconscious, there is no commonplace term to describe what the How module does. What word encompasses the activities of "kicking a ball," "swatting a mosquito," and "kissing your partner on the lips"? All three activities are handled by the How module dynamic, and yet, we have difficulty putting this dynamic into words. On the other hand, can you think of a word that encompasses the actions "watching a movie," "scrutinizing a cell under a microscope," and "enjoying fireworks"? There's a cavalcade of common words that describe *all* these activities: observing, viewing, looking, regarding, surveying, inspecting, beholding . . .

Chapter Twelve: Tortoise Mind

163 **the What module:** The What module is based on Grossberg's Adaptive Resonance Theory (ART) and LAMINART models. ART is the most important model within Grossberg's unified theory of mind. ART is a "cognitive and neural theory of how the brain autonomously learns to categorize, recognize, and predict objects and events in a changing world" that is "filled with unexpected events" (Grossberg, 2013). Central to ART is the idea that top-down expectations are matched with bottom-up sensory inputs and that this match can result in "synchronous resonances" between multiple cortical and subcortical areas. These resonances are key to "stable, autonomous learning in real time of huge amounts of data from a changing environment that can be filled with unexpected events." The LAMINART model describes how ART theories and computations are embodied in the layers of the neocortex to enable sophisticated visual perception, recognition, and learning (Raizada & Grossberg, 2003).

166 **one intriguing experiment:** Biologist Anna Wilkinson at the University of Lincoln in the UK is one of the world's leading authorities on how red-footed tortoises think. She is the author of the excellent paper "Cold-Blooded Cognition: How to Get a Turtle out of Its Shell." She's done much to demonstrate the unappreciated mental abilities of the turtle.

173 **rather than an average:** A natural question that might be asked of the What module's memory: *How else is a prototype formed except by frequency of exposure, which is necessarily statistical?* This is where Grossberg's model differs sharply from statistical models of object recognition. The stored prototypes are not averages of exemplars. Rather, they are "actively selected critical feature patterns." The formation of these feature patterns are controlled by "a gain control process, called vigilance control, which can be influenced by environmental feedback or internal volition." This allows for the flexible learning of abstract prototypes, or even individual exemplars, as the situation requires (Grossberg, 2013).

 In addition, there are multiple prototypes stored for each object category, corresponding to different views of the object, in order to solve the many-to-one mapping problem.

173 **will not trigger resonance:** The blueness will, however, resonate within the Where module, providing the conscious experience of *seeing* blue.

174 **nonconscious How module:** Another highly influential form of complementary thinking in module minds is the union of the How (VAM) and What (ART) modules. Gaudiano & Grossberg (1991): "VAM models and Adaptive Resonance Theory (ART) models exhibit complcmcntary matching, lcarning, and pcrformancc propcrtics that togcthcr provide a foun dation for designing a total sensory-cognitive and cognitive-motor autonomous system."

 Grossberg (2013): "This way of thinking provides a mechanistic reason why declarative memories (or 'learning that'), which are the sort of memories learned by ART, may be conscious, whereas procedural memories (or 'learning how'), which are the sort of memories that control spatial orienting and action, are not conscious."

 The union of inhibitory and excitatory neurons is a kind of neuron-level complementary thinking that parallels the conscious match-based learning of ART and the unconscious mismatch-based learning of VAM.

175 **memories of similar objects:** The brain must learn quickly from each new perceptual experience and do so without forgetting any knowledge acquired in the past. Grossberg calls this challenge the *stability-plasticity dilemma* (Grossberg, 1982). Grossberg points out that this problem particularly bedevils thinking mechanisms that do not incorporate active feedback. The powerful deep-learning algorithms behind many impressive advances in automated vision, text, and speech recognition usually fail to address this problem (Grossberg, 2020).

175 **"extra degree of freedom":** Grossberg (2017):

> If occluding and partially occluded objects can be recognized using boundary and surface representations that are completed in V2, then what functions are enabled by conscious 3D percepts of the unoccluded parts of opaque surfaces that are proposed to occur in V4? A synthesis of the FACADE [Visual Scene module] and ARTSCAN models [Where and What modules] proposes that a surface-shroud resonance with V4 [conscious *seeing*] provides an extra degree of freedom, namely conscious visibility, with which to distinguish directly reachable surfaces from non-reachable ones. The need for such an extra degree of freedom seems to have arisen because of the urgent need during evolution for more powerful object recognition capabilities, notably the ability to complete the boundaries and surfaces of partially occluded objects behind their occluders so that they can be recognized. . . . If the completed parts of these partially occluded objects could also be seen, then great confusion could occur in the planning of reaching behaviors, since all occluders would look transparent, and it would seem natural to reach directly through occluding objects to the occluded objects behind them. In brief, then, there is a design tension during evolution between the requirements of recognition and reaching. Conscious visibility enables the unoccluded parts of many surfaces to appear opaque, and thus good targets for reaching, without eliminating the ability of the visual cortex to correctly represent surfaces that are, in fact, transparent.

Chapter Thirteen: Rat Mind

180 **"Where's Waldo problem":** The Where module (and its interactions with the What module) is based on Grossberg's ARTSCAN and ARSTSCAN Search models. These models predict how the brain combines processing in multiple pathways to enable the recognition, learning, and localization of physical objects within three-dimensional scenes. One processing stream determines the "positional representations of the world and controls actions to acquire objects in it, but does not represent detailed properties of the objects themselves." This stream is complemented by another processing stream that allows for "object learning, recognition, and prediction," including learning view-invariant representations of objects (Chang et al., 2014).

In Britain, the *Where's Waldo?* puzzle books are titled *Where's Wally?* The books and character were created by English illustrator Martin Handford.

181 **neural architecture of the Where module:** The What module employs a fixed and rectangular conception of space, which is why our intuition tells us that we are living inside a three-dimensional grid. This is sometimes known as a Newtonian conception of the universe, after Isaac Newton's original model of a grid-like universe, which, we can now appreciate, was a direct consequence of the spatial mapping dynamics in the human Where module. Our minds could just as easily have been designed to conceive of the world as curvilinear, like the conception of the universe described by Einstein's theory of relativity, but such a conception would not have been adaptive at the scale of module minds. The grid-like dynamics of the Where module help us appreciate why for countless generations human beings assumed that the Earth was flat. It wasn't simply because the Earth *looks* flat—indeed,

it doesn't look flat at all if you're on a boat or climbing a tall hill. Rather, our Where module naturally represents everything around us as flat, whether it is or not.

182 **This synchronizes, amplifies, and prolongs:** There is evidence of ART match dynamics (on-center off-surround feedback loops) synchronizing and amplifying activity in rats.

Grossberg (2021): "Simona Temereanca and Daniel Simons (2001) produced evidence for an on-center off-surround organization that can amplify and synchronize responses in this system, and an additional 2008 article with Emory Brown further demonstrated 'millisecond by millisecond changes in thalamic near-synchronous firing . . . that may ensure transmission of preferred sensory information in local thalamocortical circuits during whisking and active touch.'"

189 **At the network level:** A good example of complementary thinking at the level of the module mind is the collaborative songwriting of Jo$ Lennon and Paul McCartney. Paul tended toward the sunny and optimistic, Jo$ tended toward the darker and pessimistic, but working together they produced arguably the greatest songs in the history of popular music. (It is inarguable that their collaborative songs have sold more and attained more critical acclaim than any they wrote individually.) Perhaps the height of this complementary module-mind thinking can be found in "A Day in the Life," which fuses together two distinct songs that each man wrote separately into a single compelling record that many consider the Beatles' finest.

Chapter Fourteen: Bird Mind

192 **c. c. cummings:** If we had included a third epigraph for this chapter, it would have been drawn from this remarkably prescient conception of the When module, posited in 1951 by eminent psychologist Karl Lashley:

> Certainly language presents in a most striking form the integrative functions that are characteristic of the cerebral cortex and that reach their highest development in human thought processes. Temporal integration is not found exclusively in language; the coordination of leg movements in insects, the song of birds, the control of trotting and pacing in a gaited horse, the rat running the maze, the architect designing a house, and the carpenter sawing a board present a problem of sequences of action which cannot be explained in terms of successions of external stimuli. In spite of the ubiquity of the problem, there have been almost no attempts to develop physiological theories to meet it.

193 **birdbrains follow after ratbrains:** Saini (2021): "European and South American researchers studying two-dozen species found that, while birds' brains may be relatively tiny, the cells within them can be more densely packed than those of rodents and some primates. Parrots and songbirds have some of the most surprising brains of all."

193 **the When module:** The When module is based on Grossberg's STORE model, which was further expanded in his LISTPARSE model. One key problem in the perception of auditory inputs is the conversion of a transient, temporal sequence into a unified mental percept. To accomplish this, neural mechanisms must account for the order in which inputs arrive and store them just long enough to allow for the recognition of the entire sequence of inputs. STORE models how temporary storage of sequential patterns, known as "working memory," can encode both the specific input (i.e., an event) that occurred as well as the order of multiple events. Grossberg asserts that such a working memory supports our capacity to "think, plan, execute and evaluate sequences of events" (Bradski et al., 1994). These events are not just limited to the auditory domain. The storage, recognition, and rehearsal of sequences support a diverse array of mental activities. "Whether we learn to understand and speak

a language, solve a mathematics problem, cook an elaborate meal, or merely dial a phone number, multiple events in a specific temporal order must somehow be stored temporarily in working memory" (Grossberg & Pearson, 2008).

196 **most studied bird mind:** There are several excellent models of song learning and production in zebra finches that were developed independently of Grossberg's previously published work, though these models are largely congruent with Grossberg's LISTPARSE family of models describing sequential learning and production.

Swaddle (2019): "There are two forebrain neural circuits associated with male singing, one related to song learning and the other to song production. . . . Males appear to learn their songs in a sequential unit structure, with introductory elements, song phrases, and the terminal distance call apparently learned in 'chunks' from the tutor. Hence, there may be a functional link between how songs are learned and how they are expressed as adults."

Bertram and colleagues (2014):

We focus on the central control of birdsong and review the recent discovery that zebra finch song is under dual premotor control. Distinct forebrain pathways for structured (theme) and unstructured (variation) singing not only raise new questions about mechanisms of sensory-motor integration, but also provide a fascinating new research opportunity. A cortical locus for a motor memory of the learned song is now firmly established, meaning that anatomical, physiological, and computational approaches are poised to reveal the neural mechanisms used by the brain to compose the songs of birds. . . . Similar to human infants, juvenile male zebra finches learn to imitate a paternal vocal pattern in a two phase process that proceeds with little or no requirement for external reinforcement. The initial "sensory" phase involves the formation of an auditory memory of the paternal vocal pattern. Notably, the memory contains only the product of the paternal vocal behavior—the acoustic structure and sequence of vocal sounds. As with human speech there is minimal transmission of information about how to produce the sounds. . . . Interestingly, the variable structure of subsong appears to be a purposeful exploration of the dynamic range of the vocal organ, and perhaps provides a period of associative learning where relationships between different vocal gestures and the sounds those gestures produce are discovered.

198 **"cocktail party effect":** Grossberg (2017):

Auditory resonances occur in cortical regions that support conscious percepts of auditory sounds as parts of auditory streams. The process that forms auditory streams is called auditory scene analysis. Auditory scene analysis enables the brain to separate multiple sound sources, such as voices or instruments, into trackable auditory streams, even when these sources may contain harmonics that overlap and are degraded by environmental noise. Auditory scene analysis hereby helps to solve the so-called cocktail party problem, which occurs when listening to a friend at a noisy cocktail party or other social occasion. By separating overlapping frequencies that are due to different acoustic sources into distinct streams, higher auditory processes, such as those involved in speech and language, can learn to recognize the meaning of these streams.

Chapter Fifteen: Monkey Mind

207 **the Why module:** The Why module is based on the CogEM model first described by Grossberg in 1971 and elaborated in numerous subsequent articles and models. This model embodies the idea that complementary cognitive and emotional dynamics support conscious feelings, conscious knowledge of these feelings, and the ability to respond appropriately to these feelings. CogEM is a model of *valuation* in module minds. CogEM also describes how object

categories computed in sensory cortical regions can interact with other cortical regions to "focus motivated attention on valued object and object-value representations." This in turn can help enable "conditioned reinforcer learning" (such as learning that pressing a button gets you a reward) and "incentive motivational learning" (such as learning about places to find food when feeling hungry) (Grossberg, 1971, 2018).

210 **an outsize role in the Who system:** We could also think of the primate Who system as the early stage of a supermind-level thinking element devoted to supermind thinking, a thinking element that is especially developed in *Homo sapiens* brains. The main developmental force shaping the Who system is supermind mental dynamics. (The language stack is another supermind-enabling module.)

210 **source of autism:** Grossberg & Seidman (2006).

The CogEM model [Why module] proposes that emotional centers like the amygdala can experience such a reduced overall output, which can thereby cause the types of problems with motivated attention and action that has just been summarized. This condition is referred to as a form of "emotional depression." Although emotionally-charged outputs from amygdala may be reduced, such reduction is not herein identified with the full-blown clinical disease of depression. How does such emotional depression arise in the model? How this occurs can be seen by noting that emotional centers are often organized into opponent affective processes, such as fear and relief. . . . The response amplitude and sensitivity to external and internal inputs of these opponent-processing emotional circuits are calibrated by an arousal level and chemical transmitters that slowly inactivate, or habituate, in an activity-dependent way. These opponent processes exhibit an Inverted-U whereby their outputs may become depressed if the arousal level is chosen too large or too small.

213 **"orienting reflex":** Kastner (2016): "Selective attention refers to the ability to voluntarily prioritize information that is chosen from competing inputs and is therefore different from the reflexive orienting to salient events in the environment. While orienting responses are mediated by the midbrain and are present in many species outside the mammalian line including reptiles, amphibians, and birds, the neural machinery needed for selective processing recruits a large-scale thalamocortical network."

221 **taking her own life:** The writing of *The Bell Jar* was supported by the Eugene F. Saxton fellowship, a grant sponsored by the publisher Harper & Row. Harper decided not to publish Plath's book, calling it "disappointing, juvenile and overwrought." One of Plath's working titles for her roman à clef was *Diary of a Suicide*.

Chapter Sixteen: Chimpanzee Mind

225 **transmogrifies into consciousness:** This chapter is based on Grossberg's treatment of "CLEARS" dynamics: Consciousness, Learning, Expectation, Attention, Resonance, and Synchrony (Grossberg, 2007) and especially his integrated model of conscious experience (Grossberg, 2017). Stephen Grossberg asserts that what we perceive as conscious events are embodied within resonances in cortical and subcortical regions of the brain.

226 **"Song of Myself":** Whitman may have been the first human to poetically describe supermind consciousness. Several passages from "Song of Myself" demonstrate Whitman's pioneering use of the first-person perspective to represent not his own personal Self but all Selves. For instance, in the opening lines:

I celebrate myself, and sing myself,
And what I assume you shall assume,
For every atom belonging to me as good belongs to you.

Holloway (1926): "The key to the understanding of this poem, as of all of Leaves of Grass, is the concept of self (typified by Walt Whitman) as both individual and universal."

Mason (1973): "The view of Whitman as a passive absorber easily explains the poet's use of the catalogues, but the poet of 'Song of Myself' is not entirely or even essentially passive. He adopts a relaxed, even passive role at the beginning of the poem, but in the course of the poem he consciously expands the self into inclusive consciousness."

226   **more than thirty models:** Here is an incomplete list of the models that support Grossberg's unified theory of mind: ART, ARTMAP (perceptual and cognitive learning); SMART (spiking dynamics); BCS/FCS, FACADE, 3D LAMINART, ARTSCAN, ARTSCENE, 3D FORMOTION (visual perception); SPINET, ARTSTREAM, NormNet, PHONET, ARTPHONE, ARTWORD, cARTWORD (auditory perception); VITE, FLETE, VITEWRITE, DIRECT, VAM, TELOS, lisTELOS (motor control); CogEM, MOTIVATOR (cognitive-emotional dynamics and reinforcement learning); pART (decision-making and executive function); START (adaptive timing); SOVEREIGN, STARS, ViSTARS, GridPlaceMap (visual and spatial navigation); STORE, LISTPARSE (working memory); SpaN (numerical estimation).

228   **governed by a consciousness cartel:** Grossberg has never used the term "consciousness cartel" or employed a similar metaphor to describe his model of consciousness. In correspondence with us (2020), Grossberg wrote: "We typically are synchronizing multiple resonances at the same time, as when we are looking at someone who is talking to us, and having feelings about what is being said based upon what we know about them and the topic of conversation."

228   **distinct type of resonance:** Grossberg describes resonance as a "dynamical state during which neuronal firings across a brain network are amplified and synchronized when they interact via reciprocal excitatory feedback signals during a matching process that occurs between bottom-up and top-down pathways." Such resonances can trigger fast learning, allowing for both learning and recognition to occur in parallel. In addition, such resonant dynamics ensure that sensory features irrelevant to the task at hand do not corrupt previously learned memories. Events and episodes that are unexpected and unfamiliar (which can prove to be of considerable significance) are remembered using a complementary reset mechanism.

228   **categories of conscious experience:** The six types of qualia-generating resonant states described in Grossberg (2017) and the models that account for each state:

Surface-shroud resonance (see a visual object or scene): ARTSCAN
Feature-category resonance (recognize a visual object or scene): ART
Stream-shroud resonance (hear an auditory object or stream): ARTSTREAM
Spectral-pitch-and-timbre resonance (recognize an auditory object or stream): SPINET, ARTSTREAM
Item-list resonance (recognize speech and language): PHONET, ARTPHONE, ARTWORD
Cognitive-emotional resonance (feel an emotion and know its source): CoGEM, MOTIVATOR

228   *I hear a sound:* Grossberg (2017):

A surface-shroud resonance has been proposed to support percepts of conscious visual qualia. . . . An analogous process of stream-shroud resonance is proposed to support percepts of conscious auditory qualia. . . . A surface-shroud resonance enables sustained spatial attention to focus on a visual object and generate its conscious qualia, and that there is a close link between problems with sustained visual spatial attention and unilateral visual neglect. A stream-shroud resonance is proposed to play a similar role in sustaining auditory spatial attention and conscious auditory quality, and problems with sustained auditory attention, say due to a parietal lesion, can cause unilateral auditory neglect.

233   **the entire "media cartel":** It's no coincidence that asking the Five W's and One H (Who, What, Where, When, Why, and How) is a fundamental tenet of journalism drilled into every journalist's head. These questions embody the same kinds of information that minds seek out and integrate through resonance into a consensual understanding of what is worth paying attention to. It also clarifies why the media is so focused on negative stories, like disasters, tragedies, and crimes—just like an individual mind, the supermind needs to nimbly focus on threats and problems rather than dwell on things that are going right. If it bleeds, it leads, in both media and minds. The crucial importance of the media for generating supermind consciousness is reflected in the fact that access to the supermind's attention, in the form of commercials, is sold by the second.

237   **"recurrent" dynamics:** Aso et al. (2014): "Thus, MBONs may modify the activity of the DANs that modulate their own activity and plasticity, resulting in a recurrent loop. This could provide positive or negative feedback to a specific compartment. Positive feedback might enhance learning to particularly salient stimuli, whereas negative feedback might suppress dopamine release once the correct response has been learned."

Hulse et al. (2020):

> The context-dependent initiation and control of many such behaviors is thought to depend on a highly conserved insect brain region called the central complex (CX).
>
> In Drosophila, this highly recurrent central brain region, which is composed of ~3000 identified neurons, enables flies to modulate their locomotor activity by time of day, maintain an arbitrary heading when flying and walking, form short- and long-term visual memories that aid in spatial navigation, use internal models of their body size when performing motor tasks, track sleep need and induce sleep, and consolidate memories during sleep.

In other words, these recurrent dynamics serve as working memory.

240   **stage of thinking:** Each stage in the journey of Mind is defined by a new layer of connectivity linking together the top-level thinking elements of the previous stage. Each layer of connectivity arose as the solution to a new version of the coordination problem.

The first stage of thinking is embodied within molecule minds, from archaea to amoebas. These minds contain molecular thinking elements that do not possess any adaptive properties on their own. An isolated molecule is not capable of learning, though a network of molecules can adapt to its environment. Communication between molecular thinking elements is often indirect and probabilistic. This renders molecular thinking slow, unstable, and inconsistent. There are no perceptions or representations in a molecule mind. When molecule minds begin to establish connectivity with one another through the exchange of molecular signals it opens the door for neuron minds.

The second stage is neuron minds, from hydras to insects. These minds contain cellular thinking elements, most notably the neuron. Each individual neuron is itself a self-contained molecule mind containing its own internal network of molecular thinking elements. Thus, each neuron functions as an independent adaptive unit. Networks of neurons begin to perform collective mental functions, including sensing, thinking, and doing, though these networks do not yet form stable and unambiguous representations.

When segregated networks of neurons begin to communicate with one another through representations it opens the door for the third stage of thinking, module minds, running from fish to nonhuman primates. The top-level thinking element is the module, a specialized network of neurons. Modules communicate with one another through stable and disambiguated representations. Module minds solve the attention dilemma by generating resonant states, which lead to conscious experience, though module minds are not able to generate a personal Self.

After module minds began to communicate with one another through complex *external* representations, it opened the door for the emergence of language and superminds.

The Darkness

247   **into a zombie:** Among professional philosophers who ponder the mind there is a prominent thought experiment known as the Zombie Problem (or Hard Problem), though a more apt moniker might be the Zombie Fantasy. This metaphysical reverie invites us to imagine that there are people who look and behave like you with bodies made of the same bones and blood and brains as you, but within these "zombies" the sacred light of consciousness is somehow absent. They can talk, they can fight, they can weep, and they will even profess, with the same urgent fervor as yourself, that they are self-aware, capable of love and pain, and equipped with free will. And yet, according to this mental flight of fantasy, these pseudo-humans have no inner life—there is no ghost in the machine, no "soul" experiencing the fragrant rainbow of reality, at least, no soul more real than the one inside Apple's Siri.

The mere fact that advocates of the Zombie Fantasy can *conceive* of the existence of human doppelgängers who possess a fully functioning brain but no sentience is held up as evidence that zombies *could* exist, and therefore must be accounted for in any explanation of consciousness. But there is no need to indulge in such extravagant fantasy to explore the implications of this conviction. Eightscore years ago, the presumption that certain human beings were mindless simulacra was not merely commonplace, it was the law of the land—scrivened into the Constitution of the United States of America. For almost a full century, in both intention and action, the US government declared that dark-skinned human beings who lived south of the Mason-Dixon line were, and should be treated as, zombies.

Christopher Hitchens offers a succinct reason to dismiss any claim that the Zombie Fantasy or "Hard Problem" purportedly holds on models of consciousness, a reason dubbed "Hitchens's razor": "What can be asserted without evidence can be dismissed without evidence."

Chapter Seventeen: Human Mind

255   **the fellowship of humankind:** In the 1960s, another nonscientist conducted the same experiment as Akbar. A girl who came to be known as Genie was raised by a psychotic father who kept her tied up in a sleeping bag in a metal-covered crib for 12 years, enforcing strict silence and forbidding anyone from interacting with her. She was not exposed to any human vocalizations other than the canine snarls her father used to intimidate her. Like the hapless subjects of Akbar's research, Genie was never able to develop fluent language and almost never engaged in spontaneous acts of speech. Even more tragic, Genie was unable to live independently and has spent her entire adult life in a private facility for mentally underdeveloped adults. Now in her sixties, Genie reportedly communicates through gestures.

Another child raised without language was Dina Sanichar, who was abandoned in India's Uttar Pradesh jungle in the 1860s and discovered in a cave when he was about six years old. He was raised in Sikandra Mission Orphanage, but never acquired language or the ability to communicate through gestures, though he eventually learned to dress himself and smoke cigarettes. Popularized in the Indian and Western press as the "Wolf Boy," Sanichar served as the inspiration for the character Mowgli in Rudyard Kipling's *Jungle Book*.

256   **power of language:** The modular mechanisms described in the language chapter (including the language stack) are *inspired* by Grossberg's cARTWORD models (which incorporate ART learning mechanisms and STORE working memory mechanisms), though our language models may *not* accurately reflect Grossberg's own opinions, theories, or models. cARTWORD describes how speech and language are "rapidly learned, stably remembered, flexibly performed, and consciously heard" (Grossberg, 2021). Speech perception involves the integration of transient, "temporally distributed phonemic information into the coherent representations of syllables and words." cARTWORD models how such temporal informa-

tion is converted into spatial patterns using working memory dynamics, which operate at multiple scales. ART dynamics act upon working memory representations, producing resonances that embody the conscious recognition of sounds, phonemes, words, sentences, and *meanings* (Grossberg, 2000).

259 **shared attention**: Arbib (2011) offers another plausible timeline for the development of shared attention that proceeds through the following stages:

1. Simple imitation (great apes)
2. Complex imitation (unique to hominin lines)
3. Emergence of pantomime (direct copy of gesture)
4. Emergence of protosign (the gesture is not a direct map of the object/thing/event)
5. Emergence of protospeech

Arbib writes, "We argue that Oldowan tool making corresponds to simple imitation and ape gestural communication and Acheulean tool making corresponds to complex imitation and protolanguage, whereas the explosion of innovations in tool making and social organization of the past 100,000 years correlates with the emergence of language."

263 **capable of eye gazing**: Human infants seek out symmetrical objects in their environment (which might be faces) as soon as they open their eyes, and closely observe faces before switching their attention to eyes. Few sources of stimuli in an infant's environment are as useful as a parent's face.

264 **mental innovation**: *imitation*: Primates *emulate* one another, but humans *imitate* one another. If a chimpanzee observes another chimpanzee slap a latched box five times before unlatching it to get a treat, the chimpanzee will quickly unlatch an identical box. But if a human observes another human slap a box five times before unlatching it, the human will do the exact same thing: namely, slap the box five times, then unlatch it. The chimpanzee emulated his companion, the human imitated (Horner & Whiten, 2005).

265 **arbitrary visual symbol**: Many philosophers have pondered how a symbol can get linked to a meaning in the first place—how Mind (or any physical entity) can link arbitrary sensory patterns with arbitrary conscious meanings. This puzzle is known as the symbol grounding problem. Grossberg (2017):

> Another kind of complementary ignorance is overcome through resonance within the attentional system, and thereby solves the symbol grounding problem: When a [bottom-up] feature pattern is activated [in the What module], its individual features have no meaning. They become meaningful only as part of the spatial pattern of features to which they belong. A [top-down] recognition category can selectively respond to this feature pattern, but does not know what these features are. When [the bottom-up representation and the top-down representation] resonate due to mutual positive feedback, the attended [bottom-up] features . . . are coherently bound together by the top-down expectation that is read out by the active [top-down] category. The attended pattern of [bottom-up] features, in turn, maintains activation of the recognition category . . . via the bottom-up adaptive filter.

265 **they don't have hands**: Perhaps elephants are closer to developing language than we might think?
Gibson (2011):

> Elephants remember and recognize by olfactory and visual means numerous conspecifics and classify them into social groups. They also sometimes cooperate to achieve joint goals and seem to understand others' intentions and emotions. Moreover, elephants have highly manipulative trunks, use tools for varied purposes, may recognize themselves in mirrors, and may have a stronger numerical sense than non-

human primates. Finally, they have elaborate vocal, olfactory, tactile, and gestural communication systems and can imitate some sounds. Given their broad cognitive and sensorimotor skills, elephants appear to have great promise as potential proto-language learners. So far, however, they have not been subject to language-training experiments.

Chapter Eighteen: Sapiens Supermind I

283   **long-term memory:** Candia et al. (2019) analyze the mental dynamics of supermind memory:

> Collective memory and attention are sustained by two channels: oral communica-tion (communicative memory) and the physical recording of information (cultural memory). . . . We use data on the citation of academic articles and patents, and on the online attention received by songs, movies and biographies, to describe the temporal decay of the attention received by cultural products. . . . Our results reveal that biog-raphies remain in our communicative memory the longest (20–30 years) and music the shortest (about 5.6 years). These findings show that the average attention received by cultural products decays following a universal biexponential function.

283   **its people-neurons:** One of the earliest *physical* instances of supermind memory are the history houses of Çatalhöyük. From Newitz (2021):

> Ian Hodder describes a curious practice that hints at one way people memorialized their nonfamily bonds. Many of Dido's neighbors built what he calls "history houses." These houses had a larger-than-average number of plastered bull heads, paintings, and bones in their walls. And these dwellings were carefully rebuilt with the exact same dimensions many times over centuries. People would even exhume and rebury the skeletons that had been placed in the floor of the earlier house, and archaeologists sometimes find dozens of skeletons in the floor of a history house. Like museums or libraries, history houses were places where the people of Çatalhöyük maintained a shared repository of cultural memories.

285   **"intergroup bias":** Hewstone et al. (2002): "Self-categorization as an intergroup member entails assimilation of the self to the intergroup category prototype and enhanced simi-larity to other intergroup members; and the intergroup is cognitively included in the self. Trust is extended to fellow intergroup, but not out-group, members, based on group living as a fundamental survival strategy. The extension of trust, positive regard, cooperation, and empathy to intergroup, but not out-group, members is an initial form of discrimination, based solely on intergroup favoritism, which must be distinguished from bias that entails an active component of aggression and out-group derogation."

287   **blistering pace:** The advent of a bird supermind united through birdsong appears to have accelerated bird mind evolution, as predicted by the metropolis principle. (This is one reason we see greater biological diversity in tropical rainforests and coral reefs, and why human tec$ological advances come at a steadily faster clip.)

From Cooney (2016): "The third and perhaps most striking conclusion is that the com-mon ancestor of all modern songbird species is likely to have lived just over 30m years ago. In evolutionary terms, this is surprisingly recent, especially compared to the probable age of the ancestor of all birds (about 95m years). When you consider that songbirds account for over half of all bird species on Earth (over 5,000 species), the relatively recent origins of song-birds mean they evolved into new species at an even faster rate than previously thought."

289   **aware of time:** Today, supermind management of time is even more important, as many superminds span multiple time zones, such as the United States, which stretches across six

time zones. Coordinating neuron-people activity across widely varying local conceptions of time demands sophisticated methods of synchronizing clocks—and, for GPS, methods of handling relativistic effects on time. The human mind is wired to naturally conceive of a Newtonian universe, but the sapiens supermind is developing the wiring to naturally conceive of an Einsteinian universe.

293   **consciousness in a nation-state:** Iain Banks, the rare novelist who found equal acclaim writing genre sci-fi and highbrow literary novels, wrote a series of books known as the Culture novels about our galaxy at a point far in the future, when it's home to a large number of competing interplanetary civilizations. Many civilizations, after hundreds of thousands of years of existence, decide to wave goodbye to the mundane world and enter another universe known as the Sublime. Banks' last novel, *The Hydrogen Sonata*, written as he was dying, is the story of one civilization's journey into the Sublime. Only a highly advanced civilization can survive the transfer into the Sublime, according to Banks, because only dynamics with an extremely high level of complexity, stability, and hierarchical sophistication can endure in the Sublime.

298   **The internet is a particularly powerful:** Some wonder whether the internet will ever become conscious. It already is. Or, more precisely, it functions as the necessary connective tissue facilitating the operation of the supermind consciousness cartel, much like the white matter in our cerebral cortex. The internet has become an essential part of generating supermind qualia, and thus it is part of the game of consciousness, though the internet itself will never wake up and realize, "Hey, I'm the internet!" any more than our white matter will wake up and say, "Hey, I'm cortical connective tissue, how about that!"

Chapter Nineteen: Sapiens Supermind II

304   **most sophisticated** *neuron:* Granato and De Giorgio (2014):

> "Even if neocortical PNs [pyramidal neurons] were homogeneous across cortical areas and layers, nonetheless each of them would represent the most complex neuron of the mammalian brain. Let us consider, for example, the L5 PN. Its apical dendrite extends through most of cortical thickness and is thus ideally suited for translaminar integration. In addition, the long, apparently homogeneous dendritic arbor of these neurons features specific functional properties: basal dendrites and the apical tufts are dominated by NMDA spikes, while Ca^{2+} spikes sustained by voltage-gated channels prevail in the distal apical trunk. Finally, dendritic, axon, and somatic domains of L5 PNs are targeted by different types of inhibitory interneurons. In summary, even the single PN is a complex world itself, able to integrate feedforward ascending input and feedback connections to generate the cognitive performance.

305   **more informed than any earthborn organism:** These facts are drawn from Hans Rosling's *Factfulness* and Steven Pinker's *The Better Angels of Our Nature*.

306   **is a cancer:** You are a god of cells. You are formed of cells, yet you can choose to annihilate any of them based upon any capricious whim that strikes you. Kill a few of your own cells, you won't even notice it. Keep on killing them, and at some point, your own welfare will be threatened. Even gods have limits.

306   **humans are willful beings:** Free will—or what neuroscientists term *volition*—is largely governed by the basal ganglia. Grossberg (2021):

> I noted the role of volitional signals in converting a modulatory top-down expectation into a driving one, and how visual imagery, thinking, and planning depend upon this property. A similar role for volition, again modulated by the basal ganglia, is predicted to control when sequences of objects or events (e.g., telephone numbers)

are temporarily stored in short-term working memory in the prefrontal cortex, and how such storage will be reset.... Let me summarize this line of causation because it clarifies how some of our most prized human endowments can be a double-edged sword: The ability to imagine, think, and plan arises from the solution of the stability-plasticity dilemma; namely, the ability to learn quickly throughout life without experiencing catastrophic forgetting. First and foremost, the ART Matching Rule, and its top-down modulatory on-center, dynamically stabilizes learned memories. The discovery during evolution that this modulatory signal *can be volitionally controlled* brought with it huge evolutionary advantages of being able to imagine visual imagery, and to quietly think and plan. But these precious consequences of our ability to learn rapidly and stably about our changing world, which is a prerequisite for developing a sense of self and civilized societies, also carry with them the risk of going crazy when volitional control goes awry. This risk is just part of the human condition. It is a risk that evolution has maintained because of its very powerful positive consequences for most of us most of the time. [Emphasis added.]

308   **sophisticated psychological speculations:** The effect of reading on individual brains is an example of the transfer of benefits acquired at a higher level (the supermind) to a lower level (individual human mind), as predicted by the metropolis principle.

Kolodny and Edelman (2018): "Interestingly, in modern humans, learning to read—a task that came into existence recently in human evolution and is unlikely to have had time to significantly influence human cognition via selection on genetic variants—causes, over a period of just several months, a significant shift in the pattern of functional connectivity in the brain."

309   **Every word in the paragraph you are currently reading:** We cheated a bit. Every word in this paragraph was generated by the DaVinci model of OpenAI's Generative Pre-trained Transformer 3, *except* the phrase "Every word in the paragraph you are currently reading," which was added by the authors for clarity.

Here is a prompt we provided to the GPT-3 to generate text: "The Self takes shape out of the endlessly circulating verbiage travelling from mind to mind and represents the front line of the conflict between your mental activity and the mental activity of the supermind. In simple terms, this means that the Self is constructed out of words. Doubt this? Then let me ask you a question. Who am I? That's right, me. Your tour guide on this mental voyage through the mind . . ."

309   **conceit of the fiction writer:** Novelists are reflecting the Self back at the Self, but they are also creating the Self. By generating new verbal concepts and models of the experience of being human, novelists and other writers are creating qualia that can empower the Self to expand itself. The history of literature is the history of the human Self: the oldest written story, *Gilgamesh*, focuses on the hero's deeds and actions, rather than his feelings or inner life. The first blossoming of human literature occurred in ancient Greece, but these stories and dramas focus on fate and predetermination, rather than individual agency or personality. Self-awareness consisted of understanding one's fixed place in society and the universe. A great flowering of the Self occurred during the Renaissance, expressed by one of the greatest writers of the age, William Shakespeare. He produced the first great literary characters who are consumed by inner contradictions and who have vivid, unique identities. The early twentieth century witnessed an explosion of the examination and complexification of the Self in literature, and the man who led the way was the Austrian psychiatrist and prolific author Sigmund Freud, followed by literary explorers of consciousness such as Virginia Woolf, James Joyce, and William Faulkner. Their books told us that we have a rich inner life and precious personal agency. And now the book in your hands suggests that your inner life (the mental dynamics of your Self) is inextricable from the mental dynamics of your supermind and the mental dynamics of your modules, neurons, and molecules.

310   **9 quintillion qualia:** Kaneda and Haub (2018) estimate that 117 billion humans have lived

over the past 192,000 years. Tseng and Poppenk (2020) estimate that humans have an average of 6,200 conscious thoughts a day. Though thoughts usually involve multiple qualia, let's assign one per thought. Wikipedia estimates that prehistoric humans lived an average of thirty-five years. Let's use that number for all humans, even though life spans are much longer today and more humans have lived in the past two centuries than in the previous two hundred centuries. This gives us 117 billion people × 6,200 thoughts/person/day × 35 years of thinking × 365 days in a year = 9.26×10^{18} thoughts = a minimum of 9.26×10^{18} human qualia. And counting. Fast.

313   **No computer program can emulate consciousness of consciousness:** Most contemporary AI designs won't ever become conscious no matter how advanced they become because they are not sufficiently anchored to physical reality. They are not sufficiently *embodied*. Specifically, most AI designs do not possess the necessary real-time hierarchical and recurrent dynamics embodied across distinctive levels of physical activity. The human mind has reciprocal dynamics operating holistically across the molecular, neural, modular, and supermodular levels. A self-driving car is closer to consciousness than the GPT-3, despite the latter's vastly superior memory, data complexity, and processing power.

   More simply, for an artificial mind to attain consciousness, it must contain analog elements that continuously respond to its physical environment in real time.

321   **That's what it's like to be you:** The Self is like the Simorq: you are each perspective and all perspectives. You are each molecular, neural, and modular thinking element participating in the dynamics of self-awareness simultaneously, bound together by the pattern of the dynamics itself, in the same way that all ten players are bound together by the game.

   In *QED: The Strange Theory of Light and Matter*, Richard Feynman writes, "The theory of quantum electrodynamics describes Nature as absurd from the point of view of common sense. And it agrees fully with experiment. So I hope you can accept Nature as She is— absurd."

   He shared this assertion to account for a phenomenon which has interesting parallels for thinking about consciousness—the fact that light appears to reflect off a mirror in a predictably symmetric fashion. Feynman points out that light actually reflects off the mirror in every possible direction—an infinite number of directions—but that most of these reflections "cancel each other out" and that the photons left over after all the cancelations happen is a precise, symmetric reflection where the angle of incidence equals the angle of reflection:

   So the theory of quantum electrodynamics gave the right answer . . . but this correct result came out at the expense of believing that light reflects all over the mirror, and having to add a bunch of little arrows together whose sole purpose was to cancel out. All that might seem to you to be a waste of time—some silly game for mathematicians only. After all, it doesn't seem like 'real physics' to have something there that only cancels out!

322   **if you believe it exists:** Beauty is another thing that only exists if you believe it exists. And if you believe it exists, you can bring more beauty into existence. Money is another: if we all stopped believing in money, it would cease to exist. (This has happened many times before: nobody believes in East German money, Confederate money, or wampum, so they have ceased to exist.) Subatomic particles, too: we *thought* into existence the Higgs boson (the ridiculously named God particle). We believed it existed, and rather than finding it lying around somewhere, we willed it into existence, like pulling a real rabbit out of an imaginary hat.

   This is where we cash out the full promise of the embodied thinking principle: The physical body and physical environment are part of the mind, after all, and therefore our thoughts are part of physical reality. All the dynamics engendering things that exist only if you believe they exist are dynamics of Mind.

323   **claim free will:** Grossberg (2021): "But these precious consequences of our ability to learn rapidly and stably about our changing world, which is a prerequisite for developing a sense of

self and civilized societies, also carry with them the risk of going crazy when volitional control goes awry. This risk is just part of the human condition. It is a risk that evolution has maintained because of its very powerful positive consequences for most of us most of the time."

The Light

331   **Frederick Douglass:** Frederick Douglass chose the name Frederick Douglass from the heroine Ellen Douglas of Walter Scott's narrative poem *The Lady of the Lake*: "I gave Mr. Jo$son the privilege of choosing me a name, but told him he must not take from me the name of 'Frederick.' I must hold on to that, to preserve a sense of my identity. Mr. Jo$son had just been reading the *Lady of the Lake*, and at once suggested that my name be 'Douglass.'"
       It is deeply ironic that, decades later, the Ku Klux Klan would adopt much of their symbolism and ritual—including the burning cross—from *The Lady of the Lake*.

332   **preserved at all costs:** Douglass (1855/2014): "About four years ago, upon a reconsideration of the whole subject, I became convinced that there was no necessity for dissolving the 'union between the northern and southern states;' that to seek this dissolution was no part of my duty as an abolitionist; that to abstain from voting, was to refuse to exercise a legitimate and powerful means for abolishing slavery; and that the constitution of the United States not only contained no guarantees in favor of slavery, but, on the contrary, it is, in its letter and spirit, an anti-slavery instrument, demanding the abolition of slavery as a condition of its own existence, as the supreme law of the land."

The Tandava

336   **one direction:** There is one remarkable implication concerning the nature of time that we can extract from the dynamics of mind. It is this: no matter the ultimate nature of our reality—whether we are a simulation running on a prodigious computer or a brain in a demonic vat—time is real. Without time moving forward in one direction, minds and consciousness could not exist.

337   **hidden structure of chaos:** Randomness demands that thinking elements form coalitions, because by joining together they share the burden of not knowing what's coming and working together to prepare for the unknown. If there was no randomness, there would be no need to cooperate with anyone else.
       Sporns (2016): "From a developmental perspective, randomness and specificity fulfill complementary roles as they jointly shape the generation and maintenance of synaptic connectivity."
       Miller (2016):

> Complex systems often have some inherent degree of randomness tied to the behavior of the agents or the structure of interactions. Perhaps surprisingly, such randomness can be useful. We often dread randomness in systems. Indeed, a key dictate in modern business management is to seek quality by removing all sources of randomness from any process. Given such imperatives, it is easy to think of randomness as a foe to be fought rather than as an opportunity to be embraced. The study of complexity suggests otherwise. Randomness is fundamental to Darwin's theory of evolution, which relies on the notion that errors (variations) during reproduction will provide grist for the mill of selection and result in "endless forms most beautiful and most wonderful."

337   **probes for loopholes:** Grossberg summarizes his thoughts on why the mind is a singular entity in the universe (Anderson & Rosenfeld, 2000):

I'll build my answer on some thoughts about why the brain is special. The following anecdote may help to make my point. Richard Feynman came into the field because he was interested in vision. When he realized that the retina is inverted, with the photodetectors behind all the other retinal layers, so that light has to go through all those layers before reaching them, he got out of the field. He couldn't figure out what kind of rational heuristics could be consistent with such a strange fact. So here we see one of the very greatest quantum mechanics admitting that brain dynamics are not just an easy application of quantum mechanics. On the other hand, the brain is tuned to the quantum level. You can see with just a few photons. The sensitivity of hearing is adjusted just above the level of thermal noise. So the brain is a quantum-sensitive measuring device. Moreover, the brain is a universal measuring device. It takes data from all the senses—vision, sound, pressure, temperature, biochemical sensors—and builds them into unified moments of resonant consciousness. But then why isn't the brain just another application of garden-variety quantum mechanics? What's different? My claim is that what's different is the brain's self-organizing capabilities. The critical thing is that we develop and learn on a very fast time scale relative to the evolution of matter. The revolution is in understanding universal quantum-sensitive rapidly self-organizing measurement devices. . . . Theories of mind and brain, in contrast, are really theories of measuring devices which happen to be self-organizing in order to adapt to an evolving world. Understanding such measurement devices would be a very big step for science. So why has it taken so long for such theories to get born?

337   **the intervention of Mind:** The journey of Mind is much like the trail of an ant over a windswept beach. The complex but purposeless structure of the sandy beach determines which paths are possible, but the ant purposefully chooses which path to take. Where he ends up is both predictable and random, the consequence of Mind and chaos interacting through time. And so is the cosmic journey of Mind, which chooses its path through spacetime from the pathways offered up by chaos.

337   **The ultimate complementary thinking:** The Tandava may also be the ultimate example of the embodied thinking principle. The dynamics of Mind cannot exist without being embodied within the dynamics of matter. But perhaps the dynamics of matter cannot exist without the dynamics of Mind, which supply matter with the arrow of time: the only variable that the dynamics of Mind and matter share in common is time.

338   **solar-system-spanning hypermind:** The qualia of truth in an individual human module mind is *art*. The qualia of truth in an individual supermind is *science*. Will a hypermind consider sapiens supermind science to be *art*?

338   **Initial value problems:** An initial value problem is the challenge of figuring out the initial conditions of a dynamic system. To conceive of the origin of the universe as an initial value problem is to conceive of the universe as fundamentally mathematical. The mere act of viewing the origin of the universe as a math problem reflects dynamics of Mind that exist outside the dynamics of matter.

***   What if the universe in its entirety is conscious? Let's say minds keep expanding, forming hyperminds, ultraminds, and so on, until the dynamics of the entire universe fall under purposeful control. Like all minds, the universe mind will want to keep existing . . . so perhaps it ignites a new Big Bang to pass as much of its mental dynamics as possible into a new universe that will eventually become self-aware and realize it needs to ignite a new Big Bang to perpetuate itself . . .

Bibliography

Initial Value Problem

Attar, F. U., Darbandi, A., & Davis, D. (1984). *The conference of the birds* (Re-issue ed.). Penguin Classics.

Bosman, C. A., & Aboitiz, F. (2015). Functional constraints in the evolution of brain circuits. *Frontiers in Neuroscience, 9*, 303.

Braitenberg, V. (1986). *Vehicles: Experiments in synthetic psychology.* MIT Press.

Breakspear, M. (2017). Dynamic models of large-scale brain activity. *Nature Neuroscience, 20*(3), 340–352.

Carroll, S. B. (2006). *Endless forms most beautiful: The new science of evo devo* (Reprint ed.). W. W. Norton.

Chirikov, B. V. (1991). Patterns in chaos. In *Chaos, order, and patterns* (pp. 109–134). Springer.

Churchland, P. S., & Sejnowski, T. J. (2016). *The computational brain* (Anniversary ed.). Bradford.

Eliav, T., Geva-Sagiv, M., Yartsev, M. M., Finkelstein, A., Rubin, A., Las, L., & Ulanovsky, N. (2018). Nonoscillatory phase coding and synchronization in the bat hippocampal formation. *Cell, 175*(4), 1119–1130.

Fornito, A., Zalesky, A., & Bullmore, E. (2016). *Fundamentals of brain network analysis* (1st ed.). Academic Press.

Grossberg, S. (2021). *Conscious mind, resonant brain: How each brain makes a mind.* Oxford University Press.

Hiesinger, P. R., & Hassan, B. A. (2018). The evolution of variability and robustness in neural development. *Trends in Neurosciences, 41*(9), 577–586.

Kaas, J. H. (2016). *Evolution of nervous systems* (2nd ed.). Academic Press.

Kaas, J. H. (2020). *Evolutionary neuroscience* (2nd ed.). Academic Press.

Lefebvre, L., Reader, S. M., & Sol, D. (2004). Brains, innovations and evolution in birds and primates. *Brain, Behavior and Evolution, 63*(4), 233–246.

Levine, D. S. (2018). *Introduction to neural and cognitive modeling* (3rd ed.). Routledge.

Lovett-Barron, M., Andalman, A. S., Allen, W. E., Vesuna, S., Kauvar, I., Burns, V. M., & Deisseroth, K. (2017). Ancestral circuits for the coordinated modulation of brain state. *Cell, 171*(6), 1411–1423.

Martinez, P., & Sprecher, S. G. (2020). Of circuits and brains: The origin and diversification of neural architectures. *Frontiers in Ecology and Evolution, 8*, 82.

Meadows, D. H., & Wright, D. (2008). *Thinking in systems: A primer.* Chelsea Green.

Moroz, L. L. (2009). On the independent origins of complex brains and neurons. *Brain, Behavior and Evolution, 74*(3), 177–190.

Murray, E., Wise, S., & Graham, K. (2017). *The evolution of memory systems: Ancestors, anatomy, and adaptations* (Reprint ed.). Oxford University Press.

Okobi, D. E., Banerjee, A., Matheson, A. M., Phelps, S. M., & Long, M. A. (2019). Motor cortical control of vocal interaction in neotropical singing mice. *Science, 363*(6430), 983–988.

Papke, R. T., & Gogarten, J. P. (2012). How bacterial lineages emerge. *Science, 336*(6077), 45–46.

Pyster, A., Hutchison, N., & Henry, D. (2018). *The paradoxical mindset of systems engineers: Uncommon minds, skills, and careers.* Jo$ Wiley & Sons.

Rabinovich, M., Friston, K. J., Varona, P., Deco, G., Jirsa, V. K., Menon, V., Haynes, J., Graben, P. B., Potthast, R., Kiebel, S. J., Bick, C., Bazhenov, M., Makeig, S., Vakorin, V. A., McIntosh, R., Huys, R., Pillai, A., Perdikis, D., Woodman, M., . . . Afraimovich, V. S. (Eds.). (2012). *Principles of brain dynamics: Global state interactions* (1st ed.). MIT Press.

Roth, G. (2015). *The long evolution of brains and minds* (1st ed.). Springer.

Shigeno, S., Murakami, Y., & Nomura, T. (2017). *Brain evolution by design: From neural origin to cognitive architecture* (1st ed.). Springer.

Sporns, O. (2016). *Networks of the brain* (Reprint ed.). MIT Press.

Sterling, P., & Laughlin, S. (2015). *Principles of neural design* (1st ed.). MIT Press.

Striedter, G. F. (2004). *Principles of brain evolution* (1st ed.). Oxford University Press.

Urushihara, H., & Muramoto, T. (2006). Genes involved in Dictyostelium discoideum sexual reproduction. *European Journal of Cell Biology, 85*(9–10), 961–968.

Wang, R., & Pan, X. (2018). *Advances in cognitive neurodynamics (V): Proceedings of the Fifth International Conference on Cognitive Neurodynamics—2015* (1st ed.). Springer.

Watanabe, S., Hofman, M. A., & Shimizu, T. (2017). *Evolution of the brain, cognition, and emotion in vertebrates* (1st ed.). Springer.

Yartsev, M. M., & Ulanovsky, N. (2013). Representation of three-dimensional space in the hippocampus of flying bats. *Science, 340*(6130), 367–372.

Zhang, W., & Yartsev, M. M. (2019). Correlated neural activity across the brains of socially interacting bats. *Cell, 178*(2), 413–428.

Chapter One: First Mind

Alberts, B., Jo$son, A. D., Lewis, J., Raff, M., Roberts, K., & Walter, P. (2002). *Molecular biology of the cell* (4th ed.). W. W. Norton.

Banno, B., Ickenstein, L. M., Chiu, G. N., Bally, M. B., Thewalt, J., Brief, E., & Wasan, E. K. (2010). The functional roles of poly (ethylene glycol)-lipid and lysolipid in the drug retention and release from lysolipid-containing thermosensitive liposomes in vitro and in vivo. *Journal of Pharmaceutical Sciences, 99*(5), 2295–2308.

Barrett, L. (2016). Why brains are not computers, why behaviorism is not satanism, and why dolphins are not aquatic apes. *The Behavior Analyst, 39*(1), 9–23.

Budin, I., & Szostak, J. W. (2010). Expanding roles for diverse physical phenomena during the origin of life. *Annual Review of Biophysics, 39*, 245–263.

Caforio, A., Siliakus, M. F., Exterkate, M., Jain, S., Jumde, V. R., Andringa, R. L., Kengen, S. W., Minnaard, A. J., Driessen, A. J., & van der Oost, J. (2018). Converting Escherichia coli into an archaebacterium with a hybrid heterochiral membrane. *Proceedings of the National Academy of Sciences, 115*(14), 3704–3709.

Cornell, C., Black, A., Xue, M., Litz, E., Ramsay, A., Gordon, M., Mileant, A., Cohen, R., Williams, A., Lee, K., & Drobny, P. (2019). Prebiotic amino acids bind to and stabilize prebiotic fatty acid membranes. *Proceedings of the National Academy of Sciences, 116*(35), 17239–17244.

Crick, F. C., & Koch, C. (2005). What is the function of the claustrum? *Philosophical Transactions of the Royal Society B: Biological Sciences, 360*(1458), 1271–1279.

Deamer, D. (2016). Membranes and the origin of life: A century of conjecture. *Journal of Molecular Evolution, 83*(5), 159–168.

Farhi, E., Guth, A. H., & Guven, J. (1990). Is it possible to create a universe in the laboratory by quantum tunneling? *Nuclear Physics B, 339*(2), 417–490.

Folse III, H. J., & Roughgarden, J. (2010). What is an individual organism? A multilevel selection perspective. *The Quarterly Review of Biology, 85*(4), 447–472.

Hoshika, S., Leal, A., Kim, J., Kim, S., Karalkar, B., Kim, J., Bates, M., Watkins, E., SantaLucia, A.,

Meyer, J., & DasGupta, S. (2019). Hachimoji DNA and RNA: A genetic system with eight building blocks. *Science, 363*(6429), 884–887.

Kiverstein, J., & Miller, M. (2015). The embodied brain: Towards a radical embodied cognitive neuroscience. *Frontiers in Human Neuroscience, 9*, 237.

Loeb, A. (2019, April 22). When lab experiments carry theological implications. *Scientific American Blog Network.* https://blogs.scientificamerican.com/observations/when-lab-experiments -carry-theological-implications/

Quammen, D. (2019). *The tangled tree: A radical new history of life* (Reprint ed.). Simon & Schuster.

Rios, A. C., & Copper, G. W. (2019). *Finding evidence of a prebiotic pyruvate reaction network in meteorites.* 50th Lunar and Planetary Science Conference in The Woodlands, Texas.

Schopf, J. W., Kitajima, K., Spicuzza, M. J., Kudryavtsev, A. B., & Valley, J. W. (2018). SIMS analyses of the oldest known assemblage of microfossils document their taxon-correlated carbon isotope compositions. *Proceedings of the National Academy of Sciences, 115*(1), 53–58.

Schrum, J. P., Zhu, T. F., & Szostak, J. W. (2010). The origins of cellular life. *Cold Spring Harbor Perspectives in Biology, 2*(9).

Shapiro, L. (2019). *Embodied cognition.* Routledge.

Simon, H. A. (2019). *The sciences of the artificial.* MIT press. (Original work published 1969)

Sporns, O. (2016). *Networks of the brain* (Reprint ed.). MIT Press.

Szostak, J. W. (2017). The narrow road to the deep past: In search of the chemistry of the origin of life. *Angewandte Chemie International Edition, 56*(37), 11037–11043.

Wallace, R. (2022). *Consciousness, cognition, and crosstalk.* Springer.

Ward, P., & Kirschvink, J. (2015). *A new history of life: The radical new discoveries about the origins and evolution of life on Earth.* Bloomsbury.

Wilson, A. D., & Golonka, S. (2013). Embodied cognition is not what you think it is. *Frontiers in Psychology, 4*, 58.

Chapter Two: Archaea Mind

Albers, S. V., & Jarrell, K. F. (2015). The archaellum: How archaea swim. *Frontiers in Microbiology, 6*, 23.

Albers, S. V., & Jarrell, K. F. (2018). The archaellum: An update on the unique archaeal motility structure. *Trends in Microbiology, 26*(4), 351–362.

Armitage, J. P., & Hellingwerf, K. J. (2005). Light-induced behavioral responses ("phototaxis") in prokaryotes. *Discoveries in Photosynthesis*, 985–995.

Beeby, M., Ferreira, J. L., Tripp, P., Albers, S. V., & Mitchell, D. R. (2020). Propulsive nanomachines: The convergent evolution of archaella, flagella and cilia. *FEMS Microbiology Reviews, 44*(3), 253–304.

Béja, O., Aravind, L., Koonin, E. V., Suzuki, M. T., Hadd, A., Nguyen, L. P., Jovanovich, S. B., Gates, C. M., Feldman, R. A., Spudich, J. L., & Spudich, E. N. (2000). Bacterial rhodopsin: Evidence for a new type of phototrophy in the sea. *Science, 289*(5486), 1902–1906.

Beznosov, S., Pyatibratov, M., Veluri, P., Mitra, S., & Fedorov, O. (2013). A way to identify archaellins in Halobacterium salinarum archaella by FLAG-tagging. *Open Life Sciences, 8*(9), 828–834.

Bollschweiler, D. (2015). *Study of the archaeal motility system of Halobacterium salinarum by cryo-electron tomography* [Doctoral dissertation, Tecsische Universität München].

Chaudhury, P., Quax, T. E., & Albers, S. V. (2018). Versatile cell surface structures of archaea. *Molecular Microbiology, 107*(3), 298–311.

Christian, B., & Griffiths, T. (2016). *Algorithms to live by: The computer science of human decisions.* Macmillan.

Fenchel, T., Blackburn, H., King, G. M., & Blackburn, T. H. (2012). *Bacterial biogeochemistry: The ecophysiology of mineral cycling.* Academic Press.

Gaines, B. R. (1972). Axioms for adaptive behaviour. *International Journal of Man-Machine Studies, 4*(2), 169–199.

Gaudet, P., Williams, J. G., Fey, P., & Chisholm, R. L. (2008). An anatomy ontology to represent biological knowledge in Dictyostelium discoideum. *BMC Genomics*, 9(1), 1–12.

Genç, B., Jara, J. H., Lagrimas, A. K., Pytel, P., Roos, R. P., Mesulam, M. M., Geula, C., Bigio, E. H., & Özdinler, P. H. (2017). Apical dendrite degeneration, a novel cellular pathology for Betz cells in ALS. *Scientific Reports*, 7(1), 1–10.

Grossberg, S. (2019). A half century of progress toward a unified neural theory of mind and brain with applications to autonomous adaptive agents and mental disorders. In Kozma, R., Alippi, C., Choe, Y., & Morabito, F. C. (Eds.), *Artificial intelligence in the age of neural networks and brain computing*. Academic Press.

Hellingwerf, K. J. (2002). The molecular basis of sensing and responding to light in microorganisms. *Antonie Van Leeuwenhoek*, 81(1), 51–59.

Hoff, W. D., Jung, K. H., & Spudich, J. L. (1997). Molecular mechanism of photosignaling by archaeal sensory rhodopsins. *Annual Review of Biophysics and Biomolecular Structure*, 26(1), 223–258.

Holland, J. H. (2006). Studying complex adaptive systems. *Journal of Systems Science and Complexity*, 19(1), 1–8.

Huber, H., Ho$, M. J., Rachel, R., Fuchs, T., Wimmer, V. C., & Stetter, K. O. (2002). A new phylum of Archaea represented by a nanosized hyperthermophilic symbiont. *Nature*, 417(6884), 63–67.

Jonas, E., & Kording, K. P. (2017). Could a neuroscientist understand a microprocessor? *PLoS Computational Biology*, 13(1), e1005268.

Kandori, H. (2015). Ion-pumping microbial rhodopsins. *Frontiers in Molecular Biosciences*, 2, 52.

Kemp, B. L., Tabish, E. M., Wolford, A. J., Jones, D. L., Butler, J. K., & Baxter, B. K. (2018). The biogeography of Great Salt Lake halophilic archaea: Testing the hypothesis of avian mechanical carriers. *Diversity*, 10(4), 124.

Khan, S., & Scholey, J. M. (2018). Assembly, functions and evolution of archaella, flagella and cilia. *Current Biology*, 28(6), R278–R292.

Koga, Y. (2012). Thermal adaptation of the archaeal and bacterial lipid membranes. *Archaea*, 2012.

Kubitschek, H. E. (1990). Cell volume increase in Escherichia coli after shifts to richer media. *Journal of Bacteriology*, 172(1), 94–101.

Kurihara, M., & Sudo, Y. (2015). Microbial rhodopsins: Wide distribution, rich diversity and great potential. *Biophysics and Physicobiology*, 12, 121–129.

Li, Z., Rodriguez-Franco, M., Albers, S. V., & Quax, T. E. (2020). The switch complex ArlCDE connects the chemotaxis system and the archaellum. *Molecular Microbiology*, 114(3), 468.

Luecke, H., Schobert, B., Lanyi, J. K., Spudich, E. N., & Spudich, J. L. (2001). Crystal structure of sensory rhodopsin II at 2.4 angstroms: Insights into color tuning and transducer interaction. *Science*, 293(5534), 1499–1503.

Nair, P. (2012). Woese and Fox: Life, rearranged. *Proceedings of the National Academy of Sciences*, 109(4), 1019–1021.

National Research Council. (1999). Correlates of smallest sizes for microorganisms. In *Size limits of very small microorganisms: Proceedings of a workshop*. National Academies Press (US).

Nutsch, T., Marwan, W., Oesterhelt, D., & Gilles, E. D. (2003). Signal processing and flagellar motor switching during phototaxis of Halobacterium salinarum. *Genome Research*, 13(11), 2406–2412.

Nutsch, T., Oesterhelt, D., Gilles, E. D., & Marwan, W. (2005). A quantitative model of the switch cycle of an archaeal flagellar motor and its sensory control. *Biophysical Journal*, 89(4), 2307–2323.

Pace, N. R., Sapp, J., & Goldenfeld, N. (2012). Phylogeny and beyond: Scientific, historical, and conceptual significance of the first tree of life. *Proceedings of the National Academy of Sciences*, 109(4), 1011–1018.

Perazzona, B., & Spudich, J. L. (1999). Identification of methylation sites and effects of phototaxis stimuli on transducer methylation in Halobacterium salinarum. *Journal of Bacteriology*, 181(18), 5676–5683.

Quax, T. E., Albers, S. V., & Pfeiffer, F. (2018). Taxis in archaea. *Emerging Topics in Life Sciences*, 2(4), 535–546.

Riley, M. (1999, October). Correlates of smallest sizes for microorganisms. In National Research Council (Ed.), *Size limits of very small microorganisms: Proceedings of a workshop* (Vol. 3, p. 21).

Schlesner, M. (2008). The Halobacterium salinarum taxis signal transduction network: A protein-protein interaction study. *Ludwig–Maximilians–Universität München.*

Sherwood, C. C., Lee, P. W., Rivara, C. B., Holloway, R. L., Gilissen, E. P., Simmons, R. M., Hakeem, A., Allman, J. M., Erwin, J. M., & Hof, P. R. (2003). Evolution of specialized pyramidal neurons in primate visual and motor cortex. *Brain, Behavior and Evolution, 61*(1), 28–44.

Sporns, O. (2016). *Networks of the brain* (Reprint ed.). MIT Press.

Sterling, P., & Laughlin, S. (2015). *Principles of neural design* (1st ed.). MIT Press.

Syutkin, A. S., Pyatibratov, M. G., & Fedorov, O. V. (2014). Flagella of halophilic archaea: Differences in supramolecular organization. *Biochemistry (Moscow), 79*(13), 1470–1482.

Thornton, K. L. (2018). *An investigation into the motility of environmental haloarchaeal species* [Doctoral dissertation, University of York].

Tien, H. T., & Ottova, A. L. (2001). The lipid bilayer concept and its experimental realization: From soap bubbles, kitchen sink, to bilayer lipid membranes. *Journal of Membrane Science, 189*(1), 83–117.

Tourte, M., Schaeffer, P., Grossi, V., & Oger, P. M. (2020). Functionalized membrane domains: An ancestral feature of archaea? *Frontiers in Microbiology, 11*, 526.

Trachtenberg, S., & Cohen-Krausz, S. (2006). The archaeabacterial flagellar filament: A bacterial propeller with a pilus-like structure. *Journal of Molecular Microbiology and Biotec$ology, 11*(3–5), 208–220.

Wilde, A., & Mullineaux, C. W. (2017). Light-controlled motility in prokaryotes and the problem of directional light perception. *FEMS Microbiology Reviews, 41*(6), 900–922.

Wirth, R. (2012). Response to Jarrell and Albers: Seven letters less does not say more. *Trends in Microbiology, 20*(11), 511–512.

Woese, C. R., & Fox, G. E. (1977). Phylogenetic structure of the prokaryotic domain: The primary kingdoms. *Proceedings of the National Academy of Sciences, 74*(11), 5088–5090.

Chapter Three: Bacteria Mind I

Addicott, M. A., Pearson, J. M., Sweitzer, M. M., Barack, D. L., & Platt, M. L. (2017). A primer on foraging and the explore/exploit trade-off for psychiatry research. *Neuropsychopharmacology, 42*(10), 1931–1939.

Alvarez, L., Friedrich, B. M., Gompper, G., & Kaupp, U. B. (2014). The computational sperm cell. *Trends in Cell Biology, 24*(3), 198–207.

Armitage, J. P. (1997). Three hundred years of bacterial motility. *Foundations of Modern Biochemistry, 3*, 107–171.

Badman, R. P., Hills, T. T., & Akaishi, R. (2020). Multiscale computation and dynamic attention in biological and artificial intelligence. *Brain Sciences, 10*(6), 396.

Berg, H. C., & Brown, D. A. (1972). Chemotaxis in Escherichia coli analysed by three-dimensional tracking. *Nature, 239*(5374), 500–504.

Berger-Tal, O., Nathan, J., Meron, E., & Saltz, D. (2014). The exploration-exploitation dilemma: A multidisciplinary framework. *PLoS One, 9*(4).

Bi, S., Jin, F., & Sourjik, V. (2018). Inverted signaling by bacterial chemotaxis receptors. *Nature Communications, 9*(1), 1–13.

Blanchard, T. C., & Gershman, S. J. (2018). Pure correlates of exploration and exploitation in the human brain. *Cognitive, Affective, & Behavioral Neuroscience, 18*(1), 117–126.

Cogliati Dezza, I., Cleeremans, A., & Alexander, W. (2019). Should we control? The interplay between cognitive control and information integration in the resolution of the exploration-exploitation dilemma. *Journal of Experimental Psychology: General, 148*(6), 977.

Costa, V. D., Mitz, A. R., & Averbeck, B. B. (2019). Subcortical substrates of explore-exploit decisions in primates. *Neuron, 103*(3), 533–545.

Ehinger, B. V., Kaufhold, L., & König, P. (2018). Probing the temporal dynamics of the exploration-exploitation dilemma of eye movements. *Journal of Vision, 18*(3), 6–6.

Faumont, S., Lindsay, T. H., & Lockery, S. R. (2012). Neuronal microcircuits for decision making in C. elegans. *Current Opinion in Neurobiology, 22*(4), 580–591.

Finn, S., Condell, O., McClure, P., Amézquita, A., & Fanning, S. (2013). Mechanisms of survival, responses and sources of Salmonella in low-moisture environments. *Frontiers in Microbiology, 4,* 331.

Giannella, R. A. (1996). Salmonella. In S. Baron (Ed.), *Medical microbiology.* University of Texas Medical Branch at Galveston.

Grossberg, S. (2021). *Conscious mind, resonant brain: How each brain makes a mind.* Oxford University Press.

Guthrie, R. K. (1991). *Salmonella* (1st ed.). CRC Press.

Hamadi, F., Latrache, H., Zahir, H., Elghmari, A., Timinouni, M., & Ellouali, M. (2008). The relation between Escherichia coli surface functional groups' composition and their physicochemical properties. *Brazilian Journal of Microbiology, 39*(1), 10–15.

Jajere, S. M. (2019). A review of Salmonella enterica with particular focus on the pathogenicity and virulence factors, host specificity and antimicrobial resistance including multidrug resistance. *Veterinary World, 12*(4), 504.

Keerthirat$e, T. P., Ross, K., Fallowfield, H., & Whiley, H. (2016). A review of temperature, pH, and other factors that influence the survival of Salmonella in mayonnaise and other raw egg products. *Pathogens, 5*(4), 63.

Keerthirat$e, T. P., Ross, K., Fallowfield, H., & Whiley, H. (2019). The combined effect of pH and temperature on the survival of Salmonella enterica serovar Typhimurium and implications for the preparation of raw egg mayonnaise. *Pathogens, 8*(4), 218.

Krummel, M. F., Bartumeus, F., & Gérard, A. (2016). T cell migration, search strategies and mechanisms. *Nature Reviews Immunology, 16*(3), 193.

Lazova, M. D., Butler, M. T., Shimizu, T. S., & Harshey, R. M. (2012). Salmonella chemoreceptors McpB and McpC mediate a repellent response to L-cystine: A potential mechanism to avoid oxidative conditions. *Molecular Microbiology, 84*(4), 697–711.

Lyon, P. (2015). The cognitive cell: Bacterial behavior reconsidered. *Frontiers in Microbiology, 6,* 264.

Macnab, R. M., & Koshland, D. E. (1972). The gradient-sensing mechanism in bacterial chemotaxis. *Proceedings of the National Academy of Sciences, 69*(9), 2509–2512.

Mauriello, E. M., Mignot, T., Yang, Z., & Zusman, D. R. (2010). Gliding motility revisited: How do the myxobacteria move without flagella? *Microbiology and Molecular Biology Reviews, 74*(2), 229–249.

Olsen, J. E., Hoegh-Andersen, K. H., Casadesús, J., & Thomsen, L. E. (2012). The importance of motility and chemotaxis for extra-animal survival of Salmonella enterica serovar Typhimurium and Dublin. *Journal of Applied Microbiology, 113*(3), 560–568.

Ramanathan, S., & Broach, J. R. (2007). Do cells think? *Cellular and Molecular Life Sciences, 64*(14), 1801–1804.

Schatten, H., Eisenstark, A., & Eisenstark, A. (Eds.). (2007). *Salmonella: Methods and protocols.* Springer Science & Business Media.

Thomas, M. A., Kleist, A. B., & Volkman, B. F. (2018). Decoding the chemotactic signal. *Journal of Leukocyte Biology, 104*(2), 359–374.

Veresoglou, S. D., Wang, D., Andrade-Linares, D. R., Hempel, S., & Rillig, M. C. (2018). Fungal decision to exploit or explore depends on growth rate. *Microbial Ecology, 75*(2), 289–292.

Webre, D. J., Wolanin, P. M., & Stock, J. B. (2003). Bacterial chemotaxis. *Current Biology, 13*(2), R47–R49.

Chapter Four: Bacteria Mind II

Bi, S., Jin, F., & Sourjik, V. (2018). Inverted signaling by bacterial chemotaxis receptors. *Nature Communications, 9*(1), 1–13.

Faumont, S., Lindsay, T. H., & Lockery, S. R. (2012). Neuronal microcircuits for decision making in C. elegans. *Current Opinion in Neurobiology, 22*(4), 580–591.

Finn, S., Condell, O., McClure, P., Amézquita, A., & Fanning, S. (2013). Mechanisms of survival, responses and sources of Salmonella in low-moisture environments. *Frontiers in Microbiology*, 4, 331.

Frankel, G., & Ron, E. Z. (Eds.). (2018). *Escherichia Coli, a versatile pathogen* (Vol. 416). Springer.

Hamadi, F., Latrache, H., Zahir, H., Elghmari, A., Timinouni, M., & Ellouali, M. (2008). The relation between Escherichia coli surface functional groups' composition and their physicochemical properties. *Brazilian Journal of Microbiology*, 39(1), 10–15.

Keerthirat$e, T. P., Ross, K., Fallowfield, H., & Whiley, H. (2019). The combined effect of pH and temperature on the survival of Salmonella enterica serovar Typhimurium and implications for the preparation of raw egg mayonnaise. *Pathogens*, 8(4), 218.

Keerthirat$e, T. P., Ross, K., Fallowfield, H., & Whiley, H. (2016). A review of temperature, pH, and other factors that influence the survival of Salmonella in mayonnaise and other raw egg products. *Pathogens*, 5(4), 63.

Kubitschek, H. E. (1990). Cell volume increase in Escherichia coli after shifts to richer media. *Journal of Bacteriology*, 172(1), 94–101.

Lan, G., & Tu, Y. (2016). Information processing in bacteria: Memory, computation, and statistical physics: A key issues review. *Reports on Progress in Physics*, 79(5), 052601.

Lazova, M. D., Butler, M. T., Shimizu, T. S., & Harshey, R. M. (2012). Salmonella chemoreceptors McpB and McpC mediate a repellent response to L-cystine: A potential mechanism to avoid oxidative conditions. *Molecular Microbiology*, 84(4), 697–711.

Lyon, P. (2015). The cognitive cell: Bacterial behavior reconsidered. *Frontiers in Microbiology*, 6, 264.

Mears, P. J., Koirala, S., Rao, C. V., Golding, I., & Chemla, Y. R. (2014). Escherichia coli swimming is robust against variations in flagellar number. *eLife*, 3, e01916.

Metris, A., George, S. M., Mulholland, F., Carter, A. T., & Baranyi, J. (2014). Metabolic shift of Escherichia coli under salt stress in the presence of glycine betaine. *Applied and Environmental Microbiology*, 80(15), 4745–4756.

Perlova, T., Gruebele, M., & Chemla, Y. R. (2019). Blue light is a universal signal for Escherichia coli chemoreceptors. *Journal of Bacteriology*, 201(11).

Sagawa, T., Kikuchi, Y., Inoue, Y., Takahashi, H., Muraoka, T., Kinbara, K., Ishijima, A., & Fukuoka, H. (2014). Single-cell E. coli response to an instantaneously applied chemotactic signal. *Biophysical Journal*, 107(3), 730–739.

Sterling, P., & Laughlin, S. (2015). *Principles of neural design* (1st ed.). MIT Press.

Stock, J. B., & Baker, M. D. (2009). Chemotaxis. *Encyclopedia of microbiology*. Elsevier.

Van Duijn, M., Keijzer, F., & Franken, D. (2006). Principles of minimal cognition: Casting cognition as sensorimotor coordination. *Adaptive Behavior*, 14(2), 157–170.

Vladimirov, N., Løvdok, L., Lebiedz, D., & Sourjik, V. (2008). Dependence of bacterial chemotaxis on gradient shape and adaptation rate. *PLoS Computational Biology*, 4(12), e1000242.

Vladimirov, N., & Sourjik, V. (2009). Chemotaxis: How bacteria use memory. *Biological Chemistry*, 390, 1097–1104.

Waite, A. J., Frankel, N. W., Dufour, Y. S., Jo$ston, J. F., Long, J., & Emonet, T. (2016). Non-genetic diversity modulates population performance. *Molecular Systems Biology*, 12(12), 895.

Waite, A. J., Frankel, N. W., & Emonet, T. (2018). Behavioral variability and phenotypic diversity in bacterial chemotaxis. *Annual Review of Biophysics*, 47, 595–616.

Chapter Five: Amoeba Mind

Berg, H. C., & Brown, D. A. (1972). Chemotaxis in Escherichia coli analysed by three-dimensional tracking. *Nature*, 239(5374), 500–504.

Bloomfield, G., Skelton, J., Ivens, A., Tanaka, Y., & Kay, R. R. (2010). Sex determination in the social amoeba Dictyostelium discoideum. *Science*, 330(6010), 1533–1536.

Bonner, J. T. (2009). *The social amoebae: The biology of cellular slime molds* (1st ed.). Princeton University Press.

Brock, D. A., Douglas, T. E., Queller, D. C., & Strassmann, J. E. (2011). Primitive agriculture in a social amoeba. *Nature, 469*(7330), 393–396.

De Palo, G., Yi, D., & Endres, R. G. (2017). A critical-like collective state leads to long-range cell communication in Dictyostelium discoideum aggregation. *PLoS Biology, 15*(4), e1002602.

DeYoung, G., Monk, P. B., & Othmer, H. G. (1988). Pacemakers in aggregation fields of Dictyostelium discoideum: Does a single cell suffice? *Journal of Mathematical Biology, 26*(5), 487–517.

Gaudet, P., Fey, P., & Chisholm, R. (2008). Dictyostelium discoideum: The social ameba. *Cold Spring Harbor Protocols, 12*, pdb-emo109.

Gaudet, P., Williams, J. G., Fey, P., & Chisholm, R. L. (2008). An anatomy ontology to represent biological knowledge in Dictyostelium discoideum. *BMC Genomics, 9*(1), 1–12.

Hashimura, H., Morimoto, Y. V., Yasui, M., & Ueda, M. (2019). Collective cell migration of Dictyostelium without cAMP oscillations at multicellular stages. *Communications Biology, 2*(1), 1–15.

Hirose, S., Benabentos, R., Ho, H. I., Kuspa, A., & Shaulsky, G. (2011). Self-recognition in social amoebae is mediated by allelic pairs of tiger genes. *Science, 333*(6041), 467–470.

Katoh, M., Chen, G., Roberge, E., Shaulsky, G., & Kuspa, A. (2007). Developmental commitment in Dictyostelium discoideum. *Eukaryotic Cell, 6*(11), 2038–2045.

LeDoux, J. (2020). *The deep history of ourselves: The four-billion-year story of how we got conscious brains.* Penguin Books.

Loomis, W. F. (2014). Cell signaling during development of Dictyostelium. *Developmental Biology, 391*(1), 1–16.

Marée, A. F., & Hogeweg, P. (2001). How amoeboids self-organize into a fruiting body: Multicellular coordination in Dictyostelium discoideum. *Proceedings of the National Academy of Sciences, 98*(7), 3879–3883.

Meece, S. (2006). A bird's eye view—of a leopard's spots: The Çatalhöyük "map" and the development of cartographic representation in prehistory. *Anatolian Studies*, January, 1–16.

Mora Van Cauwelaert, E., Arias Del Angel, J. A., Benítez, M., & Azpeitia, E. M. (2015). Development of cell differentiation in the transition to multicellularity: A dynamical modeling approach. *Frontiers in Microbiology, 6*, 603.

Newitz, A. (2021). *Four lost cities: A secret history of the urban age.* W. W. Norton.

Noorbakhsh, J., Schwab, D. J., Sgro, A. E., Gregor, T., & Mehta, P. (2015). Modeling oscillations and spiral waves in Dictyostelium populations. *Physical Review E, 91*(6), 062711.

Palsson, E., & Othmer, H. G. (2000). A model for individual and collective cell movement in Dictyostelium discoideum. *Proceedings of the National Academy of Sciences, 97*(19), 10448–10453.

Rafiq, M., & Thompson, E. (2014). The shape-shifting superhero: Dictyostelium discoideum. *Microbiology Today, 41*(1), 16–19.

Rossine, F. W., Martinez-Garcia, R., Sgro, A. E., Gregor, T., & Tarnita, C. E. (2020). Eco-evolutionary significance of "loners." *PLoS Biology, 18*(3), e3000642.

Schaap, P. (2011a). Evolution of developmental cyclic adenosine monophosphate signaling in the Dictyostelia from an amoebozoan stress response. *Development, Growth & Differentiation, 53*(4), 452–462.

Schaap, P. (2011b). Evolutionary crossroads in developmental biology: Dictyostelium discoideum. *Development, 138*(3), 387–396.

Schmidt, M. C. S. (2017). Locomotion pattern and pace of free-living amoebae—a microscopic study. In A. Méndez-Vilas (Ed.), *Microscopy and imaging science: Practical approaches to applied research and education.* Formatex Research Center.

Shaulsky, G., & Kessin, R. H. (2007). The cold war of the social amoebae. *Current Biology, 17*(16), R684–R692.

Singer, G., Araki, T., & Weijer, C. J. (2019). Oscillatory cAMP cell-cell signalling persists during multicellular Dictyostelium development. *Communications Biology, 2*(1), 1–12.

Sperelakis, N. (Ed.). (1995). *Cell physiology sourcebook: A molecular approach.* Elsevier.

Strassmann, J. E., Zhu, Y., & Queller, D. C. (2000). Altruism and social cheating in the social amoeba Dictyostelium discoideum. *Nature, 408*(6815), 965–967.

Thomas, M. A., Kleist, A. B., & Volkman, B. F. (2018). Decoding the chemotactic signal. *Journal of Leukocyte Biology, 104*(2), 359–374.

Van Haastert, P. J. (2010). Chemotaxis: Insights from the extending pseudopod. *Journal of Cell Science, 123*(18), 3031–3037.

Van Haastert, P. J., Bishop, J. D., & Gomer, R. H. (1996). The cell density factor CMF regulates the chemoattractant receptor cAR1 in Dictyostelium. *Journal of Cell Biology, 134*(6), 1543–1549.

Ward, P., & Kirschvink, J. (2015). *A new history of life: The radical new discoveries about the origins and evolution of life on Earth.* Bloomsbury.

The Metropolis Principle

Brown, P. (2010). *Indian architecture (Buddhist and Hindu period)* (2nd ed.). Read Books.

Chauhan, P. R. (2009). The lower Paleolithic of the Indian subcontinent. *Evolutionary Anthropology: Issues, News, and Reviews, 18*(2), 62–78.

Dalal, K. F., & Raghavan, G. R. (Eds.). (2020). *Explorations in Maharashtra: The Proceedings of the Fourth Workshop (16th September 2017)* (1st ed.). INSTUCEN Trust.

Enthoven, R. E. (1990). *The tribes and castes of Bombay* (Vol. 1). Asian Educational Services.

Fernandes, N. (2013). *City adrift: A short biography of Bombay.* Aleph Book Company.

Glaeser, E. L. (1998). Are cities dying? *Journal of Economic Perspectives, 12*(2), 139–160.

Hicks, D., & Stevenson, A. (Eds.). (2013). *World archaeology at the Pitt Rivers Museum: A characterization.* Archaeopress.

Hunt, M. R., & Stern, P. J. (2021). Bombay: The genealogy of a global imperial city. *Urban History*, 1–18.

Hunt, T. (2014). *Ten cities that made an empire.* Penguin UK.

Joshi, P. B. (1902). *A short sketch of the early history of the town & island of Bombay, Hindu period.* Times of India Press.

Marathe, A. (2006, June 10). Acheulian cave at Susrondi, Konkan, Maharashtra. *Current Science*, 1538–1544.

McCarthy, N. (2014, September 3). Bollywood: India's film industry by the numbers. *Forbes.*

McGowan, A. (2016). Domestic modern: Redecorating homes in Bombay in the 1930s. *Journal of the Society of Architectural Historians, 75*(4), 424–446.

Mumbai Fire Brigade. (2012, April 4). Maharashtra Fire Services. https://mahafireservice.gov.in/news/MFB 125_Yrs.pdf

Mumbai—History. (2016, June 27). *Encyclopedia Britannica.* https://www.britannica.com/place/Mumbai/History

Pandit, S. (2020). *Mumbai beyond Bombay.* Aprant.

Pinto, J., & Fernandes, N. (Eds.). (2003). *Bombay, meri jaan: Writings on Mumbai.* Penguin Books India.

Polo, M., & Cliff, N. (2016). *The travels.* Penguin Classics.

Prakash, G. (2010). *Mumbai fables.* Princeton University Press.

Pusalker, A. D., & Dighe, V. G. (1949). *Bombay: Story of the island city* [Conference session]. All India Oriental Conference, Bombay.

Riding, T. (2018). "Making Bombay Island": Land reclamation and geographical conceptions of Bombay, 1661–1728. *Journal of Historical Geography, 59*, 27–39.

Sporns, O. (2016). *Networks of the brain* (Reprint ed.). MIT Press.

Sterling, P., & Laughlin, S. (2015). *Principles of neural design* (1st ed.). MIT Press.

Tindall, G. (1992). *City of gold: The biography of Bombay.* Penguin Books India.

Vicziany, M., & Bapat, J. (2009). Mumb-dev- and the other mother goddesses in Mumbai. *Modern Asian Studies*, March 1, 511–541.

Ward, P., & Kirschvink, J. (2015). *A new history of life: The radical new discoveries about the origins and evolution of life on Earth.* Bloomsbury.

West, G. (2018). *Scale: The universal laws of life, growth, and death in organisms, cities, and companies* (Reprint ed.). Penguin.

Chapter Six: Hydra Mind

Alié, A., & Manuel, M. (2010). The backbone of the post-synaptic density originated in a unicellular ancestor of choanoflagellates and metazoans. *BMC Evolutionary Biology, 10*(1), 1–10.

Arendt, D., Bertucci, P. Y., Achim, K., & Musser, J. M. (2019). Evolution of neuronal types and families. *Current Opinion in Neurobiology, 56,* 144–152.

Arendt, D., Tosches, M. A., & Marlow, H. (2016). From nerve net to nerve ring, nerve cord and brain: Evolution of the nervous system. *Nature Reviews Neuroscience, 17*(1), 61.

Berg-Sørensen, K., & Flyvbjerg, H. (2005). The colour of thermal noise in classical Brownian motion: A feasibility study of direct experimental observation. *New Journal of Physics, 7*(1), 38.

Bucher, D., & Anderson, P. A. (2015). Evolution of the first nervous systems: What can we surmise? *The Journal of Experimental Biology, 218,* 501–503.

Budd, G. E. (2015). Early animal evolution and the origins of nervous systems. *Philosophical Transactions of the Royal Society B: Biological Sciences, 370*(1684), 20150037.

Burkhardt, P., Stegmann, C. M., Cooper, B., Kloepper, T. H., Imig, C., Varoqueaux, F., Wahl, M. C., & Fasshauer, D. (2011). Primordial neurosecretory apparatus identified in the choanoflagellate *Monosiga brevicollis. Proceedings of the National Academy of Sciences, 108*(37), 15264–15269.

Cai, X. (2008). Unicellular Ca2+ signaling "toolkit" at the origin of metazoa. *Molecular Biology and Evolution, 25*(7), 1357–1361.

Carter, J. A., Hyland, C., Steele, R. E., & Collins, E. M. S. (2016). Dynamics of mouth opening in Hydra. *Biophysical Journal, 110*(5), 1191–1201.

Churchland, P. S., & Sejnowski, T. J. (2016). *The computational brain* (Anniversary ed.). Bradford.

Dunn, A. R. (2016). How Hydra eats. *Biophysical Journal, 110*(7), 1467.

Dupre, C., & Yuste, R. (2017). Non-overlapping neural networks in Hydra vulgaris. *Current Biology, 27*(8), 1085–1097.

Ginsburg, S., & Jablonka, E. (2010). The evolution of associative learning: A factor in the Cambrian explosion. *Journal of Theoretical Biology, 266*(1), 11–20.

Han, S., Taralova, E., Dupre, C., & Yuste, R. (2018). Comprehensive machine learning analysis of Hydra behavior reveals a stable basal behavioral repertoire. *eLife, 7,* e32605.

Holland, N. D. (2003). Early central nervous system evolution: An era of skin brains? *Nature Reviews Neuroscience, 4*(8), 617–627.

Hu, G., Li, J., & Wang, G. Z. (2020). Significant evolutionary constraints on neuron cells revealed by single-cell transcriptomics. *Genome Biology and Evolution, 12*(4), 300–308.

Izhikevich, E. M. (2007). *Dynamical systems in neuroscience.* MIT Press.

Jékely, G. (2011). Origin and early evolution of neural circuits for the control of ciliary locomotion. *Proceedings of the Royal Society B: Biological Sciences, 278*(1707), 914–922.

Kass-Simon, G., & Scappaticci, A., Jr. (2002). The behavioral and developmental physiology of nematocysts. *Canadian Journal of Zoology, 80*(10), 1772–1794.

Kinnamon, J. C., & Westfall, J. A. (1981). A three dimensional serial reconstruction of neuronal distributions in the hypostome of a Hydra. *Journal of Morphology, 168*(3), 321–329.

Klimovich, A. V., & Bosch, T. C. (2018). Rethinking the role of the nervous system: Lessons from the Hydra holobiont. *BioEssays, 40*(9), 1800060.

Koizumi, O., Sato, N., & Goto, C. (2004). Chemical anatomy of Hydra nervous system using antibodies against Hydra neuropeptides: A review. *Hydrobiologia, 530*(1), 41–47.

Mackie, G. O. (2002). What's new in cnidarian biology? *Canadian Journal of Zoology, 80*(10), 1649–1653.

Martín-Durán, J. M., Pang, K., Børve, A., Lê, H. S., Furu, A., Cannon, J. T., Jondelius, U., & Hejnol, A. (2018). Convergent evolution of bilaterian nerve cords. *Nature, 553*(7686), 45–50.

Moroz, L. L., Kocot, K. M., Citarella, M. R., Dosung, S., Norekian, T. P., Povolotskaya, I. S., Grigorenko, A. P., Dailey, C., Berezikov, E., Buckley, K. M., & Ptitsyn, A. (2014). The ctenophore genome and the evolutionary origins of neural systems. *Nature, 510*(7503), 109–114.

Moroz, L. L., & Ko$, A. B. (2016). Independent origins of neurons and synapses: Insights from ctenophores. *Philosophical Transactions of the Royal Society B: Biological Sciences, 371*(1685), 20150041.

Murillo-Rincon, A. P., Klimovich, A., Pemöller, E., Taubenheim, J., Mortzfeld, B., Augustin, R., & Bosch, T. C. (2017). Spontaneous body contractions are modulated by the microbiome of Hydra. *Scientific Reports, 7*(1), 1–9.

Noro, Y., Yum, S., Nishimiya-Fujisawa, C., Busse, C., Shimizu, H., Mineta, K., Zhang, X., Holstein, T. W., David, C. N., Gojobori, T., & Fujisawa, T. (2019). Regionalized nervous system in Hydra and the mechanism of its development. *Gene Expression Patterns, 31,* 42–59.

Passano, L. M. (1963). Primitive nervous systems. *Proceedings of the National Academy of Sciences of the United States of America, 50*(2), 306.

Polilov, A. A. (2012). The smallest insects evolve anucleate neurons. *Arthropod Structure & Development, 41,* 29–34.

Siebert, S., Farrell, J. A., Cazet, J. F., Abeykoon, Y., Primack, A. S., Sc$itzler, C. E., & Juliano, C. E. (2019). Stem cell differentiation trajectories in Hydra resolved at single-cell resolution. *Science, 365*(6451).

Szymanski, J. R. (2018). *Calcium imaging of the entire muscle system of Hydra reveals extensive cellular multifunctionality* [Doctoral dissertation, Columbia University].

Szymanski, J. R., & Yuste, R. (2019). Mapping the whole-body muscle activity of Hydra vulgaris. *Current Biology, 29*(11), 1807–1817.

Varoqueaux, F., Williams, E. A., Grandemange, S., Truscello, L., Kamm, K., Schierwater, B., Jékely, G., & Fasshauer, D. (2018). High cell diversity and complex peptidergic signaling underlie placozoan behavior. *Current Biology, 28*(21), 3495–3501.

Vogt, N. (2019). Looking at Hydra cells one at a time. *Nature Methods, 16*(9), 801–801.

Weir, K., Dupre, C., van Giesen, L., Lee, A. S., & Bellono, N. W. (2020). A molecular filter for the cnidarian stinging response. *eLife, 9,* e57578.

Whelan, N. V., Kocot, K. M., Moroz, L. L., & Halanych, K. M. (2015). Error, signal, and the placement of Ctenophora sister to all other animals. *Proceedings of the National Academy of Sciences, 112*(18), 5773–5778.

Chapter Seven: Roundworm Mind

Anderson, F. E., Williams, B. W., Horn, K. M., Erséus, C., Halanych, K. M., Santos, S. R., & James, S. W. (2017). Phylogenomic analyses of Crassiclitellata support major Northern and Southern Hemisphere clades and a Pangaean origin for earthworms. *BMC Evolutionary Biology, 17*(1), 1–18.

Ardiel, E. L., & Rankin, C. H. (2010). An elegant mind: Learning and memory in Caenorhabditis elegans. *Learning & Memory, 17*(4), 191–201.

Arendt, D., Bertucci, P. Y., Achim, K., & Musser, J. M. (2019). Evolution of neuronal types and families. *Current Opinion in Neurobiology, 56,* 144–152.

Arendt, D., Musser, J. M., Baker, C. V., Bergman, A., Cepko, C., Erwin, D. H., . . . Wagner, G. P. (2016). The origin and evolution of cell types. *Nature Reviews Genetics, 17*(12), 744–757.

Bargmann, C. I., Hartwieg, E., & Horvitz, H. R. (1993). Odorant-selective genes and neurons mediate olfaction in C. elegans. *Cell, 74*(3), 515–527.

Battaglia, D., Karagiannis, A., Gallopin, T., Gutch, H. W., & Cauli, B. (2013). Beyond the frontiers of neuronal types. *Frontiers in Neural Circuits, 7,* 13.

Bono, M. D., & Villu Maricq, A. (2005). Neuronal substrates of complex behaviors in C. elegans. *Annual Review of Neuroscience, 28,* 451–501.

Boyle, J. H., Jo$son, S., & Dehghani-Sanij, A. A. (2012). Adaptive undulatory locomotion of a C. elegans inspired robot. *IEEE/ASME Transactions on Mechatronics, 18*(2), 439–448.

Brown, A. E., Yemini, E. I., Grundy, L. J., Jucikas, T., & Schafer, W. R. (2013). A dictionary of behavioral motifs reveals clusters of genes affecting Caenorhabditis elegans locomotion. *Proceedings of the National Academy of Sciences, 110*(2), 791–796.

Bryden, J., & Cohen, N. (2008). Neural control of Caenorhabditis elegans forward locomotion: The role of sensory feedback. *Biological Cybernetics, 98*(4), 339–351.

Calhoun, A. J., & Murthy, M. (2017). Quantifying behavior to solve sensorimotor transformations: Advances from worms and flies. *Current Opinion in Neurobiology, 46*, 90–98.

Chen, W. L., Ko, H., Chuang, H. S., Bau, H. H., & Raizen, D. (2019). Caenorhabditis elegans exhibits positive gravitaxis. *bioRxiv*, 658229.

Cohen, N., & Sanders, T. (2014). Nematode locomotion: Dissecting the neuronal–environmental loop. *Current Opinion in Neurobiology, 25*, 99–106.

Colen, B. (2014). Turning science on its head. *The Harvard Gazette*, April 18, 2014.

Cook, S. J., Jarrell, T. A., Brittin, C. A., Wang, Y., Bloniarz, A. E., Yakovlev, M. A., Nguyen, K. C., Tang, L. T. H., Bayer, E. A., Duerr, J. S., & Bülow, H. E. (2019). Whole-animal connectomes of both Caenorhabditis elegans sexes. *Nature, 571*(7763), 63–71.

Deng, X., Ren, Q., Du, Y., Wang, G., Wu, R., & Si, X. (2014, June). Modeling the undulatory locomotion of C. elegans based on the proprioceptive mechanism. In B. Bendib, F. Krim, H. Belmili, M. F. Almi, & S. Bolouma (Eds.), *2014 IEEE 23rd International Symposium on Industrial Electronics (ISIE)* (pp. 1560–1565). IEEE.

Deng, X., Xu, J. X., Wang, J., Wang, G. Y., & Chen, Q. S. (2016). Biological modeling of the undulatory locomotion of C. elegans using dynamic neural network approach. *Neurocomputing, 186*, 207–217.

Fang-Yen, C., Wyart, M., Xie, J., Kawai, R., Kodger, T., Chen, S., Wen, Q., & Samuel, A. D. (2010). Biomechanical analysis of gait adaptation in the nematode Caenorhabditis elegans. *Proceedings of the National Academy of Sciences, 107*(47), 20323–20328.

Faumont, S., Lindsay, T. H., & Lockery, S. R. (2012). Neuronal microcircuits for decision making in C. elegans. *Current Opinion in Neurobiology, 22*(4), 580–591.

Félix, M. A., & Braendle, C. (2010). The natural history of Caenorhabditis elegans. *Current Biology, 20*(22), R965–R969.

Fieseler, C., Kunert-Graf, J., & Kutz, J. N. (2018). The control structure of the nematode Caenorhabditis elegans: Neuro-sensory integration and proprioceptive feedback. *Journal of Biomechanics, 74*, 1–8.

Gjorgjieva, J., Biron, D., & Haspel, G. (2014). Neurobiology of Caenorhabditis elegans locomotion: Where do we stand? *Bioscience, 64*(6), 476–486.

Ho, B., Barys$ikova, A., & Brown, G. W. (2018). Unification of protein abundance datasets yields a quantitative Saccharomyces cerevisiae proteome. *Cell Systems, 6*(2), 192–205.

Holbrook, R. I., & Mortimer, B. (2018). Vibration sensitivity found in Caenorhabditis elegans. *Journal of Experimental Biology, 221*(15).

Izquierdo, E. J., & Beer, R. D. (2016). The whole worm: Brain–body–environment models of C. elegans. *Current Opinion in Neurobiology, 40*, 23–30.

Jabr, F. (2012, October 2). The connectome debate: Is mapping the mind of a worm worth it? *Scientific American*.

Jonas, E., & Kording, K. (2015). Automatic discovery of cell types and microcircuitry from neural connectomics. *Elife, 4*, e04250.

Kosinski, R. A., & Zaremba, M. (2007). Dynamics of the model of the Caenorhabditis elegans neural network. *Acta Physica Polonica B, 38*(6).

Lofting, H. (1988). *The story of Doctor Dolittle* (Reprint ed.). Yearling.

Northcutt, R. G. (2012). Evolution of centralized nervous systems: Two schools of evolutionary thought. *Proceedings of the National Academy of Sciences, 109*(Supplement 1), 10626–10633.

Poinar, G. O. (2011). *The evolutionary history of nematodes: As revealed in stone, amber and mummies* (Vol. 9). Brill.

Schafer, W. R. (2005). Deciphering the neural and molecular mechanisms of C. elegans behavior. *Current Biology, 15*(17), R723–R729.

Sengupta, P., & Samuel, A. D. (2009). Caenorhabditis elegans: A model system for systems neuroscience. *Current Opinion in Neurobiology, 19*(6), 637–643.

Sharpee, T. O. (2014). Toward functional classification of neuronal types. *Neuron, 83*(6), 1329–1334.

Shepherd, G. M., Marenco, L., Hines, M. L., Migliore, M., McDougal, R. A., Carnevale, N. T., Newton, A. J., Surles-Zeigler, M., & Ascoli, G. A. (2019). Neuron names: A gene- and property-based name format, with special reference to cortical neurons. *Frontiers in Neuroanatomy, 13*, 25.

Sporns, O. (2016). *Networks of the brain* (Reprint ed.). MIT Press.

Sterling, P., & Laughlin, S. (2015). *Principles of neural design* (1st ed.). MIT Press.

Traub, J. (2020, September 9). Doctor Dolittle's talking animals still have much to say. *New York Times.*

Vars$ey, L. R., Chen, B. L., Paniagua, E., Hall, D. H., & C€lovskii, D. B. (2011). Structural properties of the Caenorhabditis elegans neuronal network. *PLoS Computational Biology, 7*(2), e1001066.

Vogt, N. (2019). Looking at Hydra cells one at a time. *Nature Methods, 16*(9), 801–801.

White, J. G., Southgate, E., Thomson, J. N., & Brenner, S. (1986). The structure of the nervous system of the nematode Caenorhabditis elegans. *Philosophical Transactions of the Royal Society B: Biological Sciences, 314*(1165), 1–340.

Zeng, H., & Sanes, J. R. (2017). Neuronal cell-type classification: Challenges, opportunities and the path forward. *Nature Reviews Neuroscience, 18*(9), 530–546.

Zhen, M., & Samuel, A. D. (2015). C. elegans locomotion: Small circuits, complex functions. *Current Opinion in Neurobiology, 33*, 117–126.

Chapter Eight: Flatworm Mind

Agata, K., Soejima, Y., Kato, K., Kobayashi, C., Umesono, Y., & Watanabe, K. (1998). Structure of the planarian central nervous system (CNS) revealed by neuronal cell markers. *Zoological Science, 15*(3), 433–440.

Akiyama, Y., Agata, K., & Inoue, T. (2018). Coordination between binocular field and spontaneous self-motion specifies the efficiency of planarians' photo-response orientation behavior. *Communications Biology, 1*(1), 1–13.

Aoki, R., Wake, H., Sasaki, H., & Agata, K. (2009). Recording and spectrum analysis of the planarian electroencephalogram. *Neuroscience, 159*(2), 908–914.

Arenas, O. M., Zaharieva, E. E., Para, A., Vásquez-Doorman, C., Petersen, C. P., & Gallio, M. (2017). A core signaling mechanism at the origin of animal nociception. *bioRxiv*, 185405.

Deochand, N., Costello, M. S., & Deochand, M. E. (2018). Behavioral research with planaria. *Perspectives on Behavior Science, 41*(2), 447–464.

Farris, S. M. (2008). Evolutionary convergence of higher brain centers spanning the protostome-deuterostome boundary. *Brain, Behavior and Evolution, 72*(2), 106–122.

Garm, A., & Nilsson, D. E. (2014). Visual navigation in starfish: First evidence for the use of vision and eyes in starfish. *Proceedings of the Royal Society B: Biological Sciences, 281*(1777), 20133011.

Holland, L. Z., Carvalho, J. E., Escriva, H., Laudet, V., Schubert, M., Shimeld, S. M., & Yu, J. K. (2013). Evolution of bilaterian central nervous systems: A single origin? *EvoDevo, 4*(1), 1–20.

Holland, N. D. (2016). Nervous systems and scenarios for the invertebrate-to-vertebrate transition. *Philosophical Transactions of the Royal Society B: Biological Sciences, 371*(1685), 20150047.

Inoue, T. (2017). Functional specification of a primitive bilaterian brain in planarians. In *Brain Evolution by Design* (pp. 79–100). Springer.

Inoue, T., Hoshino, H., Yamashita, T., Shimoyama, S., & Agata, K. (2015). Planarian shows decision-making behavior in response to multiple stimuli by integrative brain function. *Zoological Letters, 1*(1), 1–15.

Kirwan, J. D., Bok, M. J., Smolka, J., Foster, J. J., Hernández, J. C., & Nilsson, D. E. (2018). The sea urchin Diadema africanum uses low resolution vision to find shelter and deter enemies. *Journal of Experimental Biology, 221*(14).

LeDoux, J. (2020). *The deep history of ourselves: The four-billion-year story of how we got conscious brains.* Penguin Books.

Nakazawa, M., Cebrià, F., Mineta, K., Ikeo, K., Agata, K., & Gojobori, T. (2003). Search for the evolutionary origin of a brain: Planarian brain characterized by microarray. *Molecular Biology and Evolution, 20*(5), 784–791.

Okamoto, K., Takeuchi, K., & Agata, K. (2005). Neural projections in planarian brain revealed by fluorescent dye tracing. *Zoological Science, 22*(5), 535–546.

Pagán, O. R. (2014). *The first brain: The neuroscience of planarians* (1st ed.). Oxford University Press.

Perry, C. J., Barron, A. B., & Cheng, K. (2013). Invertebrate learning and cognition: Relating phenomena to neural substrate. *Wiley Interdisciplinary Reviews: Cognitive Science, 4*(5), 561–582.

Ruiz-Trillo, I., Riutort, M., Littlewood, D. T. J., Herniou, E. A., & Baguñà, J. (1999). Acoel flatworms: Earliest extant bilaterian metazoans, not members of Platyhelminthes. *Science, 283*(5409), 1919–1923.

Sarnat, H. B., & Netsky, M. G. (1985). The brain of the planarian as the ancestor of the human brain. *Canadian Journal of Neurological Sciences, 12*(4), 296–302.

Sarnat, H. B., & Netsky, M. G. (2002). When does a ganglion become a brain? Evolutionary origin of the central nervous system. *Seminars in Pediatric Neurology, 9*(4), 240–253.

Scimone, M. L., Atabay, K. D., Fincher, C. T., Bonneau, A. R., Li, D. J., & Reddien, P. W. (2020). Muscle and neuronal guidepost-like cells facilitate planarian visual system regeneration. *Science, 368*(6498).

Shettigar, N., Joshi, A., Dalmeida, R., Gopalkris$a, R., Chakravarthy, A., Patnaik, S., Mathew, M., Palakodeti, D., & Gulyani, A. (2017). Hierarchies in light sensing and dynamic interactions between ocular and extraocular sensory networks in a flatworm. *Science Advances, 3*(7), e1603025.

Chapter Nine: Fly Mind

Akin, O., Bajar, B. T., Keles, M. F., Frye, M. A., & Zipursky, S. L. (2019). Cell-type-specific patterned stimulus-independent neuronal activity in the Drosophila visual system during synapse formation. *Neuron, 101*(5), 894–904.

Aso, Y., Hattori, D., Yu, Y., Jo$ston, R. M., Iyer, N. A., Ngo, T. T., Dionne, H., Abbott, L. F., Axel, R., Tanimoto, H., & Rubin, G. M. (2014). The neuronal architecture of the mushroom body provides a logic for associative learning. *eLife, 3*, e04577.

Avarguès-Weber, A., d'Amaro, D., Metzler, M., Finke, V., Baracchi, D., & Dyer, A. G. (2018). Does holistic processing require a large brain? Insights from honeybees and wasps in fine visual recognition tasks. *Frontiers in Psychology, 9*, 1313.

Berman, G. J., Choi, D. M., Bialek, W., & Shaevitz, J. W. (2014). Mapping the stereotyped behaviour of freely moving fruit flies. *Journal of the Royal Society Interface, 11*(99), 20140672.

Burke, C. J., Huetteroth, W., Owald, D., Perisse, E., Krashes, M. J., Das, G., Gohl, D., Silies, M., Certel, S., & Waddell, S. (2012). Layered reward signalling through octopamine and dopamine in Drosophila. *Nature, 492*(7429), 433–437.

Currier, T. A., & Nagel, K. I. (2018). Multisensory control of orientation in tethered flying Drosophila. *Current Biology, 28*(22), 3533–3546.

Dolan, M. J., Belliart-Guérin, G., Bates, A. S., Frechter, S., Lampin-Saint-Amaux, A., Aso, Y., Roberts, R. J., Schlegel, P., Wong, A., Hammad, A., & Bock, D. (2018). Communication from learned to innate olfactory processing centers is required for memory retrieval in Drosophila. *Neuron, 100*(3), 651–668.

Farris, S. M. (2013). Evolution of complex higher brain centers and behaviors: Behavioral correlates of mushroom body elaboration in insects. *Brain, Behavior and Evolution, 82*(1), 9–18.

Flood, T. F., Iguchi, S., Gorczyca, M., White, B., Ito, K., & Yoshihara, M. (2013). A single pair of interneurons commands the Drosophila feeding motor program. *Nature, 499*(7456), 83–87.

Grossberg, S. (2017). Towards solving the hard problem of consciousness: The varieties of brain resonances and the conscious experiences that they support. *Neural Networks, 87*, 38–95.

Guo, A., Lu, H., Zhang, K., Ren, Q., & Wong, Y. N. C. (2013). Visual learning and decision making in Drosophila melanogaster. In R. Menzel & P. Benjamin (Eds.), *Handbook of Behavioral Neuroscience* (Vol. 22, pp. 378–394). Elsevier.

Haberkern, H., & Jayaraman, V. (2016). Studying small brains to understand the building blocks of cognition. *Current Opinion in Neurobiology, 37*, 59–65.

Hafez, O. A., Escribano, B., Ziegler, R. L., Hirtz, J. J., Niebur, E., & Pielage, J. (2020). Dendritic signal integration in a Drosophila Mushroom Body Output Neuron (MBON) essential for learning and memory. *bioRxiv*.

Heisenberg, M. (2003). Mushroom body memoir: From maps to models. *Nature Reviews Neuroscience*, *4*(4), 266–275.

Huang, Y. C., Wang, C. T., Su, T. S., Kao, K. W., Lin, Y. J., Chuang, C. C., Chiang, A. S., & Lo, C. C. (2019). A single-cell level and connectome-derived computational model of the Drosophila brain. *Frontiers in Neuroinformatics*, *12*, 99.

Huoviala, P., Dolan, M. J., Love, F., Myers, P., Frechter, S., Namiki, S., Pettersson, L., Roberts, R. J., Turnbull, R., Mitrevica, Z., & Breads, P. (2020). Neural circuit basis of aversive odour processing in Drosophila from sensory input to descending output. *bioRxiv*, 394403.

Jhala, K. (2021, March 11). WTAF? Beeple NFT work sells for astonishing $69.3m at Christie's after flurry of last-minute bids nearly crashes website. *The Art Newspaper*.

Laissue, P. P., & Vosshall, L. B. (2009). The olfactory sensory map in Drosophila. In G. M. Tec$au (Ed.), *Brain development in Drosophila melanogaster* (Vol. 628). Springer Science & Business Media.

Li, F., Lindsey, J. W., Marin, E. C., Otto, N., Dreher, M., Dempsey, G., Stark, I., Bates, A. S., Pleijzier, M. W., Schlegel, P., & Nern, A. (2020). The connectome of the adult Drosophila mushroom body provides insights into function. *eLife*, *9*, e62576.

Li, H., Horns, F., Wu, B., Xie, Q., Li, J., Li, T., Luginbuhl, D. J., Quake, S. R., & Luo, L. (2017). Classifying Drosophila olfactory projection neuron subtypes by single-cell RNA sequencing. *Cell*, *171*(5), 1206–1220.

Li, J., Mahoney, B. D., Jacob, M. S., & Caron, S. J. C. (2020). Visual input into the Drosophila melanogaster mushroom body. *Cell Reports*, *32*(11), 108138.

Li, Q., & Liberles, S. D. (2015). Aversion and attraction through olfaction. *Current Biology*, *25*(3), R120–R129.

Martin, C. A., & Krantz, D. E. (2014). Drosophila melanogaster as a genetic model system to study neurotransmitter transporters. *Neurochemistry International*, *73*, 71–88.

Masse, N. Y., Turner, G. C., & Jefferis, G. S. (2009). Olfactory information processing in Drosophila. *Current Biology*, *19*(16), R700–R713.

Namiki, S., Dickinson, M. H., Wong, A. M., Korff, W., & Card, G. M. (2018). The functional organization of descending sensory-motor pathways in Drosophila. *Elife*, *7*, e34272.

O'Regan, G. (2008). *A brief history of computing*. Springer Science & Business Media.

Resh, V. H., & Cardé, R. T. (Eds.). (2009). *Encyclopedia of insects*. Academic Press.

Reyburn, S. (2021, March 11). JPG file sells for $69 million, as "NFT mania" gathers pace. *New York Times*.

Seki, Y., Dweck, H. K., Rybak, J., Wicher, D., Sachse, S., & Hansson, B. S. (2017). Olfactory coding from the periphery to higher brain centers in the Drosophila brain. *BMC Biology*, *15*(1), 1–20.

Shannon, C. E. (1948). A mathematical theory of communication. *The Bell System Tec$ical Journal*, *27*(3), 379–423.

Sporns, O. (2016). *Networks of the brain* (Reprint ed.). MIT Press.

Stensmyr, M. C., Dweck, H. K., Farhan, A., Ibba, I., Strutz, A., Mukunda, L., Linz, J., Grabe, V., Steck, K., Lavista-Llanos, S., & Wicher, D. (2012). A conserved dedicated olfactory circuit for detecting harmful microbes in Drosophila. *Cell*, *151*(6), 1345–1357.

Takemura, S. Y., Aso, Y., Hige, T., Wong, A., Lu, Z., Xu, C. S., Rivlin, P. K., Hess, H., Zhao, T., Parag, T., & Berg, S. (2017). A connectome of a learning and memory center in the adult Drosophila brain. *eLife*, *6*, e26975.

Turing, A. M. (1937). On computable numbers, with an application to the Entscheidungsproblem. *Proceedings of the London Mathematical Society*, *2*(1), 230–265.

Turner-Evans, D. B., & Jayaraman, V. (2016). The insect central complex. *Current Biology*, *26*(11), R453–R457.

Vogt, K., Sc$aitmann, C., Dylla, K. V., Knapek, S., Aso, Y., Rubin, G. M., & Tanimoto, H. (2014). Shared mushroom body circuits underlie visual and olfactory memories in Drosophila. *eLife*, *3*, e02395.

Waddell, S. (2013). Reinforcement signalling in Drosophila: Dopamine does it all after all. *Current Opinion in Neurobiology, 23*(3), 324–329.

Webb, B. (2012). Cognition in insects. *Philosophical Transactions of the Royal Society B: Biological Sciences, 367*(1603), 2715–2722.

Wiegmann, B. M., Trautwein, M. D., Winkler, I. S., Barr, N. B., Kim, J. W., Lambkin, C., Bertone, M. A., Cassel, B. K., Bayless, K. M., Heimberg, A. M., & Wheeler, B. M. (2011). Episodic radiations in the fly tree of life. *Proceedings of the National Academy of Sciences, 108*(14), 5690–5695.

Xu, C. S., Januszewski, M., Lu, Z., Takemura, S. Y., Hayworth, K., Huang, G., Shinomiya, K., Maitin-Shepard, J., Ackerman, D., Berg, S., & Blakely, T. (2020). A connectome of the adult Drosophila central brain. *bioRxiv*, January 1.

Zhu, Y. (2013). The Drosophila visual system: From neural circuits to behavior. *Cell Adhesion & Migration, 7*(4), 333–344.

The Unified Mathematics of the Self

Anderson, J. A., & Rosenfeld, E. (Eds.). (2000). *Talking nets: An oral history of neural networks.* MIT Press.

Boden, M. A. (2008). *Mind as machine: A history of cognitive science.* Oxford University Press.

Cao, Y., & Grossberg, S. (2019). A Laminar Cortical Model for 3D Boundary and Surface Representations of Complex Natural Scenes. In A. Adamatzky, S. Akl, & G. Sirakoulis (Eds.), *From parallel to emergent computing.* CRC Press.

Carpenter, G. A., & Grossberg, S. (2017). Adaptive resonance theory. In C. Sammut & G. I. Webb (Eds.), *Encyclopedia of machine learning and data mining.* Springer.

Christianson, G. E. (2005). *Isaac Newton.* Oxford University Press.

Engel, A. K., Fries, P., and Singer, W. (2001). Dynamic predictions: oscillations and synchrony in top–down processing. *Nature Reviews Neuroscience, 2*(10), 704–716.

Fazl, A., Grossberg, S., & Mingolla, E. (2009). View-invariant object category learning, recognition, and search: How spatial and object attention are coordinated using surface-based attentional shrouds. *Cognitive Psychology, 58*(1), 1–48.

Gaudiano, P., & Grossberg, S. (1991). Vector associative maps: Unsupervised real-time error-based learning and control of movement trajectories. *Neural Networks, 4*(2), 147–183.

Gelman, A. (2008). Objections to Bayesian statistics. *Bayesian Analysis, 3*(3), 445–449.

Gleick, J. (2004). *Isaac Newton.* Vintage.

Grossberg, S. (1968). Some physiological and biochemical consequences of psychological postulates. *Proceedings of the National Academy of Sciences of the United States of America, 60*(3), 758.

Grossberg, S. (1969). On the serial learning of lists. *Mathematical Biosciences, 4,* 201–253.

Grossberg, S. (1973). Contour enhancement, short-term memory, and constancies in reverberating neural networks. *Studies in Applied Mathematics, 52,* 213–257.

Grossberg, S. (2000). The complementary brain: Unifying brain dynamics and modularity. *Trends in Cognitive Sciences, 4*(6), 233–246.

Grossberg, S. (2007). Consciousness CLEARS the mind. *Neural Networks, 20*(9), 1040–1053.

Grossberg, S. (2013). Adaptive Resonance Theory: How a brain learns to consciously attend, learn, and recognize a changing world. *Neural Networks, 37,* 1–47.

Grossberg, S. (2016). Neural dynamics of the basal ganglia during perceptual, cognitive, and motor learning and gating. In *The basal ganglia* (pp. 457–512). Springer.

Grossberg, S. (2017a). Towards solving the hard problem of consciousness: The varieties of brain resonances and the conscious experiences that they support. *Neural Networks, 87,* 38–95.

Grossberg, S. (2017b). The visual world as illusion: The ones we know and the ones we don't. In A. G. Shapiro & D. Todorović (Eds.), *The Oxford compendium of visual illusions* (pp. 90–118). Oxford University Press.

Grossberg, S. (2018). Desirability, availability, credit assignment, category learning, and attention: Cognitive-emotional and working memory dynamics of orbitofrontal, ventrolateral, and dorsolateral prefrontal cortices. *Brain and Neuroscience Advances, 2,* 1–50.

Grossberg, S. (2019a). The embodied brain of SOVEREIGN2: From space-variant conscious percepts during visual search and navigation to learning invariant object categories and cognitive-emotional plans for acquiring valued goals. *Frontiers in Computational Neuroscience, 13*, 36.

Grossberg, S. (2019b). A half century of progress toward a unified neural theory of mind and brain with applications to autonomous adaptive agents and mental disorders. In R. Kozma, C. Alippi, Y. Choe, & F. C. Morabito (Eds.), *Artificial intelligence in the age of neural networks and brain computing.* Academic Press.

Grossberg, S. (2019c). How we see art and how artists make it. In J. L. Contreras-Vidal, D. Robleto, J. G. Cruz-Garza, J. M. Azorín, & C. S. Nam (Eds.), *Mobile brain-body imaging and the neuroscience of art, innovation and creativity* (pp. 79–99). Springer.

Grossberg, S. (2019d). The resonant brain: How attentive conscious seeing regulates action sequences that interact with attentive cognitive learning, recognition, and prediction. *Attention, Perception, & Psychophysics, 81*(7), 2237–2264.

Grossberg, S. (2020a). A path toward explainable AI and autonomous adaptive intelligence: Deep learning, adaptive resonance, and models of perception, emotion, and action. *Frontiers in Neurorobotics, 14*, 36.

Grossberg, S. (2020b). Toward autonomous adaptive intelligence: Building upon neural models of how brains make minds. *IEEE Transactions on Systems, Man, and Cybernetics: Systems,* December 30.

Grossberg, S. (2020c). Towards a unified theory of mind and brain. *Common Model of Cognition Bulletin, 1*(2), 354–358.

Grossberg, S. (2020d). A unified neural theory of conscious seeing, hearing, feeling, and knowing. *Cognitive Neuroscience,* 1–5.

Grossberg, S. (2021). *Conscious mind, resonant brain: How each brain makes a mind.* Oxford University Press.

Grossberg, S., & Huang, T. R. (2009). ARTSCENE: A neural system for natural scene classification. *Journal of Vision, 9*(4), 6–6.

Grossberg, S., & Myers, C. W. (2000). The resonant dynamics of speech perception: Interword integration and duration-dependent backward effects. *Psychological Review, 107*(4), 735.

Grossberg, S., & Seldman, D. (2006). Neural dynamics of autistic behaviors: Cognitive, emotional, and timing substrates. *Psychological Review, 113*(3), 483.

Knill, D. C., & Pouget, A. (2004). The Bayesian brain: The role of uncertainty in neural coding and computation. *TRENDS in Neurosciences, 27*(12), 712–719.

Raizada, R. D., & Grossberg, S. (2003). Towards a theory of the laminar architecture of cerebral cortex: Computational clues from the visual system. *Cerebral Cortex, 13*(1), 100–113.

Shannon, C. E. (1948). A mathematical theory of communication. *The Bell System Tec$ical Journal, 27*(3), 379–423, 623–656.

Sporns, O. (2016). *Networks of the brain* (Reprint ed.). MIT Press.

Strogatz, S. H. (2018). *Nonlinear dynamics and chaos with student solutions manual: With applications to physics, biology, chemistry, and engineering.* CRC Press.

Wunsch, D. C. II. (2019). Admiring the great mountain: A celebration special issue in honor of Stephen Grossberg's 80th birthday [Special issue]. *Neural Networks, 120*, 1–4.

Chapter Ten: Fish Mind

Abbas, F., & Meyer, M. P. (2014). Fish vision: Size selectivity in the zebrafish retinotectal pathway. *Current Biology, 24*(21), R1048–R1050.

Agrillo, C., Miletto Petrazzini, M. E., & Bisazza, A. (2016). Brightness illusion in the guppy (Poecilia reticulata). *Journal of Comparative Psychology, 130*(1), 55.

Agrillo, C., Miletto Petrazzini, M. E., & Dadda, M. (2013). Illusory patterns are fishy for fish, too. *Frontiers in Neural Circuits, 7*, 137.

Albertazzi, L., Rosa-Salva, O., Da Pos, O., & Sovrano, V. A. (2015). Fish are sensitive to expansion-contraction colour effects. *Animal Behavior and Cognition, 4*(3), 349–364.

Bielmeier, C. (2020, September 28). *An optical illusion opens a window into the brain.* Max Planck Neuroscience. https://maxplanckneuroscience.org/an-optical-illusion-opens-a-window-into-the -brain

Bilotta, J., & Saszik, S. (2001). The zebrafish as a model visual system. *International Journal of Developmental Neuroscience, 19*(7), 621–629.

Binder, M. D., Hirokawa, N., & Windhorst, U. (Eds.). (2009). *Encyclopedia of neuroscience* (Vol. 3166). Springer.

Cao, Y., & Grossberg, S. (2019). A Laminar Cortical Model for 3D Boundary and Surface Representations of Complex Natural Scenes. In A. Adamatzky, S. Akl, & G. Sirakoulis (Eds.), *From parallel to emergent computing.* CRC Press.

DeLong, C. M., Fobe, I., O'Leary, T., & Wilcox, K. T. (2018). Visual perception of planar-rotated 2D objects in goldfish (Carassius auratus). *Behavioural Processes, 157*, 263–278.

Easter, S. S. Jr., & Nicola, G. N. (1996). The development of vision in the zebrafish (Danio rerio). *Developmental Biology, 180*(2), 646–663.

Feng, L. C., Chouinard, P. A., Howell, T. J., & Bennett, P. C. (2017). Why do animals differ in their susceptibility to geometrical illusions? *Psychonomic Bulletin & Review, 24*(2), 262–276.

Froese, R., & Pauly, D. (Eds.). (n.d.). FishBase. http://www.fishbase.org/home.htm

Fuss, T., Bleckmann, H., & Schluessel, V. (2014). The brain creates illusions not just for us: Sharks (Chiloscyllium griseum) can "see the magic" as well. *Frontiers in Neural Circuits, 8*, 24.

Fuss, T., & Schluessel, V. (2017). The Ebbinghaus illusion in the gray bamboo shark (Chiloscyllium griseum) in comparison to the teleost damselfish (Chromis chromis). *Zoology, 123*, 16–29.

Gori, S., Agrillo, C., Dadda, M., & Bisazza, A. (2014). Do fish perceive illusory motion? *Scientific Reports, 4*(1), 1–6.

Grossberg, S. (1997). Cortical dynamics of three-dimensional figure–ground perception of two-dimensional pictures. *Psychological Review, 104*(3), 618.

Grossberg, S. (2000). The complementary brain: Unifying brain dynamics and modularity. *Trends in Cognitive Sciences, 4*(6), 233–246.

Grossberg, S. (2016). Cortical dynamics of figure-ground separation in response to 2D pictures and 3D scenes: How V2 combines border ownership, stereoscopic cues, and gestalt grouping rules. *Frontiers in Psychology, 6*, 2054.

Grossberg, S. (2019a). A half century of progress toward a unified neural theory of mind and brain with applications to autonomous adaptive agents and mental disorders. In R. Kozma, C. Alippi, Y. Choe, & F. C. Morabito (Eds.), *Artificial intelligence in the age of neural networks and brain computing.* Academic Press.

Grossberg, S. (2019b). How we see art and how artists make it. In J. L. Contreras-Vidal, D. Robleto, J. G. Cruz-Garza, J. M. Azorín, & C. S. Nam (Eds.), *Mobile brain-body imaging and the neuroscience of art, innovation and creativity* (pp. 79–99). Springer.

Grossberg, S. (2019c). The resonant brain: How attentive conscious seeing regulates action sequences that interact with attentive cognitive learning, recognition, and prediction. *Attention, Perception, & Psychophysics, 81*(7), 2237–2264.

Holland, N. D. (2016). Nervous systems and scenarios for the invertebrate-to-vertebrate transition. *Philosophical Transactions of the Royal Society B: Biological Sciences, 371*(1685), 20150047.

Kanizsa, G. (1974). Contours without gradients or cognitive contours? *Giornale Italiano di Psicologia,* 1, 93–113.

Kanizsa, G. (1976). Subjective contours. *Scientific American, 234*, 48–52.

Kawasaka, K., Hotta, T., & Kohda, M. (2019). Does a cichlid fish process face holistically? Evidence of the face inversion effect. *Animal Cognition, 22*(2), 153–162.

Marquez Legorreta, E. (2020). Visual learning and its underlying neural substrate in two species of teleost fish (Zebrafish and Ambon damselfish) [Doctoral dissertation, University of Queensland].

Miletto Petrazzini, M. E., Parrish, A. F., Beran, M. J., & Agrillo, C. (2018). Exploring the solitaire illusion in guppies (Poecilia reticulata). *Journal of Comparative Psychology, 132*(1), 48.

Neri, P. (2012). Feature binding in zebrafish. *Animal Behaviour, 84*(2), 485–493.

Neri, P. (2020). Complex visual analysis of ecologically relevant signals in Siamese fighting fish. *Animal Cognition, 23*(1), 41–53.

Newport, C., Wallis, G., & Siebeck, U. E. (2014). Concept learning and the use of three common psychophysical paradigms in the archerfish (Toxotes chatareus). *Frontiers in Neural Circuits, 8*, 39.

Newport, C., Wallis, G., & Siebeck, U. E. (2018). Object recognition in fish: Accurate discrimination across novel views of an unfamiliar object category (human faces). *Animal Behaviour, 145*, 39–49.

Portugues, R., & Engert, F. (2009). The neural basis of visual behaviors in the larval zebrafish. *Current Opinion in Neurobiology, 19*(6), 644–647.

Qiu, F. T., & Von Der Heydt, R. (2005). Figure and ground in the visual cortex: V2 combines stereoscopic cues with Gestalt rules. *Neuron, 47*(1), 155–166.

Rosa Salva, O., Sovrano, V. A., & Vallortigara, G. (2014). What can fish brains tell us about visual perception? *Frontiers in Neural Circuits, 8*, 119.

Sovrano, V. A., Albertazzi, L., & Salva, O. R. (2015). The Ebbinghaus illusion in a fish (Xenotoca eiseni). *Animal Cognition, 18*(2), 533–542.

Sovrano, V. A., Da Pos, O., & Albertazzi, L. (2016). The Müller-Lyer illusion in the teleost fish Xenotoca eiseni. *Animal Cognition, 19*(1), 123–132.

Wu, Y., Dal Maschio, M., Kubo, F., & Baier, H. (2020). An optical illusion pinpoints an essential circuit node for global motion processing. *Neuron, 108*(4), 722–734.

Wyzisk, K., & Neumeyer, C. (2007). Perception of illusory surfaces and contours in goldfish. *Visual Neuroscience, 24*(3), 291.

Zoccolan, D., Cox, D. D., & Benucci, A. (2015). What can simple brains teach us about how vision works. *Frontiers in Neural Circuits, 9*, 51.

Chapter Eleven: Frog Mind

Anderson, C. W. (2001). Anatomical evidence for brainstem circuits mediating feeding motor programs in the leopard frog, Rana pipiens. *Experimental Brain Research, 140*(1), 12–19.

Birinyi, A., Rácz, N., Kecskes, S., Matesz, C., & Kovalecz, G. (2018). Neural circuits underlying jaw movements for the prey-catching behavior in frog: Distribution of vestibular afferent terminals on motoneurons supplying the jaw. *Brain Structure and Function, 223*(4), 1683–1696.

Bullock, D., & Grossberg, S. (1989). VITE and FLETE: Neural modules for trajectory formation and postural control. *Advances in Psychology, 62*, 253–297.

Chipman, H., George, E. I., McCulloch, R. E., Clyde, M., Foster, D. P., & Stine, R. A. (2001). The practical implementation of Bayesian model selection. *Lecture Notes—Monograph Series*, 65 134.

Combes, P. P., & Overman, H. G. (2004). The spatial distribution of economic activities in the European Union. In *Handbook of regional and urban economics* (Vol. 4, pp. 2845–2909). Elsevier.

Corbacho, F., Nishikawa, K. C., Weerasuriya, A., Liaw, J. S., & Arbib, M. A. (2005). Schema-based learning of adaptable and flexible prey-catching in anurans: I. The basic architecture. *Biological Cybernetics, 93*(6), 391–409.

Gallwey, W. T. (2014). *The inner game of tennis: The ultimate guide to the mental side of peak performance.* Pan Macmillan.

Gaudiano, P., & Grossberg, S. (1991). Vector associative maps: Unsupervised real-time error-based learning and control of movement trajectories. *Neural Networks, 4*(2), 147–183.

Gaudiano, P., & Grossberg, S. (1992). Adaptive vector integration to endpoint: Self-organizing neural circuits for control of planned movement trajectories. *Human Movement Science, 11*(1–2), 141–155.

Grossberg, S. (2017). Towards solving the hard problem of consciousness: The varieties of brain resonances and the conscious experiences that they support. *Neural Networks, 87*, 38–95.

Kecskes, S., Matesz, C., Gaál, B., & Birinyi, A. (2016). Neural circuits underlying tongue movements for the prey-catching behavior in frog: Distribution of primary afferent terminals on motoneurons supplying the tongue. *Brain Structure and Function, 221*(3), 1533–1553.

Kovalecz, G., Kecskes, S., Birinyi, A., & Matesz, C. (2015). Possible neural network mediating jaw opening during prey-catching behavior of the frog. *Brain Research Bulletin, 119*, 19–24.

Mandal, R., & Anderson, C. W. (2009). Anatomical organization of brainstem circuits mediating feeding motor programs in the marine toad, Bufo marinus. *Brain Research, 1298*, 99–110.

Manzano, A. S., Herrel, A., Fabre, A. C., & Abdala, V. (2017). Variation in brain anatomy in frogs and its possible bearing on their locomotor ecology. *Journal of Anatomy, 231*(1), 38–58.

Monroy, J. A., & Nishikawa, K. (2011). Prey capture in frogs: Alternative strategies, biomechanical trade-offs, and hierarchical decision making. *Journal of Experimental Zoology Part A: Ecological Genetics and Physiology, 315*(2), 61–71.

Nishikawa, K. C. (1999). Neuromuscular control of prey capture in frogs. *Philosophical Transactions of the Royal Society B: Biological Sciences, 354*(1385), 941–954.

Sakes, A., van der Wiel, M., Henselmans, P. W., van Leeuwen, J. L., Dodou, D., & Breedveld, P. (2016). Shooting mechanisms in nature: A systematic review. *PLoS One, 11*(7), e0158277.

Weitzenfeld, A. (2008). From schemas to neural networks: A multi-level modelling approach to biologically-inspired autonomous robotic systems. *Robotics and Autonomous Systems, 56*(2), 177–197.

Chapter Twelve: Tortoise Mind

Carpenter, G. A., & Grossberg, S. (2017). Adaptive resonance theory. In C. Sammut & G. I. Webb (Eds.), *Encyclopedia of machine learning and data mining*. Springer.

Fournier, J., Müller, C. M., & Laurent, G. (2015). Looking for the roots of cortical sensory computation in three-layered cortices. *Current Opinion in Neurobiology, 31*, 119–126.

Fournier, J., Müller, C. M., Sc$eider, I., & Laurent, G. (2018). Spatial information in a non-retinotopic visual cortex. *Neuron, 97*(1), 164–180.

Fro$wieser, A., Pike, T. W., Murray, J. C., & Wilkinson, A. (2019). Perception of artificial conspecifics by bearded dragons (Pogona vitticeps). *Integrative Zoology, 14*(2), 214–222.

Gaudiano, P., & Grossberg, S. (1991). Vector associative maps: Unsupervised real-time error-based learning and control of movement trajectories. *Neural Networks, 4*(2), 147–183.

Glavaschi, A. and Beaumont, E. (2014). The escape behaviour of wild Greek tortoises Testudo graeca with and emphasis on geometrical shape discrimination. *Basic and Applied Herpetology, 28*, 21–33.

Grossberg, S. (1982). How does a brain build a cognitive code? *Studies of mind and brain: Neural principles of learning, perception, development, cognition, and motor control* (pp. 1–52). D. Reidel Publishing Co.

Grossberg, S. (2013). Adaptive Resonance Theory: How a brain learns to consciously attend, learn, and recognize a changing world. *Neural Networks, 37*, 1–47.

Grossberg, S. (2017). Towards solving the hard problem of consciousness: The varieties of brain resonances and the conscious experiences that they support. *Neural Networks, 87*, 38–95.

Grossberg, S. (2020). A path toward explainable AI and autonomous adaptive intelligence: Deep learning, adaptive resonance, and models of perception, emotion, and action. *Frontiers in Neurorobotics, 14*, 36.

Grossberg, S. (2021). *Conscious mind, resonant brain: How each brain makes a mind*. Oxford University Press.

Hay, J. M., Subramanian, S., Millar, C. D., Mohandesan, E., & Lambert, D. M. (2008). Rapid molecular evolution in a living fossil. *Trends in Genetics, 24*(3), 106–109.

Hoseini, M. S., Pobst, J., Wright, N. C., Clawson, W., Shew, W., & Wessel, R. (2018). The turtle visual system mediates a complex spatiotemporal transformation of visual stimuli into cortical activity. *Journal of Comparative Physiology A, 204*(2), 167–181.

Kis, A., Huber, L., & Wilkinson, A. (2015). Social learning by imitation in a reptile (Pogona vitticeps). *Animal Cognition, 18*(1), 325–331.

Northcutt, R. G. (2013). Variation in reptilian brains and cognition. *Brain, Behavior and Evolution, 82*(1), 45–54.

Raizada, R. D., & Grossberg, S. (2003). Towards a theory of the laminar architecture of cerebral cortex: Computational clues from the visual system. *Cerebral Cortex, 13*(1), 100–113.

Roth, T. C., Krochmal, A. R., & LaDage, L. D. (2019). Reptilian cognition: A more complex picture via integration of neurological mechanisms, behavioral constraints, and evolutionary context. *BioEssays, 41*(8), 1900033.

Santacà, M., Miletto Petrazzini, M. E., Agrillo, C., & Wilkinson, A. (2019). Can reptiles perceive visual illusions? Delboeuf illusion in red-footed tortoise (Chelonoidis carbonaria) and bearded dragon (Pogona vitticeps). *Journal of Comparative Psychology, 133*(4), 419.

Santacà, M., Miletto Petrazzini, M. E., Agrillo, C., & Wilkinson, A. (2020). Exploring the Müller-Lyer illusion in a nonavian reptile (Pogona vitticeps). *Journal of Comparative Psychology, 134,* 391–400.

Santacà, M., Petrazzini, M. E. M., Wilkinson, A., & Agrillo, C. (2020). Anisotropy of perceived space in non-primates? The horizontal-vertical illusion in bearded dragons (Pogona vitticeps) and red-footed tortoises (Chelonoidis carbonaria). *Behavioural Processes, 176,* 104117.

Spotila, J. R., & Tomillo, P. S. (Eds.). (2015). *The leatherback turtle: Biology and conservation.* JHU Press.

Strogatz, S. (2004). *Sync: The emerging science of spontaneous order.* Penguin UK.

Wilkinson, A., & Glass, E. (2018). 18 tortoises–cold-blooded cognition: How to get a tortoise out of its shell. In N. Bueno-Guerra & F. Amici (Eds.), *Field and laboratory methods in animal cognition: A comparative guide* (p. 401). Cambridge University Press.

Wilkinson, A., Kuenstner, K., Mueller, J., & Huber, L. (2010). Social learning in a non-social reptile (Geochelone carbonaria). *Biology Letters, 6*(5), 614–616.

Wilkinson, A., Mandl, I., Bugnyar, T., & Huber, L. (2010). Gaze following in the red-footed tortoise (Geochelone carbonaria). *Animal Cognition, 13*(5), 765–769.

Wilkinson, A., Mueller-Paul, J., & Huber, L. (2013). Picture–object recognition in the tortoise Chelonoidis carbonaria. *Animal Cognition, 16*(1), 99–107.

Wilkinson, A., Sebanz, N., Mandl, I., & Huber, L. (2011). No evidence of contagious yawning in the red-footed tortoise Geochelone carbonaria. *Current Zoology, 57*(4), 477–484.

Chapter Thirteen: Rat Mind

Aggarwal, A. (2016). The sensori-motor model of the hippocampal place cells. *Neurocomputing, 185,* 142–152.

Barrera, A., & Weitzenfeld, A. (2008). Computational modeling of spatial cognition in rats and robotic experimentation: Goal-oriented navigation and place recognition in multiple directions. In *2008 2nd IEEE RAS & EMBS International Conference on Biomedical Robotics and Biomechatronics* (pp. 789–794). IEEE.

Bevins, R. A., & Besheer, J. (2006). Object recognition in rats and mice: A one-trial non-matching-to-sample learning task to study "recognition memory." *Nature Protocols, 1*(3), 1306–1311.

Bicanski, A., & Burgess, N. (2018). A neural-level model of spatial memory and imagery. *eLife, 7,* e33752.

Buzsáki, G., & Tingley, D. (2018). Space and time: The hippocampus as a sequence generator. *Trends in Cognitive Sciences, 22*(10), 853–869.

Chang, H. C., Grossberg, S., & Cao, Y. (2014). Where's Waldo? How perceptual, cognitive, and emotional brain processes cooperate during learning to categorize and find desired objects in a cluttered scene. *Frontiers in Integrative Neuroscience, 8,* 43.

Churakov, G., Sadasivuni, M. K., Rosenbloom, K. R., Huchon, D., Brosius, J., & Schmitz, J. (2010). Rodent evolution: Back to the root. *Molecular Biology and Evolution, 27*(6), 1315–1326.

Fazl, A., Grossberg, S., & Mingolla, E. (2009). View-invariant object category learning, recognition, and search: How spatial and object attention are coordinated using surface-based attentional shrouds. *Cognitive Psychology, 58*(1), 1–48.

Geva-Sagiv, M., Las, L., Yovel, Y., & Ulanovsky, N. (2015). Spatial cognition in bats and rats: From sensory acquisition to multiscale maps and navigation. *Nature Reviews Neuroscience, 16*(2), 94–108.

Gibbs, R. A., & Pachter, L. (2004). Genome sequence of the brown Norway rat yields insights into mammalian evolution. *Nature, 428*(6982), 493–521.

Grossberg, S. (2014). How visual illusions illuminate complementary brain processes: illusory depth from brightness and apparent motion of illusory contours. *Frontiers in Human Neuroscience, 8*, 854.

Grossberg, S. (2021). *Conscious mind, resonant brain: How each brain makes a mind.* Oxford University Press.

Ito, H. T. (2018). Prefrontal–hippocampal interactions for spatial navigation. *Neuroscience Research, 129*, 2–7.

Joesch, M., Sc$ell, B., Raghu, S. V., Reiff, D. F., & Borst, A. (2010). ON and OFF pathways in Drosophila motion vision. *Nature, 468*(7321), 300–304.

Matsumoto, J., Makino, Y., Miura, H., & Yano, M. (2011). A computational model of the hippocampus that represents environmental structure and goal location, and guides movement. *Biological Cybernetics, 105*(2), 139–152.

Modlinska, K., & Pisula, W. (2020). The natural history of model organisms: The Norway rat, from an obnoxious pest to a laboratory pet. *eLife, 9*, e50651.

Purves, D., Augustine, G. J., Fitzpatrick, D., Hall, W. C., LaMantia, A., Mooney, R. D., Platt, M. L., & White, L. E. (Eds.). (2017). *Neuroscience* (6th ed.). Oxford University Press.

Rowland, D. C., Roudi, Y., Moser, M. B., & Moser, E. I. (2016). Ten years of grid cells. *Annual Review of Neuroscience, 39*, 19–40.

Schwartz, D. M., & Koyluoglu, O. O. (2019). On the organization of grid and place cells: Neural denoising via subspace learning. *Neural Computation, 31*(8), 1519–1550.

Sweetman, S., Smith, G., & Martill, D. (2017). Highly derived eutherian mammals from the earliest Cretaceous of southern Britain. *Acta Palaeontologica Polonica, 62*.

Tafazoli, S., Safaai, H., De Franceschi, G., Rosselli, F. B., Vanzella, W., Riggi, M., Buffolo, F., Panzeri, S., & Zoccolan, D. (2017). Emergence of transformation-tolerant representations of visual objects in rat lateral extrastriate cortex. *eLife, 6*, e22794.

Tejera, G., Llofriu, M., Barrera, A., & Weitzenfeld, A. (2018). Bio-inspired robotics: A spatial cognition model integrating place cells, grid cells and head direction cells. *Journal of Intelligent & Robotic Systems, 91*(1), 85–99.

Tyler, C. W., & Kontsevich, L. L. (1995). Mechanisms of stereoscopic processing: stereoattention and surface perception in depth reconstruction. *Perception, 24*(2), 127–53.

Van Wyk, M., Wässle, H., & Taylor, W. R. (2009). Receptive field properties of ON-and OFF-ganglion cells in the mouse retina. *Visual Neuroscience, 26*(3), 297.

Wansley, L., & Kennedy, M. (2021, June). *After years of detecting land mines, a heroic rat is hanging up his sniffer.* National Public Radio.

Chapter Fourteen: Bird Mind

Adam, I., & Elemans, C. P. (2019). Vocal motor performance in birdsong requires brain–body interaction. *Eneuro, 6*(3).

Armstrong, E., & Abarbanel, H. D. (2016). Model of the songbird nucleus HVC as a network of central pattern generators. *Journal of Neurophysiology, 116*(5), 2405–2419.

Bertram, R., Daou, A., Hyson, R. L., Jo$son, F., & Wu, W. (2014). Two neural streams, one voice: Pathways for theme and variation in the songbird brain. *Neuroscience, 277*, 806–817.

Bertram, R., Hyson, R. L., Brunick, A. J., Flores, D., & Jo$son, F. (2020). Network dynamics underlie learning and performance of birdsong. *Current Opinion in Neurobiology, 64*, 119–126.

Berwick, R. C., Okanoya, K., Beckers, G. J., & Bolhuis, J. J. (2011). Songs to syntax: The linguistics of birdsong. *Trends in Cognitive Sciences, 15*(3), 113–121.

Betts, M. G., Hadley, A. S., Rodenhouse, N., & Nocera, J. J. (2008). Social information trumps vegetation structure in breeding site selection by a migrant songbird. *Proceedings of the Royal Society B: Biological Sciences, 275*(1648), 2257–2263.

Bradski, G., Carpenter, G. A., & Grossberg, S. (1994). STORE working memory networks for storage and recall of arbitrary temporal sequences. *Biological Cybernetics, 71*(6), 469–480.

Chen, R., Bollu, T., & Goldberg, J. H. (2018). A stable neural code for birdsong. *Neuron, 98*(6), 1057–1059.

DeCandido, R., & Allen, D. (2010). The falcon that nests on Broadway. *Winging It, 22*(4).

Dima, G. C., Goldin, M. A., & Mindlin, B. G. (2018). Modeling temperature manipulations in a circular model of birdsong production. *Papers in Physics, 10,* 100002.

Dooling, R. J., & Prior, N. H. (2017). Do we hear what birds hear in birdsong? *Animal Behaviour, 124,* 283–289.

Engesser, S., Crane, J. M., Savage, J. L., Russell, A. F., & Townsend, S. W. (2015). Experimental evidence for phonemic contrasts in a nonhuman vocal system. *PLoS Biology, 13*(6), e1002171.

Fehér, O., Wang, H., Saar, S., Mitra, P. P., & Tchernichovski, O. (2009). De novo establishment of wild-type song culture in the zebra finch. *Nature, 459*(7246), 564–568.

Fiete, I. R., & Seung, H. S. (2007). Neural network models of birdsong production, learning, and coding. In L. Squire, T. Albright, F. Bloom, F. Gage, and N. Spitzer (Eds.), *New encyclopedia of neuroscience.* Elsevier.

Fortune, E. S., Rodríguez, C., Li, D., Ball, G. F., & Coleman, M. J. (2011). Neural mechanisms for the coordination of duet singing in wrens. *Science, 334*(6056), 666–670.

Galvis, D., Wu, W., Hyson, R. L., Jo$son, F., & Bertram, R. (2017). A distributed neural network model for the distinct roles of medial and lateral HVC in zebra finch song production. *Journal of Neurophysiology, 118*(2), 677–692.

Galvis, D., Wu, W., Hyson, R. L., Jo$son, F., & Bertram, R. (2018). Interhemispheric dominance switching in a neural network model for birdsong. *Journal of Neurophysiology, 120*(3), 1186–1197.

Grossberg, S. (2017). Towards solving the hard problem of consciousness: The varieties of brain resonances and the conscious experiences that they support. *Neural Networks, 87,* 38–95.

Grossberg, S., & Pearson, L. R. (2008). Laminar cortical dynamics of cognitive and motor working memory, sequence learning and performance: Toward a unified theory of how the cerebral cortex works. *Psychological Review, 115*(3), 677.

Hamaguchi, K., Tanaka, M., & Mooney, R. (2016). A distributed recurrent network contributes to temporally precise vocalizations. *Neuron, 91*(3), 680–693.

Jelbert, S. A., Hosking, R. J., Taylor, A. H., & Gray, R. D. (2018). Mental template matching is a potential cultural transmission mechanism for New Caledonian crow tool manufacturing traditions. *Scientific Reports, 8*(1), 1–8.

Lashley, K. S. (1951). *The problem of serial order in behavior* (Vol. 21). Bobbs-Merrill.

Mackevicius, E. L., & Fee, M. S. (2018). Building a state space for song learning. *Current Opinion in Neurobiology, 49,* 59–68.

Mindlin, G. B. (2017). Nonlinear dynamics in the study of birdsong. *Chaos: An Interdisciplinary Journal of Nonlinear Science, 27*(9), 092101.

Morinay, J., Cardoso, G. C., Doutrelant, C., & Covas, R. (2013). The evolution of birdsong on islands. *Ecology and Evolution, 3*(16), 5127–5140.

Moyle, R. G., Oliveros, C. H., Andersen, M. J., Hosner, P. A., Benz, B. W., Manthey, J. D., Travers, S. L., Brown, R. M., & Faircloth, B. C. (2016). Tectonic collision and uplift of Wallacea triggered the global songbird radiation. *Nature Communications, 7*(1), 1–7.

Nowicki, S., Searcy, W. A., Hughes, M., & Podos, J. (2001). The evolution of bird song: Male and female response to song innovation in swamp sparrows. *Animal Behaviour, 62*(6), 1189–1195.

Podos, J., Huber, S. K., & Taft, B. (2004). Bird song: The interface of evolution and mechanism. *Annual Review of Ecology, Evolution, and Systematics, 35,* 55–87.

Saini, A. (2021, April 14). *The link between birdsong and language.* BBC Earth.

Schmidt, M. F., & Goller, F. (2016). Breathtaking songs: Coordinating the neural circuits for breathing and singing. *Physiology, 31*(6), 442–451.

Sterling, P., & Laughlin, S. (2015). *Principles of neural design.* MIT Press.

Swaddle, J. P. (2019). Zebra finches. In J. Chloe (Ed.), *Encyclopedia of animal behavior.* Academic Press.

Templeton, C. N., Ríos-Chelén, A. A., Quirós-Guerrero, E., Mann, N. I., & Slater, P. J. (2013). Female happy wrens select songs to cooperate with their mates rather than confront intruders. *Biology Letters, 9*(1), 20120863.

Teşileanu, T., Ölveczky, B., & Balasubramanian, V. (2017). Rules and mechanisms for efficient two-stage learning in neural circuits. *eLife, 6,* e20944.

Xiao, L., Chattree, G., Oscos, F. G., Cao, M., Wanat, M. J., & Roberts, T. F. (2018). A basal ganglia circuit sufficient to guide birdsong learning. *Neuron, 98*(1), 208–221.

Chapter Fifteen: Monkey Mind

Arbib, M. A. (2016). Towards a computational comparative neuroprimatology: Framing the language-ready brain. *Physics of Life Reviews, 16,* 1–54.

Beul, S. F., Barbas, H., & Hilgetag, C. C. (2017). A predictive structural model of the primate connectome. *Scientific Reports, 7*(1), 1–12.

Beul, S. F., & Hilgetag, C. C. (2019). Neuron density fundamentally relates to architecture and connectivity of the primate cerebral cortex. *NeuroImage, 189,* 777–792.

Butler, D., & Suddendorf, T. (2014). Reducing the neural search space for hominid cognition: What distinguishes human and great ape brains from those of small apes? *Psychonomic Bulletin & Review, 21*(3), 590–619.

Capitanio, J. P., & Emborg, M. E. (2008). Contributions of non-human primates to neuroscience research. *The Lancet, 371*(9618), 1126–1135.

Costa, V. D., Mitz, A. R., & Averbeck, B. B. (2019). Subcortical substrates of explore-exploit decisions in primates. *Neuron, 103*(3), 533–545.

Croxson, P. L., Forkel, S. J., Cerliani, L., & Thiebaut de Schotten, M. (2018). Structural variability across the primate brain: A cross-species comparison. *Cerebral Cortex, 28*(11), 3829–3841.

De Waal, F., & Ferrari, P. F. E. (2012). *The primate mind: Built to connect with other minds.* Harvard University Press.

Donahue, C. J., Glasser, M. F., Preuss, T. M., Rilling, J. K., & Van Essen, D. C. (2018). Quantitative assessment of prefrontal cortex in humans relative to nonhuman primates. *Proceedings of the National Academy of Sciences, 115*(22), E5183–E5192.

Dranias, M. R., Grossberg, S., & Bullock, D. (2008). Dopaminergic and non-dopaminergic value systems in conditioning and outcome-specific revaluation. *Brain Research, 1238,* 239–287.

Fagot, J., Boë, L. J., Berthomier, F., Claidière, N., Malassis, R., Meguerditchian, A., Rey, A., & Montant, M. (2019). The baboon: A model for the study of language evolution. *Journal of Human Evolution, 126,* 39–50.

Franklin, D. J., & Grossberg, S. (2017). A neural model of normal and abnormal learning and memory consolidation: Adaptively timed conditioning, hippocampus, amnesia, neurotrophins, and consciousness. *Cognitive, Affective, & Behavioral Neuroscience, 17*(1), 24–76.

Grossberg, S. (1971). On the dynamics of operant conditioning. *Journal of Theoretical Biology, 33*(2), 225–255.

Grossberg, S. (2013). Adaptive Resonance Theory: How a brain learns to consciously attend, learn, and recognize a changing world. *Neural Networks, 37,* 1–47.

Grossberg, S. (2018). Desirability, availability, credit assignment, category learning, and attention: Cognitive-emotional and working memory dynamics of orbitofrontal, ventrolateral, and dorsolateral prefrontal cortices. *Brain and Neuroscience Advances, 2,* 1–50.

Grossberg, S., Bullock, D., & Dranias, M. R. (2008). Neural dynamics underlying impaired autonomic and conditioned responses following amygdala and orbitofrontal lesions. *Behavioral Neuroscience, 122*(5), 1100.

Grossberg, S., & Seidman, D. (2006). Neural dynamics of autistic behaviors: Cognitive, emotional, and timing substrates. *Psychological Review, 113*(3), 483.

Hanson, K. L., Hrvoj-Mihic, B., & Semendeferi, K. (2014). A dual comparative approach: Integrating lines of evidence from human evolutionary neuroanatomy and neurodevelopmental disorders. *Brain, Behavior and Evolution, 84*(2), 135–155.

Herculano-Houzel, S., Collins, C. E., Wong, P., & Kaas, J. H. (2007). Cellular scaling rules for primate brains. *Proceedings of the National Academy of Sciences, 104*(9), 3562–3567.

Hofman, M. A., & Falk, D. (Eds.). (2012). *Evolution of the primate brain: From neuron to behavior* (Vol. 195). Elsevier.

Hutchison, R. M., & Everling, S. (2012). Monkey in the middle: Why non-human primates are needed to bridge the gap in resting-state investigations. *Frontiers in Neuroanatomy, 6,* 29.

Iriki, A., & Sakura, O. (2008). The neuroscience of primate intellectual evolution: Natural selection and passive and intentional niche construction. *Philosophical Transactions of the Royal Society B: Biological Sciences, 363*(1500), 2229–2241.

Kaas, J. H. (2019). The origin and evolution of neocortex: From early mammals to modern humans. *Progress in Brain Research, 250,* 61–81.

Kastner, S. (2016). Evolution of visual attention in primates. In *Evolution of nervous systems* (2nd ed., pp. 237–256). Elsevier.

Plath, S. (2008). *The bell jar.* Faber & Faber.

Ravosa, M. J., & Dagosto, M. (Eds.). (2007). *Primate origins: Adaptations and evolution.* Springer Science & Business Media.

Reid, A. T., Lewis, J., Bezgin, G., Khundrakpam, B., Eickhoff, S. B., McIntosh, A. R., Bellec, P., & Evans, A. C. (2016). A cross-modal, cross-species comparison of connectivity measures in the primate brain. *Neuroimage, 125,* 311–331.

Ruppert, N., Holzner, A., & Widdig, A. (2020). Of pig-tails and palm oil: How rat-eating macaques increase oil palm sustainability. *The Science Breaker, 6*(2).

Shen, K., Bezgin, G., Schirner, M., Ritter, P., Everling, S., & McIntosh, A. R. (2019). A macaque connectome for large-scale network simulations in TheVirtualBrain. *Scientific Data, 6*(1), 1–12.

Smaers, J. B., & Vanier, D. R. (2019). Brain size expansion in primates and humans is explained by a selective modular expansion of the cortico-cerebellar system. *Cortex, 118,* 292–305.

Tardif, S. D., Coleman, K., Hobbs, T. R., & Lutz, C. (2013). IACUC review of nonhuman primate research. *ILAR Journal, 54*(2), 234–245.

Van Schaik, C. P. (2016). *The primate origins of human nature.* Jo$ Wiley & Sons.

Wilson, V. A., Kade, C., Moeller, S., Treue, S., Kagan, I., & Fischer, J. (2020). Macaque gaze responses to the primatar: A virtual macaque head for social cognition research. *Frontiers in Psychology, 11,* 1645.

Zhang, D., Guo, L., Zhu, D., Li, K., Li, L., Chen, H., Zhao, Q., Hu, X., & Liu, T. (2013). Diffusion tensor imaging reveals evolution of primate brain architectures. *Brain Structure and Function, 218*(6), 1429–1450.

Chapter Sixteen: Chimpanzee Mind

Aso, Y., Hattori, D., Yu, Y., Jo$ston, R. M., Iyer, N. A., Ngo, T. T., Dionne, H., Abbott, L. F., Axel, R., Tanimoto, H., & Rubin, G. M. (2014). The neuronal architecture of the mushroom body provides a logic for associative learning. *elife, 3,* e04577.

Berke, J. D., Hetrick, V., Breck, J., & Greene, R. W. (2008). Transient 23–30 Hz oscillations in mouse hippocampus during exploration of novel environments. *Hippocampus, 18*(5), 519–529.

Birkhead, T. (2012). *Bird sense: What it's like to be a bird.* Bloomsbury.

Buffalo, E. A., Fries, P., Landman, R., Buschman, T. J., & Desimone, R. (2011). Laminar differences in gamma and alpha coherence in the ventral stream. *Proceedings of the National Academy of Sciences, 108*(27), 11262–11267.

Buschman, T. J., & Miller, E. K. (2009). Serial, covert shifts of attention during visual search are reflected by the frontal eye fields and correlated with population oscillations. *Neuron*, *63*(3), 386–396.

Buzsaki, G. (2006). *Rhythms of the brain*. Oxford University Press.

Chalmers, D. J. (1996). *The conscious mind: In search of a fundamental theory*. Oxford Paperbacks.

Crick, F., & Koch, C. (2003). A framework for consciousness. *Nature Neuroscience*, *6*(2), 119–126.

Crick, F. C., & Koch, C. (2005). What is the function of the claustrum? *Philosophical Transactions of the Royal Society B: Biological Sciences*, *360*(1458), 1271–1279.

Dennett, D. (2000). Commentary on Chalmers "Facing backwards on the problem of consciousness." *Journal of Consciousness Studies*, *3*(1) (Special issue—part 2), 4–6.

Dennett, D. C. (2017). *From bacteria to Bach and back: The evolution of minds*. W. W. Norton.

Douglas, A. E. (2018). *Fundamentals of microbiome science: How microbes shape animal biology*. Princeton University Press.

Edelman, G. M., & Tononi, G. (2003). A universe of consciousness: How matter becomes imagination. *Contemporary Psychology*, *48*, 92.

Feinberg, T. E., & Mallatt, J. (2013). The evolutionary and genetic origins of consciousness in the Cambrian Period over 500 million years ago. *Frontiers in Psychology*, *4*, 667.

Grossberg, S. (2007). Consciousness CLEARS the mind. *Neural Networks*, *20*(9), 1040–1053.

Grossberg, S. (2013). Recurrent neural networks. *Scholarpedia*, *8*(2), 1888.

Grossberg, S. (2017). Towards solving the hard problem of consciousness: The varieties of brain resonances and the conscious experiences that they support. *Neural Networks*, *87*, 38–95.

Grossberg, S. (2019). The embodied brain of SOVEREIGN2: From space-variant conscious percepts during visual search and navigation to learning invariant object categories and cognitive-emotional plans for acquiring valued goals. *Frontiers in Computational Neuroscience*, *13*, 36.

Grossberg, S. (2021). *Conscious mind, resonant brain: How each brain makes a mind*. Oxford University Press.

Holloway, E. (1926). *Whitman: An interpretation in narrative*. A.A. Knopf.

Hopkins, W. D., Li, X., & Roberts, N. (2019). More intelligent chimpanzees (Pan troglodytes) have larger brains and increased cortical thickness. *Intelligence*, *74*, 18–24.

Hulse, B. K., Haberkern, H., Franconville, R., Turner-Evans, D. B., Takemura, S., Wolff, T., Noorman, M., Dreher, M., Dan, C., Parekh, R., & Hermundstad, A. M. (2020). A connectome of the Drosophila central complex reveals network motifs suitable for flexible navigation and context-dependent action selection. *bioRxiv*.

James, W. (1904). Does "consciousness" exist? *The Journal of Philosophy, Psychology and Scientific Methods*, *1*(18), 477–491.

Jun, J. J., Longtin, A., & Maler, L. (2016). Active sensing associated with spatial learning reveals memory-based attention in an electric fish. *Journal of Neurophysiology*, *115*(5), 2577–2592.

Krauzlis, R. J., Bogadhi, A. R., Herman, J. P., & Bollimunta, A. (2018). Selective attention without a neocortex. *Cortex*, *102*, 161–175.

Kühl, H. S., Kalan, A. K., Arandjelovic, M., Aubert, F., D'Auvergne, L., Goedmakers, A., Jones, S., Kehoe, L., Regnaut, S., Tickle, A., & Ton, E. (2016). Chimpanzee accumulative stone throwing. *Scientific Reports*, *6*(1), 1–8.

Lundqvist, M., Herman, P., Warden, M. R., Brincat, S. L., & Miller, E. K. (2018). Gamma and beta bursts during working memory readout suggest roles in its volitional control. *Nature Communications*, *9*(1), 1–12.

Magyary, I. (2019). Floating novel object recognition in adult zebrafish: A pilot study. *Cognitive Processing*, *20*(3), 359–362.

Marachlian, E., Avitan, L., Goodhill, G. J., & Sumbre, G. (2018). Principles of functional circuit connectivity: Insights from spontaneous activity in the zebrafish optic tectum. *Frontiers in Neural Circuits*, *12*, 46.

Mason, J. (1973). Walt Whitman's catalogues: Rhetorical means for two journeys in "Song of Myself." *American Literature*, *45*(1), 34–49.

Newport, C., Wallis, G., & Siebeck, U. E. (2018). Object recognition in fish: Accurate discrimination across novel views of an unfamiliar object category (human faces). *Animal Behaviour, 145*, 39–49.

Northmore, D. P. (2017). Holding visual attention for 400 million years: A model of tectum and torus longitudinalis in teleost fishes. *Vision Research, 131*, 44–56.

Parker, A. N., Wallis, G. M., Obergrussberger, R., & Siebeck, U. E. (2020). Categorical face perception in fish: How a fish brain warps reality to dissociate "same" from "different." *Journal of Comparative Neurology, 528*(17), 2919–2928.

Pruetz, J. D., & Herzog, N. M. (2017). Savanna chimpanzees at Fongoli, Senegal, navigate a fire landscape. *Current Anthropology, 58*(S16), S337–S350.

Pruetz, J. D., & LaDuke, T. C. (2010). Brief communication: Reaction to fire by savanna chimpanzees (Pan troglodytes verus) at Fongoli, Senegal: Conceptualization of "fire behavior" and the case for a chimpanzee model. *American Journal of Physical Anthropology, 141*(4), 646–650.

Rabinovich, M. I., Friston, K. J., & Varona, P. (Eds.). (2012). *Principles of brain dynamics: Global state interactions.* MIT Press.

Seth, A. K., Baars, B. J., & Edelman, D. B. (2005). Criteria for consciousness in humans and other mammals. *Consciousness and Cognition, 14*(1), 119–139.

Singer, W., Sejnowski, T. J., & Rakic, P. (Eds.). (2019). *The neocortex* (Vol. 27). MIT Press.

Strogatz, S. H. (2012). *Sync: How order emerges from chaos in the universe, nature, and daily life.* Hachette UK.

Taylor, J. G., Freeman, W., & Cleeremans, A. (2007). Introduction to the special issue on "Brain and consciousness." *Neural Networks, 20*(9), 929–931.

Temereanca, S., Brown, E. N., & Simons, D. J. (2008). Rapid changes in thalamic firing synchrony during repetitive whisker stimulation. *Journal of Neuroscience, 28*(44), 11153–11164.

Temereanca, S., & Simons, D. J. (2001). Topographic specificity in the functional effects of corticofugal feedback in the whisker/barrel system. *Society for Neuroscience Abstracts, 393*(6).

Whitman, W., & Reynolds, D. S. (2005). *Walt Whitman's Leaves of Grass* (150th anniversary ed.). Oxford University Press.

Woodruff, M. L. (2018a). The fish in the creek is sentient, even if I can't speak with it. *Trans/Form/Ação, 41*(SPE), 119–152.

Woodruff, M. L. (2018b). Sentience in fishes: More on the evidence. *Animal Sentience, 2*(13), 16.

The Darkness

Blight, D. W. (2020). *Frederick Douglass: Prophet of freedom.* Simon & Schuster.

Douglass, F. (2014). *My bondage and my freedom.* Yale University Press. (Original work published 1855)

Douglass, F. (2016). *Narrative of the life of Frederick Douglass, an American slave.* Yale University Press. (Original work published 1845)

P. Gabrielle Foreman, et al. (n.d.). Writing about slavery/teaching about slavery: This might help [Community-sourced document]. https://naacpculpeper.org/resources/writing-about-slavery-this-might-help/

US Census Bureau. (1975). *Bicentennial edition: Historical statistics of the United States, colonial times to 1970.*

Chapter Seventeen: Human Mind

Arbib, M. A. (2011). From mirror neurons to complex imitation in the evolution of language and tool use. *Annual Review of Anthropology, 40*, 257–273.

Arbib, M. A. (2018). Computational challenges of evolving the language-ready brain: 2. Building towards neurolinguistics. *Interaction Studies, 19*(1–2), 22–37.

Arbib, M. A., & Bickerton, D. (Eds.). (2010). *The emergence of protolanguage: Holophrasis vs compositionality* (Vol. 24). Jo$ Benjamins Publishing.

Arbib, M. A., Gasser, B., & Barrès, V. (2014). Language is handy but is it embodied? *Neuropsychologia, 55*, 57–70.

Barnard, A. (2016). *Language in prehistory.* Cambridge University Press.

Braaten, R. F., & Leary, J. C. (1999). Temporal induction of missing birdsong segments in European starlings. *Psychological Science, 10*(2), 162–166.

Brinkhof, T. (2021, June 8). *Dina Sanichar, the real-life "Mowgli" who was raised by wolves.* All That's Interesting. https://allthatsinteresting.com/dina-sanichar

Bugnyar, T., Stöwe, M., & Heinrich, B. (2004). Ravens, Corvus corax, follow gaze direction of humans around obstacles. *Proceedings of the Royal Society B: Biological Sciences, 271*(1546), 1331–1336.

Busnel, R. G., & Classe, A. (2013). *Whistled languages* (Vol. 13). Springer Science & Business Media.

Butler, A. M., & Wolff, W. (1995). *United States Senate election, expulsion, and censure cases, 1793–1990* (Vol. 103, No. 33). US Government Printing Office.

Byrne, R. W., & Cochet, H. (2017). Where have all the (ape) gestures gone? *Psychonomic Bulletin & Review, 24*(1), 68–71.

Carmody, R. N., & Wrangham, R. W. (2009). The energetic significance of cooking. *Journal of Human Evolution, 57*(4), 379–391.

Chamberlain, A. *Indians, North American.* (1911). *Encyclopedia Britannica* (Vol. 14).

Christiansen, M. H., & Kirby, S. E. (2003). *Language evolution.* Oxford University Press.

Corballis, M. C. (1999). The gestural origins of language: Human language may have evolved from manual gestures, which survive today as a "behavioral fossil" coupled to speech. *American Scientist, 87*(2), 138–145.

Cooney, C. (2016, August 30). *How songbirds island-hopped their way from Australia to colonise the world.* The Conversation. https://theconversation.com/how-songbirds-island-hopped-their-way-from-australia-to-colonise-the-world-64616

Coupé, C., Oh, Y. M., Dediu, D., & Pellegrino, F. (2019). Different languages, similar encoding efficiency: Comparable information rates across the human communicative niche. *Science Advances, 5*(9), eaaw2594.

Curtiss, S. (2014). *Genie: A psycholinguistic study of a modern-day wild child.* Academic Press.

Declaration of the immediate causes which induce and justify the secession of South Carolina. (1860). Louise Pettus Archives and Special Collections, Winthrop University. Accession 1256-M608 (661). https://digitalcommons.winthrop.edu/manuscriptcollection_findingaids/793/

Dove, G. (2018). Language as a disruptive tec$ology: Abstract concepts, embodiment and the flexible mind. *Philosophical Transactions of the Royal Society B: Biological Sciences, 373*(1752), 20170135.

Dovidio, J. F., & Gaertner, S. L. (2010). Intergroup bias. In S. T. Fiske, D. T. Gilbert, & G. Lindzey (Eds.), *Handbook of social psychology* (pp. 1084–1121). Jo$ Wiley & Sons.

Ellis, N. C. (2011). The emergence of language as a complex adaptive system. In J. Simpson (Ed.), *The Routledge handbook of applied linguistics* (pp. 654, 667). Taylor & Francis.

Ferdinand, V., Kirby, S., & Smith, K. (2019). The cognitive roots of regularization in language. *Cognition, 184*, 53–68.

Ferretti, F., Adornetti, I., Chiera, A., Cosentino, E., & Nicchiarelli, S. (2018). Introduction: Origin and evolution of language—an interdisciplinary perspective. *Topoi, 37*(2), 219–234.

Fiske, S. T., & North, M. S. (2015). Measures of stereotyping and prejudice: Barometers of bias. In G. J. Boyle, D. H. Saklofske, & G. Matthews (Eds.), *Measures of personality and social psychological constructs* (pp. 684–718). Academic Press.

Fitch, W. T. (2010). *The evolution of language.* Cambridge University Press.

Fitch, W. T., & Martins, M. D. (2014). Hierarchical processing in music, language, and action: Lashley revisited. *Annals of the New York Academy of Sciences, 1316*(1), 87–104.

Fromkin, V., Krashen, S., Curtiss, S., Rigler, D., & Rigler, M. (1974). The development of language in Genie: A case of language acquisition beyond the "critical period." *Brain and Language, 1*(1), 81–107.

Gibson, K. (2011). Are other animals as smart as great apes? Do others provide better models for the evolution of speech or language? In M. Tallerman & K. R. Gibson (Eds.), *The Oxford handbook of language evolution*. Oxford University Press.

Grodzinski, U., Watanabe, A., & Clayton, N. S. (2012). Peep to pilfer: What scrub-jays like to watch when observing others. *Animal Behaviour, 83*(5), 1253–1260.

Grossberg, S. (2000). The complementary brain: Unifying brain dynamics and modularity. *Trends in Cognitive Sciences, 4*(6), 233–246.

Grossberg, S. (2003). Resonant neural dynamics of speech perception. *Journal of Phonetics, 31*(3–4), 423–445.

Grossberg, S. (2017). Towards solving the hard problem of consciousness: The varieties of brain resonances and the conscious experiences that they support. *Neural Networks, 87*, 38–95.

Grossberg, S. (2021). *Conscious mind, resonant brain: How each brain makes a mind.* Oxford University Press.

Grossberg, S., & Kazerounian, S. (2011). Laminar cortical dynamics of conscious speech perception: Neural model of phonemic restoration using subsequent context in noise. *The Journal of the Acoustical Society of America, 130*(1), 440–460.

Grossberg, S., & Kazerounian, S. (2016). Phoneme restoration and empirical coverage of Interactive Activation and Adaptive Resonance models of human speech processing. *The Journal of the Acoustical Society of America, 140*(2), 1130–1153.

Grossberg, S., & Myers, C. W. (2000). The resonant dynamics of speech perception: Interword integration and duration-dependent backward effects. *Psychological Review, 107*(4), 735.

Grossberg, S., & Vladusich, T. (2010). How do children learn to follow gaze, share joint attention, imitate their teachers, and use tools during social interactions? *Neural Networks, 23*(8–9), 940–965.

Harnad, S. (2011). Symbol grounding and the origin of language: From show to tell. In M. Tallerman & K. R. Gibson (Eds.), *The Oxford handbook of language evolution*. Oxford University Press.

Horner, V., & Whiten, A. (2005). Causal knowledge and imitation/emulation switching in chimpanzees (Pan troglodytes) and children (Homo sapiens). *Animal Cognition, 8*(3), 164–181.

Hublin, J. J., Ben-Ncer, A., Bailey, S. E., Freidline, S. E., Neubauer, S., Skinner, M. M., Bergmann, I., Le Cabec, A., Benazzi, S., Harvati, K., & Gunz, P. (2017). New fossils from Jebel Irhoud, Morocco and the pan-African origin of Homo sapiens. *Nature, 546*(7657), 289–292.

Hurford, J. R. (2014). *Origins of language: A slim guide.* Oxford University Press.

Itakura, S. (2004). Gaze-following and joint visual attention in nonhuman animals. *Japanese Psychological Research, 46*(3), 216–226.

Kirby, S., Cornish, H., & Smith, K. (2008). Cumulative cultural evolution in the laboratory: An experimental approach to the origins of structure in human language. *Proceedings of the National Academy of Sciences, 105*(31), 10681–10686.

Klein, R. G. (2017). Language and human evolution. *Journal of Neurolinguistics, 43*, 204–221.

Knight, C., Studdert-Kennedy, M., & Hurford, J. (Eds.). (2000). *The evolutionary emergence of language: Social function and the origins of linguistic form.* Cambridge University Press.

Kolodny, O., & Edelman, S. (2018). The evolution of the capacity for language: The ecological context and adaptive value of a process of cognitive hijacking. *Philosophical Transactions of the Royal Society B: Biological Sciences, 373*(1743), 20170052.

Land, M. F., & Nilsson, D.-E. (2002). *Animal eyes.* Oxford University Press.

Lindzey, G. E., & Aronson, E. E. (1968). *The handbook of social psychology.* Addison-Wesley.

Lubbock, J. (1867). On the origin of civilization and the early condition of man. *British Association for the Advancement of Science: Transactions,* 120.

Luca, F., Perry, G. H., & Di Rienzo, A. (2010). Evolutionary adaptations to dietary changes. *Annual Review of Nutrition, 30*, 291–314.

MacNeilage, P. F. (2010). *The origin of speech* (No. 10). Oxford University Press.

McWhorter, J. (2016, June 29). The world's most efficient languages. *The Atlantic.*

Mooney, R. (2009). Birdsong: The neurobiology of avian vocal learning. In L. Squire (Ed.), *Encyclopedia of Neuroscience*, Academic Press.

Nogier, J. F., & Zock, M. (1992). Lexical choice as pattern matching. *Knowledge-Based Systems*, 5(3), 200–212.

Peretz, I., Vuvan, D., Lagrois, M. É., & Armony, J. L. (2015). Neural overlap in processing music and speech. *Philosophical Transactions of the Royal Society B: Biological Sciences*, 370(1664), 20140090.

Pinker, S. (1994). *The language instinct: How the mind creates language*. HarperCollins.

Pinker, S. (2003). *The blank slate: The modern denial of human nature*. Penguin.

Pinker, S. (2015). *Words and rules: The ingredients of language*. Basic Books.

Ross, A. (2017). Insect evolution: The origin of wings. *Current Biology*, 27(3), R113–R115.

Sardi, S., Vardi, R., Sheinin, A., Goldental, A., & Kanter, I. (2017). New types of experiments reveal that a neuron functions as multiple independent threshold units. *Scientific Reports*, 7(1), 1–17.

Sc$eider, C., Call, J., & Liebal, K. (2012). Onset and early use of gestural communication in non-human great apes. *American Journal of Primatology*, 74(2), 102–113.

Schoenemann, P. T. (2009). Evolution of brain and language. *Language Learning*, 59, 162–186.

Searcy, W. A. (2019). Animal communication, cognition, and the evolution of language. *Animal Behaviour*, 151, 203–205.

Simpson, J., & O'Hara, S. J. (2019). Gaze following in an asocial reptile (Eublepharis macularius). *Animal Cognition*, 22(2), 145–152.

Steel, K. (2016). Feral and isolated children from Herodotus to Akbar to Hesse: Heroes, thinkers, and friends of wolves. Presentation, April 11, CUNY Brooklyn College.

Tallerman, M., & Gibson, K. R. (Eds.). (2011). *The Oxford handbook of language evolution*. Oxford University Press.

Vail, A. L., Manica, A., & Bshary, R. (2013). Referential gestures in fish collaborative hunting. *Nature Communications*, 4(1), 1–7.

White, L., Togneri, R., Liu, W., & Bennamoun, M. (2018). *Neural representations of natural language* (Vol. 783). Springer.

Wrangham, R. (2009). *Catching fire: How cooking made us human*. Basic Books.

Wrangham, R. (2017). Control of fire in the Paleolithic: Evaluating the cooking hypothesis. *Current Anthropology*, 58(S16), S303–S313.

Chapter Eighteen: Sapiens Supermind I

Almaatouq, A., Noriega-Campero, A., Alotaibi, A., Krafft, P. M., Moussaid, M., & Pentland, A. (2020). Adaptive social networks promote the wisdom of crowds. *Proceedings of the National Academy of Sciences*, 117(21), 11379–86.

Bal, P. M., & Veltkamp, M. (2013). How does fiction reading influence empathy? An experimental investigation on the role of emotional transportation. *PLoS One*, 8(1), e55341.

Brockman, J. (Ed.). (2020). *Possible minds: Twenty-five ways of looking at AI*. Penguin.

Broushaki, F., Thomas, M. G., Link, V., López, S., van Dorp, L., Kirsanow, K., Hofmanová, Z., Diekmann, Y., Cassidy, L. M., Díez-del-Molino, D., & Kousathanas, A. (2016). Early Neolithic genomes from the eastern Fertile Crescent. *Science*, 353(6298), 499–503.

Burdick, A. (2017). *Why time flies: A mostly scientific investigation*. Simon & Schuster.

Candia, C., Jara-Figueroa, C., Rodriguez-Sickert, C., Barabási, A. L., & Hidalgo, C. A. (2019). The universal decay of collective memory and attention. *Nature Human Behaviour*, 3(1), 82–91.

Cooney, C. (2016, August 30). *How songbirds island-hopped their way from Australia to colonise the world*. The Conversation. https://theconversation.com/how-songbirds-island-hopped-their-way-from-australia-to-colonise-the-world-64616

Fekete, T., van Leeuwen, C., & Edelman, S. (2016). System, subsystem, hive: Boundary problems in computational theories of consciousness. *Frontiers in Psychology*, 7, 1041.

Hewstone, M., Rubin, M., & Willis, H. (2002). Intergroup bias. *Annual Review of Psychology*, 53(1), 575–604.

Newitz, A. (2021). *Four lost cities: A secret history of the urban age*. W. W. Norton.

Schwitzgebel, E. (2015). If materialism is true, the United States is probably conscious. *Philosophical Studies, 172*(7), 1697–1721.

Chapter Nineteen: Sapiens Supermind II

Baines, J. (1982). Interpreting Sinuhe. *The Journal of Egyptian Archaeology, 68*(1), 31–44.

Djikic, M., Oatley, K., & Moldoveanu, M. C. (2013). Reading other minds: Effects of literature on empathy. *Scientific Study of Literature, 3*(1), 28–47.

Feynman, R. P. (2014). *QED*. Princeton University Press.

Ford, M. (2018). *Architects of intelligence: The truth about AI from the people building it*. Packt.

Foster, J. L. (2001). *Ancient Egyptian literature: An anthology* (Illustrated ed.). University of Texas Press.

Galassi, F. M., Böni, T., Rühli, F. J., & Habicht, M. E. (2016). Fight-or-flight response in the ancient Egyptian novel "Sinuhe"(c. 1800 BCE). *Autonomic Neuroscience, 195*, 27–28.

Glimcher, P. W. (2005). Indeterminacy in brain and behavior. *Annual Review of Psychology, 56*.

Granato, A., & De Giorgio, A. (2014). Alterations of neocortical pyramidal neurons: Turning points in the genesis of mental retardation. *Frontiers in Pediatrics, 2*, 86.

Grossberg, S. (2021). *Conscious mind, resonant brain: How each brain makes a mind*. Oxford University Press.

Herodotus. (2015). (P. Cartledge, Ed., & T. Holland, Trans.). *The histories* (Penguin Classics Deluxe reprint ed.). Penguin Classics.

Hewstone, M., Rubin, M., & Willis, H. (2002). Intergroup bias. *Annual Review of Psychology, 53*(1), 575–604.

Hofstadter, D. R. (2007). *I am a strange loop*. Basic Books.

Jennings, C. D. (2017). I attend, therefore I am. *Aeon*. https://aeon.co/essays/what-is-the-self-if-not-that-which-pays-attention

Kaneda, T., & Haub, C. (2018). How many people have ever lived on earth. *Population Reference Bureau, 9*.

Kidd, D. C., & Castano, E. (2013). Reading literary fiction improves theory of mind. *Science, 342*(6156), 377–380.

Kidd, D., & Castano, E. (2017). Different stories: How levels of familiarity with literary and genre fiction relate to mentalizing. *Psychology of Aesthetics, Creativity, and the Arts, 11*(4), 474.

Kolodny, O., & Edelman, S. (2018). The evolution of the capacity for language: The ecological context and adaptive value of a process of cognitive hijacking. *Philosophical Transactions of the Royal Society B: Biological Sciences, 373*(1743), 20170052.

Li, P., Qian, H., & Wu, J. (2014). Environment: Accelerate research on land creation. *Nature News, 510*(7503), 29.

Mar, R. A., Oatley, K., Hirsh, J., Dela Paz, J., & Peterson, J. B. (2006). Bookworms versus nerds: Exposure to fiction versus non-fiction, divergent associations with social ability, and the simulation of fictional social worlds. *Journal of Research in Personality, 40*(5), 694–712.

Mar, R. A., Oatley, K., & Peterson, J. B. (2009). Exploring the link between reading fiction and empathy: Ruling out individual differences and examining outcomes. *Communications, 34*, 407–428.

Miller, J. H. (2016). *A crude look at the whole: The science of complex systems in business, life, and society*. Basic Books.

Mithen, S. (2011). *After the ice: A global human history, 20,000–5000 BC*. Weidenfeld & Nicolson.

Oatley, K. (2016). Fiction: Simulation of social worlds. *Trends in Cognitive Sciences, 20*(8), 618–628.

Pancro, M. E., Weisberg, D. S., Black, J., Goldstein, T. R., Barnes, J. L., Brownell, H., & Winner, E. (2016). Does reading a single passage of literary fiction really improve theory of mind? An attempt at replication. *Journal of Personality and Social Psychology, 111*(5), e46.

Pinker, S. (2012). *The better angels of our nature: Why violence has declined*. Penguin.

Renfrew, C. (2009). *Prehistory: The making of the human mind* (Modern Library Chronicles, Reprint ed.). Modern Library.

Rosling, H. (2019). *Factfulness*. Flammarion.

Sejnowski, T. J. (2018). *The deep learning revolution*. MIT Press.

Tseng, J., & Poppenk, J. (2020). Brain meta-state transitions demarcate thoughts across task contexts exposing the mental noise of trait neuroticism. *Nature Communications, 11*(1), 1–12.

Twenge, J. M., Campbell, W. K., & Gentile, B. (2013). Changes in pronoun use in American books and the rise of individualism, 1960–2008. *Journal of Cross-Cultural Psychology, 44*(3), 406–415.

Whitehouse, H., Francois, P., Savage, P. E., Currie, T. E., Feeney, K. C., Cioni, E., Purcell, R., Ross, R. M., Larson, J., Baines, J., & Ter Haar, B. (2019). Complex societies precede moralizing gods throughout world history. *Nature, 568*(7751), 226–229.

Willison, H. J., Jacobs, B. C., & van Doorn, P. A. (2016). Guillain-Barré syndrome. *The Lancet, 388*(10045), 717–727.

The Light

Blight, D. W. (2020, August 21). Barack Obama delivers a jeremiad. *New York Times*.

Blight, D. W. (2020). *Frederick Douglass: Prophet of freedom*. Simon & Schuster.

Douglass, F. (2014). *My bondage and my freedom*. Yale University Press. (Original work published 1855)

Douglass, F. (2016). *Narrative of the life of Frederick Douglass, an American slave*. Yale University Press. (Original work published 1845)

The Tandava

Anderson, J. A., & Rosenfeld, E. (Eds.). (2000). *Talking nets: An oral history of neural networks*. MIT Press.

Coomaraswamy, A. K. (1958). *The dance of Shiva*. Owen.

Holland, J. H. (2006). Studying complex adaptive systems. *Journal of Systems Science and Complexity, 19*(1), 1–8.

Mack, K. (2020b). *The end of everything: (Astrophysically speaking)* (Illustrated ed.). Scribner.

Miller, J. H. (2016). *A crude look at the whole: The science of complex systems in business, life, and society* (1st ed.). Basic Books.

Sporns, O. (2016). *Networks of the brain* (Reprint ed.). MIT Press.

Index